Recent Development in Stochastic Dynamics and Stochastic Analysis

INTERDISCIPLINARY MATHEMATICAL SCIENCES

Series Editor: Jinqiao Duan *(Illinois Inst. of Tech., USA)*

Editorial Board: Ludwig Arnold, Roberto Camassa, Peter Constantin, Charles Doering, Paul Fischer, Andrei V. Fursikov, Sergey V. Lototsky, Fred R. McMorris, Daniel Schertzer, Bjorn Schmalfuss, Yuefei Wang, Xiangdong Ye, and Jerzy Zabczyk

---

Published

Vol. 1: Global Attractors of Nonautonomous Dissipative Dynamical Systems
David N. Cheban

Vol. 2: Stochastic Differential Equations: Theory and Applications
A Volume in Honor of Professor Boris L. Rozovskii
eds. Peter H. Baxendale & Sergey V. Lototsky

Vol. 3: Amplitude Equations for Stochastic Partial Differential Equations
Dirk Blömker

Vol. 4: Mathematical Theory of Adaptive Control
Vladimir G. Sragovich

Vol. 5: The Hilbert–Huang Transform and Its Applications
Norden E. Huang & Samuel S. P. Shen

Vol. 6: Meshfree Approximation Methods with MATLAB
Gregory E. Fasshauer

Vol. 7: Variational Methods for Strongly Indefinite Problems
Yanheng Ding

Vol. 8: Recent Development in Stochastic Dynamics and Stochastic Analysis
eds. Jinqiao Duan, Shunlong Luo & Caishi Wang

Vol. 9: Perspectives in Mathematical Sciences
eds. Yisong Yang, Xinchu Fu & Jinqiao Duan

Interdisciplinary Mathematical Sciences – Vol. 8

Recent Development in Stochastic Dynamics and Stochastic Analysis

Editors

Jinqiao Duan
Illinois Institute of Technology, USA

Shunlong Luo
Chinese Academy of Sciences, China

Caishi Wang
Northwest Normal University, China

World Scientific

NEW JERSEY · LONDON · SINGAPORE · BEIJING · SHANGHAI · HONG KONG · TAIPEI · CHENNAI

Published by

World Scientific Publishing Co. Pte. Ltd.
5 Toh Tuck Link, Singapore 596224
USA office: 27 Warren Street, Suite 401-402, Hackensack, NJ 07601
UK office: 57 Shelton Street, Covent Garden, London WC2H 9HE

British Library Cataloguing-in-Publication Data
A catalogue record for this book is available from the British Library.

RECENT DEVELOPMENT IN STOCHASTIC DYNAMICS AND STOCHASTIC ANALYSIS
Interdisciplinary Mathematical Sciences — Vol. 8

Copyright © 2010 by World Scientific Publishing Co. Pte. Ltd.

All rights reserved. This book, or parts thereof, may not be reproduced in any form or by any means, electronic or mechanical, including photocopying, recording or any information storage and retrieval system now known or to be invented, without written permission from the Publisher.

For photocopying of material in this volume, please pay a copying fee through the Copyright Clearance Center, Inc., 222 Rosewood Drive, Danvers, MA 01923, USA. In this case permission to photocopy is not required from the publisher.

ISBN-13 978-981-4277-25-9
ISBN-10 981-4277-25-8

Printed in Singapore.

Professor Zhi-Yuan Huang

Editorial Foreword

This volume of *Interdisciplinary Mathematical Sciences* collects invited contributions giving timely surveys on a diverse range of topics on stochastic dynamics and stochastic analysis. This includes dynamics under random boundary conditions, decoherent information analysis, stabilization by noise, stochastic parameterization, white noise theory, self-similar processes and colored noise, non-Gaussian noise, stochastic evolutionary equations, infinite dimensional rotation group and Lie algebra, Weyl processes, random fields, Malliavin calculus, and stochastic integral.

The ordering of the chapters follows the alphabetic order of the last names of the first authors of the contributions.

We are grateful to all the authors for their contributions to this volume. We thank Dr. Chujin Li, Huazhong University of Science and Technology, Wuhan, China, for technical support. We would also like to thank Rok Ting Tan and Rajesh Babu at World Scientific Publishing, for their professional editorial assistance.

We are privileged and honored to dedicate this volume to Professor Zhi-Yuan Huang, on the occasion of his 75th birthday, in celebration of his achievements in mathematical sciences.

Jinqiao Duan, Illinois Institute of Technology, Chicago, USA
Shunlong Luo, Chinese Academy of Sciences, Beijing, China
Caishi Wang, Northwest Normal University, Lanzhou, China

September 1, 2009

Preface

This festschrift volume is dedicated to professor Zhi-Yuan Huang on the occasion of his 75th birthday.

Zhi-Yuan Huang was born on June 2, 1934, in Nanchang, the capital city of Jiangxi province of southern China. He entered Wuhan University, a prestigious university in China, in 1956 and graduated in 1960. After graduation, he remained as a faculty member at the Department of Mathematics. In 1962 he went to Zhongshan University for advanced studies on theory of stochastic processes and published his first research papers. Two years later, he returned to Wuhan University, where he spent more than thirty years as a faculty member.

From 1982 to 1983, he worked with Professor S. Orey as a senior visiting scholar at the University of Minnesota, USA. During that time, he proposed a theory of stochastic integration over a general topological measurable space, which included the well-known Itô integral as well as other existing stochastic integrals as special cases.

At the end of 1983, he returned to Wuhan University and gave lectures to graduate students on stochastic analysis. In 1984, he was appointed as associate director of the Research Institute for Mathematics at Wuhan University and the following year he was promoted to full professor. He continued his research on stochastic analysis and visited Kyoto University, Japan for one month in 1987 as visiting professor supported by JSPS.

In 1988, his book *Foundations of Stochastic Analysis* was published, which was the first monograph systematically presenting Malliavin calculus in Chinese. In fact he had got interested in Malliavin calculus earlier than 1985, when he published his paper *Malliavin calculus and its applications*. At the end of 1988, he also wrote an article introducing the quantum stochastic calculus, a new theory created by R. L. Hudson and K. R. Parthasarathy.

In 1990, he received the Natural Science Prize of Ministry of Education, China. The same year he was appointed as a doctoral advisor, a higher position at universities in China. In 1992, he moved to Huazhong University of Science and Technology (HUST), also in Wuhan, and since then he has been working there as a professor of mathematics, dean of the School of Science (1994-2000) and Vice President of Academic Committee of HUST (2000-2004). He also held a concurrent position of research fellow in the Institute of Applied Mathematics, Chinese Academy of Sciences (1996-98). Currently, he is Director of the Research Center for Stochastics at HUST.

The white noise theory initiated by T. Hida is essentially an infinite dimensional analog of Schwartz distribution theory. Zhi-Yuan Huang was the first one who noted the potential role of the white noise theory in developing the quantum stochastic calculus. In 1993, he published his celebrated research paper *Quantum white noises: White noise approach to quantum stochastic calculus*. As is suggested by the title, he applied the white noise theory to the quantum stochastic calculus and proposed the notion of quantum white noises. He showed that the quantum white noises were pointwise-defined creation and annihilation operators on the boson Fock space and could be used to extend the Hudson and Parthasarathy's quantum stochastic integral to the non-adapted situation. Following this work, together with Shunlong Luo, he further developed the Wick calculus for generalized operators and successfully applied it to quantum field theory. Professor Hida said: Huang's calculus "has become a very powerful and important tool in the theory of quantum stochastic analysis". Professor Parthasarathy also highly appraised Huang's work and wrote: "the contributions of Professor Huang are very original and have a great potential for further developments in our understanding of rigorous quantum field theory."

From 1993 to 1999, Huang was invited to give talks about his calculus at several international conferences. In 1997, he and Jia-an Yan coauthored a new book entitled *Introduction to Infinite Dimensional Stochastic Analysis*, which was the first monograph dealing with Malliavin calculus and white noise theory in a unified framework (the English version appeared in 2000). The same year he was one more time awarded the Natural Science Prize of Ministry of Education, China. In 1998, he became an editor of the journal *Infinite Dimensional Analysis, Quantum Probability and Related Topics*. In 1999, he and Jia-an Yan shared a prize awarded by the Chinese Government for their joint monograph promoting mathematical learning.

Since 2000, Huang has been devoting himself to the study of Lévy white noise as well as fractional noises. He has published a series of research papers independently or jointly with his students and much progress has been made. He, together with his Ph.D. students, gave an interacting Fock expansion of Lévy white noise functionals and developed a white noise approach to analysis for fractional Lévy processes. In 2004, he published his third book *Quantum White Noise Analysis* (in Chinese, with Caishi Wang and Guanglin Rang). In 2006 and 2009, he received the Natural Science Prize and Teaching Achievement Prize of Hubei Province, China, respectively.

Ph.D. Graduate Students of Zhi-Yuan Huang

Shunlong Luo (Wuhan University, 1995)
Zongxia Liang (Institute of Applied Mathematics CAS, 1996)
Mingli Zheng (Institute of Applied Mathematics CAS, 1996)
Qingquan Lin (Institute of Applied Mathematics CAS, 1998)
Caishi Wang (Huazhong University of Science and Technology, 1999)
Xiangjun Wang (Huazhong University of Science and Technology, 1999)
Shaopu Zhou (Huazhong University of Science and Technology, 2000)
Xiaoshan Hu (Huazhong University of Science and Technology, 2002)
Jihui Hu (Huazhong University of Science and Technology, 2002)
Guanglin Rang (Huazhong University of Science and Technology, 2003)
Ying Wu (Huazhong University of Science and Technology, 2005)
Chujin Li (Huazhong University of Science and Technology, 2005)
Guanghui Huang (Huazhong University of Science and Technology, 2006)
Peiyan Li (Huazhong University of Science and Technology, 2007)
Xuebin Lii (Huazhong University of Science and Technology, 2009)
Junjun Liao (Huazhong University of Science and Technology, 2010)

Publications of Zhi-Yuan Huang Since 1980

Books

(1) Foundations of Stochastic Analysis (in Chinese), Wuhan Univ. Press (1988); Second edition, Science Press (2001)
(2) Introduction to Infinite Dimensional Stochastic Analysis (in Chinese, with Jia-An Yan), Science Press (1997); English version, Kluwer (2000)
(3) Quantum White Noise Analysis (in Chinese, with Caishi Wang and Guanglin Rang), Hubei Sci. Tech. Publ. (2004)

Papers and Invited Lectures

(1) On the generalized sample solutions of stochastic differential equations, Wuhan Univ. J., No. 2 (1981), 11-21 (in Chinese, with M. Xu and Z. Hu)
(2) Martingale measures and stochastic integrals on metric spaces, Wuhan Univ. J. (Special issue for Math.), No. 1 (1981), 89-102
(3) Stochastic integrals on general topological measurable spaces, Z. Wahrs. verw. Gebiete, Vol. 66 (1984), 25-40
(4) On the generalized sample solutions of stochastic boundary value problems, Stochastics, Vol. 11 (1984), 237-248

(5) A comparison theorem for solutions of stochastic differential equations and its applications, Proc. Amer. Math. Soc., Vol. 91 (1984), 611-617
(6) The weak projection theory and decompositions of quasi-martingale measures, Chinese Ann. Math. (Ser. B), Vol. 6 (1985), 395-399
(7) The Malliavin calculus and its applications, J. Applied Probab. Statist., Vol. 1, No. 2 (1985), 161-172 (in Chinese)
(8) On the product martingale measure and multiple stochastic integral, Chinese Ann. Math. (Ser. B), Vol. 7 (1986), 207-210
(9) Spectral analysis for stochastic integral operators, Wuhan Univ. J., No. 4 (1986), 17-24 (in Chinese, with Y. Liao)
(10) Functional integration and partial differential equations, Adv. Math. (China), Vol. 15, No. 2 (1986), 131-174 (based on a series of lectures given by Prof. P. Malliavin in Wuhan Univ. in June 1984)
(11) Spectral analysis for stochastic integral operators, invited lecture given in Kyoto Univ. (1987)
(12) An introduction to quantum stochastic calculus, Adv. Math. (China), Vol. 17, No. 4 (1988), 360-378
(13) Quasi sure stochastic flows, Stochastics, Vol. 33 (1990), 149-157 (with J. Ren)
(14) Some recent development of stochastic calculus in China, Contemporary Math., Vol. 118 (1991), 177-185
(15) Stochastic calculus of variation on Gaussian space and white noise analysis, in "Gaussian Random Fields", K. Ito and T. Hida eds., World Scientific (1991)
(16) Quantum white noises - White noise approach to quantum stochastic calculus, Nagoya Math. J., Vol. 129 (1993), 23-42
(17) Quantum white noises analysis - a new insight into quantum stochastic calculus, invited lecture of 9th Conference on Quantum Probability and Applications, Nottingham (1993)
(18) P-adic valued white noise functionals, Quantum Prob. Related Topics, Vol. IX, (1994), 273-294 (with Khrennikov)
(19) An extension of Hida's distribution theory via analyticity spaces, in "Dirichlet Forms and Stochastic Processes", Walter de Gruyter, 1995 (with H. Song)
(20) Generalized functionals of a p-adic white noise, Doklady Mathematics, Vol. 52 (1995), 175-178 (with Khrennikov)
(21) Quantum white noises and quantum fields, 23rd SPA Conference, Singapore (1995) (with S. Luo)
(22) A model for white noise analysis in p-adic number fields, Acta Math. Sci., Vol. 16 (1996), 1-14 (with Khrennikov)
(23) Quantum white noise analysis, invited lecture of Conference on Stochastic Differential Geometry and Infinite Dimensional Analysis, Hangzhou (1996)
(24) Analytic functionals and a new distribution theory over infinite dimensional spaces, Chinese Ann. Math. (Ser. B), Vol. 17 (1996), 507-514 (with J. Ren)
(25) Quantum white noises, Wick calculus and quantum fields, invited lecture of

Conference on Infinite Dimensional Analysis and Quantum Probability, Rome (1997)

(26) The weak solution for SDE with terminal conditions, Math. Appl., Vol. 10, No. 4 (1997), 60-64 (with Q. Lin)

(27) Quantum white noises and free fields, IDAQP, Vol. 1, No. 1 (1998), 69-82 (with S. Luo)

(28) D^∞-Approximation of quadratic variations of smooth Ito processes, Chinese Ann. Math. (Ser. B), Vol. 19 (1998), 305-310 (with J. Ren)

(29) Wick calculus of generalized operators and its applications to quantum stochastic calculus, IDAQP, Vol. 1, No. 3 (1998), 455-466 (with S. Luo)

(30) Positivity-preservingness of differential second quantization, Math. Appl., Vol. 11 (1998), 31-32 (with C. Wang)

(31) Quantum integral equation of Volterra type with generalized operator-valued kernels, IDAQP, Vol. 3 (2000), 505-517 (with C. Wang and X. Wang)

(32) Quantum cable equations in terms of generalized operators, Acta Appl. Math., Vol. 63 (2000), 151-164 (with C. Wang and X. Wang)

(33) Wick tensor products in Levy white noise spaces, Second Sino-French Colloquium in Probability and Applications, Invited Lecture, Wuhan (2001)

(34) Explicit forms of Wick powers in general white noise spaces, IJMMS, Vol. 31 (2002), 413-420 (with X. Hu and X. Wang)

(35) $L^2(E^*, \mu)$-Weyl representations, IDAQP, Vol. 5 (2002), 581-592 (with X. Hu and X. Wang)

(36) A white noise approach to quantum stochastic cable equations, Acta Math. Sinica, Vol. 45 (2002), 851-862 (with C. Wang)

(37) Quantum integral equations with kernels of quantum white noise in space and time, Advances in Math. Research Vol. 1 (2002), G.Oyibo eds. NOVA SCI. PUBL., 97-108 (with C. Wang and T-S. Chew)

(38) Quadratic covariation and extended Ito formula for continuous semimartingales, Math. Appl., Vol. 15 (2002), 81-84 (with J. Hu)

(39) White noise approach to interacting quantum field theory, in "Recent Developments in Stochastic Analysis and Related Topics", eds. S. Albeverio et al., World Scientific, 2004 (with G. Rang)

(40) Quantum stochastic differential equations in terms of generalized operators, Adv. Math. (China), Vol. 32 (2003), 53-62 (with C. Wang and X. Wang)

(41) White noise approach to the construction of Φ_4^4 quantum fields, Acta Appl. Math., Vol. 77 (2003), 299-318 (with G. Rang)

(42) Generalized operators and operator-valued distribution in quantum field theory, Acta. Math. Scientia, Vol. 23(B) (2003), 145-154 (with X. Wang and C. Wang)

(43) A W-transform-based criterion for the existence of bounded extensions of E-operators, J. Math. Anal. Appl., Vol. 288 (2003), 397-410 (with C. Wang and X. Wang)

(44) The non-uniform Riemann approach to anticipating stochastic integrals, Stoch.

Anal. Appl. Vol. 22 (2004), 429-442 (with T-S. Chew and C. Wang)
(45) Generalized operators and $P(\Phi)_2$ quantum fields, Acta Math. Scientia, Vol. 24(B) (2004), 589-596 (with G. Rang)
(46) Analytic characterization for Hilbert-Schmidt operators on Fock space, Acta Math. Sin. (Engl. Ser.), Vol. 21 (2005), 787-796 (with C. Wang and X. Wang)
(47) A filtration of Wick algebra and its application to quantum SDE, Acta Math. Sin. (Engl. Ser.), Vol. 20 (2004), 999-1008 (with C. Wang)
(48) A moment characterization of B-valued generalized functionals of white noise, Acta Math. Sin. (Engl. Ser.), Vol. 22 (2006), 157-168 (with C. Wang)
(49) δ-Function of an operator: A white noise approach, Proc. Amer. Math. Soc., Vol. 133, No. 3 (2005), 891-898 (with C. Wang and X. Wang)
(50) Fractional Brownian motion and sheet as white noise functionals, Acta Math. Sin. (Engl. Ser.), Vol. 22, No. 4 (2006), 1183-1188 (with C. Li, J. Wan and Y. Wu)
(51) Interacting Fock expansion of Lévy white noise functionals, Acta Appl. Math., Vol. 82 (2004), 333-352 (with Y. Wu)
(52) Lévy white noise calculus based on interaction exponents, Acta Appl. Math., Vol. 88 (2005), 251-268 (with Y. Wu)
(53) Anisotropic fractional Brownian random fields as white noise functionals, Acta Math. Appl. Sinica, Vol. 21, No. 4 (2005), 655-660 (with C. Li)
(54) White noise approach to the construction of Φ_4^4 quantum fields (II), Acta Math. Sin. (Engl. Ser.), Vol. 23, No. 5 (2007), 895-904 (with G. Rang)
(55) On fractional stable processes and sheets: white noise approach, J. Math. Anal. Appl., Vol. 325, No. 1 (2007), 624-635 (with C. Li)
(56) Explicit forms of q-deformed Lévy-Meixner polynomials and their generating functions, Acta Math. Sin. (Engl. Ser.), Vol. 24, No. 2 (2008), 201-214 (with P. Li and Y. Wu)
(57) Generalized fractional Lévy processes: A white noise approach, Stochastics and Dynamics, Vol. 6, No. 4 (2006), 473-485 (with P. Li)
(58) Fractional generalized Lévy random fields as white noise functionals, Front. Math. China, Vol. 2, No. 2 (2007), 211-226 (with P. Li)
(59) Fractional Lévy processes on Gel'fand triple and stochastic integration, Front. Math. China, Vol. 3, No. 2 (2008), 287-303 (with X. Lü and J. Wan)
(60) Fractional noises on Gel'fand triples, Invited lecture of International Conference on Stochastic Analysis and Related Fields, Wuhan (2008) (with X. Lü).

Contents

Editorial Foreword   vii

Preface   ix

1. Hyperbolic Equations with Random Boundary Conditions   1
   Zdzisław Brzeźniak and Szymon Peszat

2. Decoherent Information of Quantum Operations   23
   Xuelian Cao, Nan Li and Shunlong Luo

3. Stabilization of Evolution Equations by Noise   43
   Tomás Caraballo and Peter E. Kloeden

4. Stochastic Quantification of Missing Mechanisms in Dynamical Systems   67
   Baohua Chen and Jinqiao Duan

5. Banach Space-Valued Functionals of White Noise   77
   Yin Chen and Caishi Wang

6. Hurst Index Estimation for Self-Similar Processes with Long-Memory   91
   Alexandra Chronopoulou and Frederi G. Viens

7. Modeling Colored Noise by Fractional Brownian Motion   119
   Jinqiao Duan, Chujin Li and Xiangjun Wang

8. A Sufficient Condition for Non-Explosion for a Class of
   Stochastic Partial Differential Equations ... 131
   Hongbo Fu, Daomin Cao and Jinqiao Duan

9. The Influence of Transaction Costs on Optimal Control for an
   Insurance Company with a New Value Function ... 143
   Lin He, Zongxia Liang and Fei Xing

10. Limit Theorems for p-Variations of Solutions of SDEs Driven by
    Additive Stable Lévy Noise and Model Selection for
    Paleo-Climatic Data ... 161
    Claudia Hein, Peter Imkeller and Ilya Pavlyukevich

11. Class II Semi-Subgroups of the Infinite Dimensional Rotation
    Group and Associated Lie Algebra ... 177
    Takeyuki Hida and Si Si

12. Stopping Weyl Processes ... 185
    Robin L. Hudson

13. Karhunen-Loéve Expansion for Stochastic Convolution of
    Cylindrical Fractional Brownian Motions ... 195
    Zongxia Liang

14. Stein's Method Meets Malliavin Calculus:
    A Short Survey With New Estimates ... 207
    Ivan Nourdin and Giovanni Peccati

15. On Stochastic Integrals with Respect to an Infinite Number of
    Poisson Point Process and Its Applications ... 237
    Guanglin Rang, Qing Li and Sheng You

16. Lévy White Noise, Elliptic SPDEs and Euclidean Random Fields ... 251
    Jiang-Lun Wu

17. A Short Presentation of Choquet Integral ... 269
    Jia-An Yan

Chapter 1

Hyperbolic Equations with Random Boundary Conditions

Zdzisław Brzeźniak and Szymon Peszat

Department of Mathematics, University of York, York, YO10 5DD, UK,
zb500@york.ac.uk
and
Institute of Mathematics, Polish Academy of Sciences, Św. Tomasza 30/7, 31-027
Kraków, Poland, e-mail: napeszat@cyf-kr.edu.pl

Following Lasiecka and Triggiani an abstract hyperbolic equation with random boundary conditions is formulated. As examples wave and transport equations are studied.

Contents

1 Introduction 1
   1.1 The Wave equation 2
   1.2 The Transport equation 3
2 Abstract formulation 3
3 The Wave equation - introduction 4
4 Weak solution to the wave equation 6
5 Mild formulations 7
6 Scales of Hilbert spaces 8
   6.1 Application to the boundary value problem 10
7 Equivalence of weak and mild solutions 11
8 The Fundamental Solution 13
9 Applications 14
10 The Transport equation 16
References 20

1. Introduction

Assume that $\mathcal{A}$ is the generator of a C_0-group $U = (U(t))_{t \in \mathbb{R}}$ of bounded linear operators on a Hilbert space $\mathcal{H}$. Let $\mathcal{U}$ be another Hilbert space and let $u \in L^2_{\text{loc}}(0, +\infty; \mathcal{U})$. Typical examples of $\mathcal{H}$ and $\mathcal{U}$ will be spaces $L^2(\mathcal{O})$ and $L^2(\partial \mathcal{O})$.

Research of the second named author was supported by Polish Ministry of Science and Higher Education Grant PO3A 034 29 "Stochastic evolution equations driven by Lévy noise". Research of the first named author was supported by an EPSRC grant number EP/E01822X/1. In addition the authors acknowledge the support of EC FP6 Marie Curie ToK programme SPADE2, MTKD-CT-2004-014508 and Polish MNiSW SPB-M

Lasiecka and Triggiani, see Refs. 14–17, 4 and 18, discovered that for a large class of boundary operators τ, the hyperbolic initial value problem

$$\frac{d}{dt}X(t) = \mathcal{A}X(t), \qquad t \geq 0, \qquad X(0) = X_0,$$

considered with a non-homogeneous boundary condition

$$\tau(X(t)) = u(t), \qquad t \geq 0,$$

can be written in the form of the homogeneous boundary problem

$$\frac{d}{dt}X(t) = \mathcal{A}X(t) + (\lambda - \mathcal{A})Eu(t), \qquad X(0) = X_0, \qquad (1.1)$$

by choosing properly the space $\mathcal{U}$, an operator $E \in L(\mathcal{U}, \mathcal{H})$ and a scalar λ from the resolvent set $\rho(\mathcal{A})$ of $\mathcal{A}$.

In this chapter u will be the time derivative of a $\mathcal{U}$-valued càdlàg process ξ. We will discuss the existence and regularity of a solution to the problem (1.1). The abstract framework will be illustrated by the wave and the transport equations.

Let us describe briefly the history of the problem studied in this chapter. To our (very limited) knowledge, the first paper which studied evolution problems with boundary noise was a paper[3] by Balakrishnan. The equation studied in that paper was first order in time and fourth order in space with Dirichlet boundary noise. Later Sowers[28] investigated general reaction diffusion equation with Neumann type boundary noise. Da Prato and Zabczyk in their second monograph,[12] see also Ref. 11, have explained the difference between the problems with Dirichlet and Neumann boundary noises. In particular, the solution to the former is less regular in space than the solution to the latter. Maslowski[23] studied some basic questions such as exponential stability in the mean of the solutions and the existence and uniqueness of an invariant measure. Other related works for parabolic problems with boundary noise are E. Alòs and S. Bonaccorsi[1,2] and Brzeźniak et. al.[5] Similar question have also been investigated in the case of hyperbolic SPDEs with the Neumann boundary conditions, see for instance Mao and Markus,[22] Dalang and Lévêque.[7–9,19] Moreover, some authours, see for example Chueshov, Duan and Schmalfuss[6,13] have studied problems in which deterministic partial differential equations are coupled to stochastic ones by some sort of boundary conditions. To our best knowledge, our paper is the first one in which the hyperbolic SPDEs with Dirichlet boundary conditions are studied.

1.1. *The Wave equation*

Let Δ denote the Laplace operator, τ be a boundary operator, $\mathcal{O}$ be a domain in $\mathbb{R}^d$ with smooth boundary $\partial\mathcal{O}$, and let $\mathcal{S}'(\partial\mathcal{O})$ denote the space of distributions on $\partial\mathcal{O}$. We assume that ξ take values in $\mathcal{S}'(\partial\mathcal{O})$. In the first part of the paper, see

Sections from 3 to 9 we will be concerned with the following initial value problem

$$\begin{cases} \dfrac{\partial^2 u}{\partial t^2} = \Delta u & \text{on } (0,\infty) \times \mathcal{O}, \\ \tau u = \dfrac{d\xi}{dt} & \text{on } (0,\infty) \times \partial \mathcal{O}, \\ u(0,\cdot) = u_0 & \text{on } \mathcal{O}, \\ \dfrac{\partial u}{\partial t}(0,\cdot) = u_{0,1} & \text{on } \mathcal{O}. \end{cases} \quad (1.2)$$

In fact, we will only consider the Dirichlet and the Neumann boundary conditions. In the case of the Dirichlet boundary conditions we put $\tau = \tau_D$, where $\tau_D \psi(x) = \psi(x)$ for $x \in \partial \mathcal{O}$, whereas in case of the Neumann boundary conditions we put $\tau = \tau_N$, where $\tau_N \psi(x) = \frac{\partial \psi}{\partial \mathbf{n}}(x)$, $x \in \partial \mathcal{O}$ and $\mathbf{n}$ is the exterior unit normal vector field on $\partial \mathcal{O}$.

1.2. *The Transport equation*

In Section 10 we will consider the following stochastic generalization of the boundary value problem associated to the following simple transport equation introduced in Ref. 4 [Example 4.1, p. 466],

$$\begin{cases} \dfrac{\partial u}{\partial t} = \dfrac{\partial u}{\partial x} & \text{on } (0,\infty) \times (0,2\pi), \\ u(0,\cdot) = u_0 & \text{on } (0,2\pi), \\ \tau(u(t,\cdot)) = \dfrac{d\xi}{dt}(t) & \text{for } t \in (0,\infty), \end{cases} \quad (1.3)$$

where $\tau(\psi) = \psi(2\pi) - \psi(0)$.

This paper is organized as follows. The next section is devoted to an abstract framework. This framework is adapted from works by Lasiecka and Triggiani and also the book by Bensoussan et al.[4] In the following section we study the wave equation with the Dirichlet and the Neumann boundary conditions. In particular we will investigate the concepts of weak and mild solutions, their equivalence, their relation with the abstract framework and their regularity. The final section (Section 10) is devoted to the transport equation.

2. Abstract formulation

We will identify the Hilbert space $\mathcal{H}$ with its dual $\mathcal{H}'$. Therefore the adjoint operator $\mathcal{A}^*$ is bounded from its domain $\mathrm{D}(\mathcal{A}^*)$, equipped with the graph norm, into $\mathcal{H}$ and hence it has a bounded dual operator $(\mathcal{A}^*)' \colon \mathcal{H} \mapsto (\mathrm{D}(\mathcal{A}^*))'$. It is easy to see that the latter operator is a bounded linear extension of a linear map $\mathcal{A} \colon \mathrm{D}(\mathcal{A}) \mapsto \mathcal{H}$. Since $\mathcal{A}$ is a closed operator, $\mathrm{D}(\mathcal{A})$ endowed with a graph norm is a Hilbert space.

Alternatively, we can take $\kappa \in \rho(\mathcal{A})$ and endow $D(\mathcal{A})$ with the norm $\|(\kappa - \mathcal{A})f\|_{\mathcal{H}}$. Clearly these two norms are equivalent.

We assume that for any $T > 0$ there exists a constant $K > 0$ such that

$$\int_0^T \|E^*\mathcal{A}^*U(t)^*f\|_{\mathcal{H}}^2 dt \leq K \|f\|_{\mathcal{H}}^2, \qquad f \in D(\mathcal{A}^*). \tag{2.1}$$

Assume that X is a mild solution of (1.1) with u being the weak time derivative of ξ, i.e.

$$X(t) = U(t)X_0 + \int_0^t U(t-r)(\lambda - \mathcal{A})E\frac{d\xi(r)}{dr}, \qquad t \geq 0. \tag{2.2}$$

Then, integrating by parts we see that the mild form of (1.1) is

$$\begin{aligned} X(t) = U(t)X_0 &+ (\lambda - (\mathcal{A}^*)')\left[E\xi(t) - U(t)E\xi(0)\right] \\ &+ \int_0^t (\mathcal{A}^*)'U(t-r)(\lambda - \mathcal{A})E\xi(r)dr, \qquad t \geq 0. \end{aligned} \tag{2.3}$$

In other words,

$$X(t) = U(t)x_0 + (\lambda - (\mathcal{A}^*)')\left[E\xi(t) - U(t)E\xi(0)\right] + (\mathcal{A}^*)'Y(t), \qquad t \geq 0,$$

where

$$Y(t) := \int_0^t U(t-r)(\lambda - \mathcal{A})E\xi(r)dr, \qquad t \geq 0.$$

Since the trajectories of the process ξ are càdlàg, they are also locally bounded and hence locally square integrable. Hence, by[4] [Proposition 4.1], the process Y has trajectories in $C(\mathbb{R}_+; \mathcal{H})$. Since $(\mathcal{A}^*)'$ is a bounded operator from $\mathcal{H}$ to $(D(\mathcal{A}^*))'$ we have the following result.

Theorem 2.1. *Assume (2.1). If ξ is an $\mathcal{U}$-valued càdlàg process, then the process X defined by (2.3) has càdlàg trajectories in $(D(\mathcal{A}^*))'$.*

3. The Wave equation - introduction

In the next section we will introduce a concept of weak, in the PDEs sense, solution to the boundary problem for wave equation (1.2). Next, denoting by τ the appropriate boundary operator, by D_τ the boundary map with certain parameter κ and by Δ_τ the Laplace operator with homogeneous boundary conditions, in Section 5 we will show in a heuristic way that problem (1.2) can be written as follows

$$\begin{aligned} dX &= \mathcal{A}_\tau X dt + (\kappa - \mathcal{A}_\tau^2)\mathbf{D}_\tau d\xi, \\ &= \mathcal{A}_\tau X dt + ((\kappa - \Delta_\tau)D_\tau)^\dagger d\xi, \end{aligned} \tag{3.1}$$

where

$$X = \begin{pmatrix} u \\ \frac{\partial u}{\partial t} \end{pmatrix}, \quad \mathcal{A}_\tau = \begin{bmatrix} 0 & I \\ \Delta_\tau & 0 \end{bmatrix}, \quad \mathbf{D}_\tau = \begin{pmatrix} 0 \\ D_\tau \end{pmatrix}, \qquad (3.2)$$

$$((\kappa - \Delta_\tau) D_\tau)^\dagger = \begin{pmatrix} 0 \\ (\kappa - \Delta_\tau) D_\tau \end{pmatrix}.$$

Note that

$$(\kappa - \mathcal{A}_\tau^2)\mathbf{D}_\tau = \left(\begin{bmatrix} \kappa & 0 \\ 0 & \kappa \end{bmatrix} - \begin{bmatrix} 0 & I \\ \Delta_\tau & 0 \end{bmatrix} \begin{bmatrix} 0 & I \\ \Delta_\tau & 0 \end{bmatrix} \right) \begin{pmatrix} 0 \\ D_\tau \end{pmatrix}$$

$$= \begin{bmatrix} \kappa - \Delta_\tau & 0 \\ 0 & \kappa - \Delta_\tau \end{bmatrix} \begin{pmatrix} 0 \\ D_\tau \end{pmatrix} = \begin{pmatrix} 0 \\ (\kappa - \Delta_\tau) D_\tau \end{pmatrix} \qquad (3.3)$$

$$= ((\kappa - \Delta_\tau) D_\tau)^\dagger.$$

Let U_τ be the semigroup generated by the operator $\mathcal{A}_\tau$. In Section 7 we will show that a weak solution to problem (1.2) exists and moreover, it is given by the following formula

$$X(t) = U_\tau(t)X(0) + \int_0^t U_\tau(t - r)\left(\kappa - \mathcal{A}_\tau^2\right)\mathbf{D}_\tau \mathrm{d}\xi(r), \qquad t \geq 0. \qquad (3.4)$$

In other words, a weak solution to problem (1.2) is the mild solution to problem (3.1). In (3.4), the integrals are defined by integration by parts. Thus

$$\int_0^t U_\tau(t - r)(\kappa - \mathcal{A}_\tau^2)\mathbf{D}_\tau \mathrm{d}\xi(r)$$
$$= (\kappa - \mathcal{A}_\tau^2)\mathbf{D}_\tau \xi(t) - U_\tau(t)(\kappa - \mathcal{A}_\tau^2)\mathbf{D}_\tau \xi(0) \qquad (3.5)$$
$$+ \int_0^t \mathcal{A}_\tau U_\tau(t - r)\left(\kappa - \mathcal{A}_\tau^2\right)\mathbf{D}_\tau \xi(r)\mathrm{d}r, \qquad t \geq 0.$$

From now on we will assume that $\kappa > 0$ is such that both κ and $\sqrt{\kappa}$ belong to the resolvent set of $\mathcal{A}_\tau$. Then

$$(\sqrt{\kappa} - \mathcal{A}_\tau)\begin{pmatrix} D_\tau \\ \sqrt{\kappa}D_\tau \end{pmatrix} = \begin{bmatrix} \sqrt{\kappa} & -I \\ -\Delta_\tau & \sqrt{\kappa} \end{bmatrix}\begin{pmatrix} D_\tau \\ \sqrt{\kappa}D_\tau \end{pmatrix}$$

$$= \begin{pmatrix} 0 \\ (\kappa - \Delta_\tau) D_\tau \end{pmatrix} = ((\kappa - \Delta_\tau) D_\tau)^\dagger,$$

and hence we see that our case fits into the abstract framework with $\lambda = \sqrt{\kappa}$, and

$$E = \begin{pmatrix} D_\tau \\ \sqrt{\kappa}D_\tau \end{pmatrix}.$$

Since condition (2.1) is satisfied[4], we get the following corollary to Theorem 2.1.

Proposition 3.1. *The process X defined by formula (3.4), is $(D(\mathcal{A}^*))'$-valued càdlàg.*

Alternatively, one can give a proper meaning of the term $(\kappa - \mathcal{A}_\tau^2)\mathbf{D}_\tau$ by using the scales of Hilbert spaces

$$H_s^\tau = \mathrm{D}(\kappa - \Delta_\tau)^{s/2}, \qquad \mathcal{H}_s^\tau := \begin{pmatrix} H_{s+1}^\tau \\ \times \\ H_s^\tau \end{pmatrix}, \qquad s \in \mathbb{R},$$

where κ belongs to the resolvent set of Δ_τ, see Section 6 for more details. It turns out that if ξ is sufficiently regular in space variable, then $D_\tau \xi$ takes values in the domain of Δ_τ considered on $H_{\tau,-s}$ for s large enough, so the term appearing in the utmost right hand side of (3.1) is well defined.

4. Weak solution to the wave equation

We will introduce a notion of weak solution to the wave equation and we will discuss their uniqueness.

By $\mathcal{S}(\overline{\mathcal{O}})$ we will denote the class of restrictions of all test function $\varphi \in \mathcal{S}(\mathbb{R}^d)$ to $\overline{\mathcal{O}}$ and we will denote by $(\cdot, \cdot)$ the duality forms on both $\mathcal{S}'(\overline{\mathcal{O}}) \times \mathcal{S}(\overline{\mathcal{O}})$ and $\mathcal{S}'(\partial \mathcal{O}) \times \mathcal{S}(\partial \mathcal{O})$. We will always assume that any $\mathcal{S}'(\overline{\mathcal{O}})$-valued process v is weakly measurable, that is

$$\Omega \times [0, \infty) \ni (t, \omega) \mapsto (v(t)(\omega), \varphi) \in \mathbb{R}$$

is measurable for any $\varphi \in \mathcal{S}(\overline{\mathcal{O}})$.

Assume now that ξ is an $\mathcal{S}'(\partial \mathcal{O})$-valued process. Taking into account the Green formula, see e.g. the monograph[20] by Lions and Magenes, we arrive at the following definitions of a weak solution.

Definition 4.1. We will say that an $\mathcal{S}'(\overline{\mathcal{O}}) \times \mathcal{S}'(\overline{\mathcal{O}})$-valued process (u, v) is a *weak solution* to (1.2) considered with the Dirichlet boundary condition, i.e. $\tau = \tau_D$, iff (aD) for all $t > 0$,

$$(u(t), \varphi) = (u_0, \varphi) + \int_0^t (v(r), \varphi) dr, \qquad \mathbb{P}-a.s. \quad \forall \varphi \in \mathcal{S}(\overline{\mathcal{O}}), \tag{4.1}$$

and (bD) for all $t \geq 0$, $\mathbb{P}$-a.s. for all $\psi \in \mathcal{S}(\overline{\mathcal{O}})$ satisfying $\psi = 0$ on $\partial \mathcal{O}$,

$$(v(t), \psi) = (u_{0,1}, \psi) + \int_0^t (u(r), \Delta \psi) dr + \left(\xi(t) - \xi(0), \frac{\partial \psi}{\partial \mathbf{n}} \right). \tag{4.2}$$

We will call an $\mathcal{S}'(\overline{\mathcal{O}}) \times \mathcal{S}'(\overline{\mathcal{O}})$-valued process (u, v) a *weak solution* to (1.2) considered with the Neumann boundary condition, i.e. $\tau = \tau_N$, iff (aN) equality (4.1) holds and (bN) for all $t > 0$ and $\psi \in \mathcal{S}(\overline{\mathcal{O}})$ satisfying $\frac{\partial \psi}{\partial \mathbf{n}} = 0$ on $\partial \mathcal{O}$, $\mathbb{P}$-a.s.

$$(u(t), \psi) = (u_0, \psi) + \int_0^t (u(r), \Delta \psi) dr - (\xi(t) - \xi(0), \psi). \tag{4.3}$$

Let $\kappa \geq 0$. For the future consideration we will need also a concept of a weak solution to the (deterministic) elliptic problem

$$\begin{aligned} \Delta u(x) &= \kappa u(x), & x \in \mathcal{O}, \\ \tau u(x) &= \gamma(x), & x \in \partial\mathcal{O}. \end{aligned} \tag{4.4}$$

Definition 4.2. Let $\gamma \in \mathcal{S}'(\partial\mathcal{O})$. We call $u \in \mathcal{S}'(\overline{\mathcal{O}})$ the *weak solution* to (4.4) considered with the Dirichlet boundary condition $\tau = \tau_D$ iff

$$(u, \Delta\psi) + \left(\gamma, \frac{\partial\psi}{\partial \mathbf{n}}\right) = \kappa(u, \psi), \qquad \forall \psi \in \mathcal{S}(\overline{\mathcal{O}}): \ \psi = 0 \text{ on } \partial\mathcal{O}. \tag{4.5}$$

We will call $u \in \mathcal{S}'(\overline{\mathcal{O}})$ a *weak solution* to (4.4) considered with the Neumann boundary condition ($\tau = \tau_N$) iff

$$(u, \Delta\psi) - (\gamma, \psi) = \kappa(u, \psi), \qquad \forall \psi \in \mathcal{S}(\overline{\mathcal{O}}): \ \frac{\partial\psi}{\partial \mathbf{n}} = 0 \text{ on } \partial\mathcal{O}. \tag{4.6}$$

For the completeness of our presentation we present the following result on the uniqueness of solutions.[5]

Proposition 4.2. *(i) For any u_0, $u_{0,1}$ and γ problem (1.2) considered with Dirichlet or Neumann boundary conditions has at most one solution.*
(ii) For any $\gamma \in \mathcal{S}'(\partial\mathcal{O})$ and $\kappa \geq 0$ problem (4.4) with Dirichlet boundary condition has at most one solution.
(iii) For any $\gamma \in \mathcal{S}'(\partial\mathcal{O})$ and $\kappa > 0$ problem (4.4) with Neumann boundary condition has at most one solution.

We will denote by $D_D\gamma$ and $D_N\gamma$ the solution to (4.4) with Dirichlet and Neumann boundary conditions. We call D_D and D_N the *Dirichlet and Neumann boundary maps*. Note that both these maps depend on the parameter κ and hence should be denoted by D_D^κ and D_N^κ. However, we have decided to use less cumbersome notation.

5. Mild formulations

In this section we will heuristically derive a mild formulation of the solution to the stochastic nonhomogeneous boundary value problems to the wave equation. In Section 7 we will show that a mild solution is in fact a weak solution.

Assume now that a process u solves wave problem (1.2). As in Ref. 11 we consider a new process $y := u - D_\tau \frac{\partial \xi}{\partial t}$. Clearly $\tau y(t) = 0$ for $t > 0$ and

$$y(0) = u_0 - D_\tau \frac{\partial \xi}{\partial t}(0), \qquad \frac{\partial y}{\partial t}(0) = u_{0,1} - \frac{\partial}{\partial t} D_\tau \frac{\partial \xi}{\partial t}(0). \tag{5.1}$$

Next, by the definition of the map D_τ, we have

$$\frac{\partial^2 y}{\partial t^2} = \Delta y + \kappa D_\tau \frac{\partial \xi}{\partial t} - \frac{\partial^2}{\partial t^2} D_\tau \frac{\partial \xi}{\partial t}.$$

Let $(U_\tau(t))_{t\in\mathbb{R}}$ be the group generated by the operator $\mathcal{A}_\tau$ defined by equality (3.2). Let us put $z = \frac{\partial y}{\partial t}$. Then, for $t \geq 0$,

$$\begin{pmatrix} y \\ z \end{pmatrix}(t) = U_\tau(t) \begin{pmatrix} y(0) \\ z(0) \end{pmatrix}$$

$$+ \int_0^t U_\tau(t-r) \begin{pmatrix} 0 \\ \kappa D_\tau \frac{\partial \xi}{\partial r}(r) - \frac{\partial^2}{\partial r^2} D_\tau \frac{\partial \xi}{\partial r}(r) \end{pmatrix} \mathrm{d}r.$$

On the other hand, by the integration by parts formula, we have

$$-\int_0^t U_\tau(t-r) \begin{pmatrix} 0 \\ \frac{\partial^2}{\partial r^2} D_\tau \frac{\partial \xi}{\partial r}(r) \end{pmatrix} \mathrm{d}r = -\begin{pmatrix} 0 \\ \frac{\partial}{\partial t} D_\tau \frac{\partial \xi}{\partial t}(t) \end{pmatrix}$$

$$+ U_\tau(t) \begin{pmatrix} 0 \\ \frac{\partial}{\partial t} D_\tau \frac{\partial \xi}{\partial t}(0) \end{pmatrix} - \int_0^t \mathcal{A}_\tau U_\tau(t-r) \begin{pmatrix} 0 \\ \frac{\partial}{\partial r} D_\tau \frac{\partial \xi}{\partial r}(r) \end{pmatrix} \mathrm{d}r.$$

Set

$$Z(t) := -U_\tau(t) \begin{pmatrix} D_\tau \frac{\partial \xi}{\partial t}(0) \\ 0 \end{pmatrix} + \begin{pmatrix} D_\tau \frac{\partial \xi}{\partial t}(t) \\ 0 \end{pmatrix}.$$

Then, for $t \geq 0$,

$$\int_0^t \mathcal{A}_\tau U(t-r) \begin{pmatrix} 0 \\ \frac{\partial}{\partial r} D_\tau \frac{\partial \xi}{\partial r}(r) \end{pmatrix} \mathrm{d}r = \int_0^t U_\tau(t-r) \begin{pmatrix} \frac{\partial}{\partial r} D_\tau \frac{\partial \xi}{\partial r}(r) \\ 0 \end{pmatrix} \mathrm{d}r$$

$$= Z(t) + \int_0^t \mathcal{A}_\tau U_\tau(t-r) \begin{pmatrix} D_\tau \frac{\partial \xi}{\partial r}(r) \\ 0 \end{pmatrix} \mathrm{d}r$$

$$= Z(t) + \int_0^t \mathcal{A}_\tau^2 U_\tau(t-r) \begin{pmatrix} 0 \\ D_\tau \frac{\partial \xi}{\partial r}(r) \end{pmatrix} \mathrm{d}r.$$

Hence, in view of the equalities appearing in (5.1), we arrive at the following identity

$$\begin{pmatrix} y \\ z \end{pmatrix}(t) = U_\tau(t) \begin{pmatrix} u_0 \\ u_{0,1} \end{pmatrix} - \begin{pmatrix} D_t \frac{\partial \xi}{\partial t}(t) \\ \frac{\partial}{\partial t} D_\tau \frac{\partial \xi}{\partial t}(t) \end{pmatrix}$$

$$+ \int_0^t (\kappa - \mathcal{A}_\tau^2) U_\tau(t-r) \begin{pmatrix} 0 \\ D_\tau \end{pmatrix} \frac{\partial \xi}{\partial r}(r) \mathrm{d}r, \qquad t \geq 0.$$

Putting $v(t) := \frac{\partial u(t)}{\partial t}$ we observe that

$$v(t) = \frac{\partial y(t)}{\partial t} + \frac{\partial}{\partial t} D_\tau \frac{\partial \xi}{\partial t}(t) = z(t) + \frac{\partial}{\partial t} D_\tau \frac{\partial \xi}{\partial t}(t),$$

and hence we obtain (3.4).

6. Scales of Hilbert spaces

Let $(A, \mathrm{D}(A))$ be the infinitesimal generator of an analytic semigroup S on a real separable Hilbert space H. Let $\tilde{\kappa}$ belongs to the resolvent set of A. Then, see e.g. Ref. 21, the fractional power operators $(\tilde{\kappa} - A)^s$, $s \in \mathbb{R}$, are well defined.

In particular, for $s < 0$, $(\tilde{\kappa} - A)^s$ is a bounded linear operator and for $s > 0$, $(\tilde{\kappa} - A)^s := ((\tilde{\kappa} - A)^{-s})^{-1}$. For any $s \geq 0$ we set $H_s := D(\tilde{\kappa} - A)^{s/2} - R(\tilde{\kappa} - A)^{-s/2}$. We equip the space H_s with the norm

$$\|f\|_{H_s} := \|(\tilde{\kappa} - A)^{s/2} f\|_H, \qquad f \in H_s.$$

Note that for all $s, r \geq 0$, $(\tilde{\kappa} - A)^{r/2}: H_{s+r} \mapsto H_s$ is an isometric isomorphism.

We also introduce the spaces H_s for $s < 0$. To do this let us fix $s < 0$. Note that the operator $(\tilde{\kappa} - A)^{s/2}: H \mapsto H_s$ is an isometric isomorphism. Hence we can define H_s as the completion of H with respect to the norm $\|f\|_{H_s} := \|(\tilde{\kappa} - A)^{s/2} f\|_H$. We have

$$\|(\tilde{\kappa} - A)^{-s/2} f\|_{H_s} = \|f\|_H \qquad \text{for } f \in H_{-s}.$$

Thus, since H_{-s} is dense in H, $(\tilde{\kappa} - A)^{-s/2}$ can be uniquely extended to the linear isometry denoted also by $(\tilde{\kappa} - A)^{-s/2}$ between H_s and H.

Assume now that A is a self-adjoint non-positive definite linear operator in H. It is well known (and easy to see) that A considered on any H_s with $s < 0$ is essentially self-adjoint. We denote by A_s its unique self-adjoint extension. Note that $D(A_s) = H_{s+2}$. Finally, for any $s \geq 0$, the restriction A_s of A to H_{s+2} is a self-adjoint operator on H_s. The spaces H_s, $s < 0$, can be chosen in such a way that

$$H_s \hookrightarrow H_r \hookrightarrow H \hookrightarrow H_{-r} \hookrightarrow H_{-s}, \qquad \forall s \geq r \geq 0,$$

with all embedding dense and continuous. Identifying H with its dual space H' we obtain

$$H_s \hookrightarrow H \equiv H' \hookrightarrow H'_s, \qquad s \geq 0.$$

Remark 6.1. Under the identification above we have $H'_s = H_{-s}$. Moreover,

$$\langle A_s f, g \rangle = \langle f, A_{-s} g \rangle, \qquad \forall f \in D(A_s), \, g \in D(A_{-s}),$$

where $\langle \cdot, \cdot \rangle$ is the bilinear form on $H_s \times H_{-s}$ whose restriction to $(H_s \cap H) \times H$ is the scalar product on H.

Given $s \in \mathbb{R}$, define

$$\mathcal{H}_s := \begin{pmatrix} H_{s+1} \\ \times \\ H_s \end{pmatrix}.$$

On $\mathcal{H}_s$ we consider an operator $\mathcal{A}_s$ defined by the following formulas

$$\mathcal{A}_s := \begin{bmatrix} 0 & I \\ A_s & 0 \end{bmatrix}, \qquad D(\mathcal{A}_s) = \mathcal{H}_{s+1}.$$

By the Lumer–Philips theorem[24], $\mathcal{A}_s$ generates an unitary group U_s on $\mathcal{H}_s$.

6.1. Application to the boundary value problem

Let $-\Delta_D$ and $-\Delta_N$ be the Laplace operators on $H = L^2(\mathcal{O})$ with the homogeneous Dirichlet and Neumann boundary conditions, respectively. The corresponding scales of Hilbert spaces will be denoted by (H_s^D) and (H_s^N) and the restriction (or, if $s < 0$, the unique self-adjoint extension) of Δ_D and Δ_N to H_s^D and H_s^N by Δ_s^D and Δ_s^N. Finally, we write

$$\mathcal{H}_s^D := \begin{matrix} H_{s+1}^D \\ \times \\ H_s^D \end{matrix} \quad \text{and} \quad \mathcal{H}_s^N := \begin{matrix} H_{s+1}^N \\ \times \\ H_s^N \end{matrix}$$

and

$$\mathcal{A}_s^D := \begin{bmatrix} 0 & I \\ \Delta_s^D & 0 \end{bmatrix}, \quad D(\mathcal{A}_s^D) = \mathcal{H}_{s+1}^D,$$

$$\mathcal{A}_s^N := \begin{bmatrix} 0 & I \\ \Delta_s^N & 0 \end{bmatrix}, \quad D(\mathcal{A}_s^N) = \mathcal{H}_{s+1}^N.$$

Example 6.1. (*i*) Let $\mathcal{O} = (0,1)$ and $\kappa = 0$. Define functions ψ_i, $i = 1,2$ by $\psi_1(x) = 1$ and $\psi_2(x) = x$ for $x \in [0,1]$. Then, see e.g. Ref. 5, $D_D \colon \partial \mathcal{O} \equiv \mathbb{R}^2 \mapsto C([0,1])$ is given by

$$D_D \begin{pmatrix} a \\ b \end{pmatrix} = a\psi_1 + (b-a)\psi_2,$$

and, by Remark 6.1, see also Ref. 5, $\Delta_s^D \psi_1 = \frac{d}{dx}\delta_0 - \frac{d}{dx}\delta_1$ and $\Delta_s^D \psi_2 = -\frac{d}{dx}\delta_1$ for $s \leq -1$. Taking into account (3.3), for $s \geq 1$ we have

$$-(\mathcal{A}_{-s}^D)^2 \mathbf{D}_D \begin{pmatrix} a \\ b \end{pmatrix} = (-\Delta_{-s}^D D_D)^\dagger \begin{pmatrix} a \\ b \end{pmatrix} = \begin{pmatrix} 0 \\ -a\frac{d}{dx}\delta_0 + b\frac{d}{dx}\delta_1 \end{pmatrix}.$$

(*ii*) For the Neumann boundary conditions on $\mathcal{O} = (0,1)$ we take $\kappa = 1$. Then

$$D_N \begin{pmatrix} a \\ b \end{pmatrix} = \frac{b - ae}{e - e^{-1}}\psi_1 + \left(a + \frac{b - ae}{e - e^{-1}}\right)\psi_2, \quad \begin{pmatrix} a \\ b \end{pmatrix} \in \mathbb{R}^2 \equiv \partial\mathcal{O},$$

where $\psi_1(x) = e^{-x}$ and $\psi_2(x) = e^x$. Then, by Remark 6.1, for any $\phi \in D(\Delta_N^1)$, and $s \geq 1$,

$$(\Delta_{-s}^N \psi_i, \phi) = (\Delta_{-1}^N \psi_i, \phi) = \int_0^1 \psi_i(x) \frac{d^2\phi}{dx^2}(x) dx.$$

Since

$$\int_0^1 \psi_1(x) \frac{d^2\phi}{dx^2}(x) dx = \int_0^1 e^{-x} \frac{d^2\phi}{dx^2}(x) dx$$

$$= e^{-1}\frac{d\phi}{dx}(1) - \frac{d\phi}{dx}(0) + e^{-1}\phi(1) - \phi(0) + \int_0^1 e^{-x}\phi(x) dx$$

and
$$\int_0^1 \psi_2(x)\frac{d^2\phi}{dx^2}(x)dx = \int_0^1 e^x \frac{d^2\phi}{dx^2}(x)dx$$
$$= e\frac{d\phi}{dx}(1) - \frac{d\phi}{dx}(0) - e\phi(1) + \phi(0) + \int_0^1 e^x \phi(x)dx,$$

and since $\frac{d\phi}{dx}(0) = \frac{d\phi}{dx}(1) = 0$, it follows
$$\Delta^N_{-s}\psi_1 = e^{-1}\delta_1 - \delta_0 + \psi_1,$$
$$\Delta^N_{-s}\psi_2 = -e\delta_1 + \delta_0 + \psi_2.$$

Consequently,
$$\left(1 - \Delta^N_{-s}\right) D_N \begin{pmatrix} a \\ b \end{pmatrix}$$
$$= \frac{b-ae}{e-e^{-1}}\left(\delta_0 - e^{-1}\delta_1\right) + \left(a + \frac{b-ae}{e-e^{-1}}\right)(e\delta_1 - \delta_0)$$
$$= -a\delta_0 + (b-ae)\delta_1,$$

and hence
$$\left(1 - (\mathcal{A}^N_{-s})^2\right) \mathbf{D}_N \begin{pmatrix} a \\ b \end{pmatrix} = \left((1 - \Delta^N_{-s}) D_N\right)^\dagger \begin{pmatrix} a \\ b \end{pmatrix}$$
$$= \begin{pmatrix} 0 \\ -a\delta_0 + (b-ae)\delta_1 \end{pmatrix}.$$

(iii) Let $\mathcal{O} = (0,\infty)$ and $\kappa = 1$. Then, see e.g. Ref. 5, $D_D \colon \partial\mathcal{O} \equiv \mathbb{R} \mapsto C([0,\infty))$ is given by $D_D a = a\psi$, where $\psi(x) = e^{-x}$. Next, again by Remark 6.1, see also Ref. 5, $\Delta^D_{-s}\psi = \frac{d}{dx}\delta_0 + \psi$, $s \geq 1$, and consequently for any $s \geq 1$,
$$\left(1 - (\mathcal{A}^D_{-s})^2\right) \mathbf{D}_D(a) = \left((1 - \Delta^D_{-s}) D_D\right)^\dagger(a) = \begin{pmatrix} 0 \\ -a\frac{d}{dx}\delta_0 \end{pmatrix}.$$

(iv) For the Neumann boundary problem on $(0,\infty)$ with $\kappa = 1$ we have $D_N a = -a\psi$, where $\psi(x) = e^{-x}$. Then, $\Delta^N_{-s}\psi = \delta_0 + \psi$. Consequently,
$$\left(1 - (\mathcal{A}^N_{-s})^2\right) \mathbf{D}_N(a) = \left((1 - \Delta^N_{-s}) D_N\right)^\dagger(a) = \begin{pmatrix} 0 \\ -a\delta_0 \end{pmatrix}.$$

7. Equivalence of weak and mild solutions

We denote by τ either the Dirichlet or the Neumann boundary condition. We assume that D_τ is a corresponding boundary map, i.e. it satisfies
$$\Delta D_\tau \psi = \kappa D_\tau \psi, \qquad \tau D_\tau \psi = \psi \quad \text{on } \partial\mathcal{O}. \tag{7.1}$$

Recall that H_s^τ is the domain of $(\tilde\kappa-\Delta_\tau)^s$, hence in particular, H_{s+1}^τ is the domain of the Laplace operator Δ_s^τ considered on H_s^τ. We denote by $\left(U_s^\tau(t)\right)_{t\in\mathbb{R}}$ the C_0-group on $\mathcal{H}_s^\tau$ generated by $\mathcal{A}_s^\tau$.

The following existence theorem is the main result of this section.

Theorem 7.1. *Assume that $D_\tau\xi$ is a càdlàg process in a space H_s^τ for some $s \in \mathbb{R}$. Then for any $X_0 := (u_0, u_{0,1})^T \in \mathcal{H}_{s-4}^\tau$ there exists a unique weak solution u to problem (1.2) whose trajectories are H_{s-3}-valued càdlàg, with the boundary condition $\tau u = \dot\xi$. Moreover, the process $X = \left(u, \frac{\partial u}{\partial t}\right)^T$ is given by*

$$U_{s-3}^\tau(t)X_0 + \int_0^t U_{s-3}^\tau(t-r)\left((\kappa-\Delta_{s-2}^\tau)D_\tau\right)^\dagger \mathrm{d}\xi(r)$$

$$= U_{s-3}^\tau(t)X_0 + \int_0^t \mathcal{A}_{s-3}^\tau U_{s-2}^\tau(t-r)\left((\kappa-\Delta_{s-2}^\tau)D_\tau\right)^\dagger \xi(r)\mathrm{d}r \qquad (7.2)$$

$$+ \left((\kappa-\Delta_{s-2}^\tau)D_\tau\right)^\dagger \xi(t) - U_{s-2}^\tau(t)\left((\kappa-\Delta_{s-2}^\tau)D_\tau\right)^\dagger \xi(0), \qquad t \geq 0.$$

Proof. Clearly the process X defined by formula (7.2) is càdlàg in $\mathcal{H}_{s-3}^\tau$. We will show that it is a weak solution to problem (1.2). By the well-known equivalence result, see e.g. Ref. 10 or Ref. 25, for any $h \in \mathrm{D}\left(\left(\mathcal{A}_{s-3}^\tau\right)^*\right)$,

$$\langle X(t), h\rangle_{\mathcal{H}_{s-3}^\tau} = \langle X(0), h\rangle_{\mathcal{H}_{s-3}^\tau} + \int_0^t \langle X(r), \left(\mathcal{A}_{s-3}^\tau\right)^* h\rangle_{\mathcal{H}_{s-3}^\tau}\mathrm{d}r \qquad (7.3)$$

$$+ \langle h, \left((\kappa-\Delta_{s-2}^\tau)D_\tau\right)^\dagger(\xi(t)-\xi(0))\rangle_{\mathcal{H}_{s-3}^\tau}, \qquad t \geq 0.$$

Clearly,

$$\left(\mathcal{A}_{s-3}^\tau\right)^* = \begin{pmatrix} 0 & \Delta_{s-3}^\tau \\ I & 0 \end{pmatrix}.$$

Let $X = (u, v)^T$, and let $\varphi \in \mathcal{S}(\overline{\mathcal{O}})$ and $\psi \in \mathcal{S}(\overline{\mathcal{O}})$ be such that $\tau\psi = 0$ on $\partial\mathcal{O}$. Note that $h_1 := (\varphi, 0)^T$, $h_2 := (0, \psi)^T \in \mathrm{D}\left(\left(\mathcal{A}_{s-3}^\tau\right)^*\right)$. Applying equality (7.3) to $h = h_1$ we obtain

$$\langle u(t), \varphi\rangle_{H_{s-2}} = \langle u_0, \varphi\rangle_{H_{s-2}} + \int_0^t \langle v(r), \varphi\rangle_{H_{s-2}}\mathrm{d}r, \qquad t \geq 0.$$

Consequently, for $\tilde\varphi := (\kappa-\Delta)^{s-2}\varphi$,

$$(u(t), \tilde\varphi) = (u_0, \tilde\varphi) + \int_0^t (v(r), \tilde\varphi)\mathrm{d}r, \qquad t \geq 0.$$

Next, for $h = h_2$,

$$\langle v(t), \psi\rangle_{H_{s-3}} = \langle u_{0,1}, \psi\rangle_{H_{s-3}} + \int_0^t \langle u(r), \Delta\psi\rangle_{H_{s-3}}\mathrm{d}r + R_\tau(t), \ t \geq 0,$$

where, for $t \geq 0$,

$$R_\tau(t) := \langle \psi, \left(\kappa-\Delta_{s-2}^\tau\right)D_\tau\xi(t)\rangle_{H_{s-3}} - \langle \psi, \left(\kappa-\Delta_{s-2}^\tau\right)D_\tau\xi(0)\rangle_{H_{s-3}}.$$

Let $\tilde{\psi} := (\tilde{\kappa} - \Delta)^{s-3}\psi$. It remains to show that in the case of the Dirichlet boundary operator,
$$R_\tau(t) = \left(\xi(t) - \xi(0), \frac{\partial \tilde{\psi}}{\partial \mathbf{n}}\right)$$
and, in the case of the Neumann boundary operator,
$$R_\tau(t) = -\left(\xi(t) - \xi(0), \tilde{\psi}\right).$$
These two identities follow from an observation that by Definition 4.2, if z is such that $D_\tau z \in H_s$, then for any $\psi \in \mathcal{S}(\overline{\mathcal{O}})$ satisfying $\tau\psi = 0$,
$$(\psi, (\lambda - \Delta^\tau_{s-2})\, D_\tau z(t)) = \begin{cases} \left(z(t), \frac{\partial \psi}{\partial \mathbf{n}}\right) & \text{if } \tau \text{ is Dirichlet,} \\ -\left(\xi(t) - \xi(0), \tilde{\psi}\right) & \text{if } \tau \text{ is Neumann.} \end{cases}$$
$\square$

8. The Fundamental Solution

Let G_τ be the fundamental solution to the Cauchy problem for $\frac{\partial^2 u}{\partial t^2} = \Delta u$ associated with the boundary operator τ. In other words, $G_\tau : (0, \infty) \times \overline{\mathcal{O}} \times \overline{\mathcal{O}} \mapsto \mathbb{R}$ satisfies $\tau G_\tau(t, x, y) = 0$ with respect to x and y variables, $G_\tau(0, x, y) = 0$ and $\frac{\partial G_\tau}{\partial t}(0, x, y) = \delta_x(y)$,
$$\frac{\partial^2 G_\tau}{\partial t^2}(t, x, y) = \Delta_x G_\tau(t, x, y) = \Delta_y G_\tau(t, x, y), \quad t > 0, \ x, u \in \mathcal{O}.$$
Then the wave semigroup is given by, for $x \in \mathcal{O}$,
$$U_\tau(t)\begin{pmatrix} u_0 \\ u_{0,1} \end{pmatrix}(x)$$
$$= \begin{pmatrix} \int_\mathcal{O} \left(\frac{\partial}{\partial t} G_\tau(t, x, y) u_0(y) + G_\tau(t, x, y) u_{0,1}(y)\right) dy \\ \int_\mathcal{O} \left(\frac{\partial^2}{\partial t^2} G_\tau(t, x, y) u_0(y) + \frac{\partial}{\partial t} G_\tau(t, x, y) u_{0,1}(y)\right) dy \end{pmatrix}.$$
Hence, for $x \in \mathcal{O}$,
$$U_\tau(t)\left((\kappa - \Delta_\tau)\, D_\tau\right)^\dagger v(x)$$
$$= \begin{pmatrix} \int_\mathcal{O} G_\tau(t, x, y)(\kappa - \Delta_\tau) D_\tau v(y) dy \\ \int_\mathcal{O} \frac{\partial}{\partial t} G_\tau(t, x, y)(\kappa - \Delta_\tau) D_\tau v(y) dy \end{pmatrix}. \tag{8.1}$$

Let us now denote by σ the surface measure on $\partial \mathcal{O}$. The following result gives the formula for the solution to wave problem (1.2) in terms of the fundamental solution.

Theorem 8.1. *Assume that u is a solution to wave problem (1.2), where for simplicity $u_0 = 0 = u_{0,1}$.*
(i) If τ is the Dirichlet boundary operator, then the solution is given by
$$u(t, x) = -\int_0^t \int_{\partial \mathcal{O}} \frac{\partial G_\tau}{\partial \mathbf{n}_y}(t - s, x, y) v(y) d\xi(s)(y) \sigma(dy), \quad t \geq 0, \ x \in \mathcal{O}.$$

(ii) If τ is the Neumann boundary operator, then
$$u(t,x) = \int_0^t \int_{\partial \mathcal{O}} G_\tau(t-s,x,y) d\xi(s)(y)\sigma(dy), \qquad t \geq 0,\ x \in \mathcal{O}.$$

Proof. Let us assume that $u_0 = u_{0,1} = 0$. Then by (8.1), the solution to wave problem (1.2) is given by
$$u(t,x) = \int_0^t \mathcal{T}_\tau(t-s) d\xi(s), \qquad t \geq 0,\ x \in \mathcal{O},$$
where
$$\mathcal{T}_\tau(t)v(x) := \int_{\mathcal{O}} G_\tau(t,x,y)(\kappa - \Delta_\tau) D_\tau v(y) dy, \qquad t \geq 0,\ x \in \mathcal{O}.$$

Note that, first by Remark 6.1, and then by the fact that G_τ satisfies the boundary condition $\tau G_\tau(t,x,y) = 0$ with respect to y-variable, we obtain, for $t \geq 0$, $x \in \mathcal{O}$,
$$\mathcal{T}_\tau(t)v(x) := \int_{\mathcal{O}} G_\tau(t,x,y)(\kappa - \Delta_\tau) D_\tau v(y) dy$$
$$= \int_{\mathcal{O}} (\kappa - \Delta_\tau)_y G_\tau(t,x,y) D_\tau v(y) dy.$$

The Green formula and the fact that $(\kappa - \Delta) D_\tau v = 0$, yield then
$$\mathcal{T}_\tau(t)v(x) = \int_{\partial\mathcal{O}} \left(-\frac{\partial G}{\partial n_y}(t,x,y) D_\tau v(y) + G(t,x,y) \frac{\partial D_\tau v}{\partial n_y}(y) \right) \sigma(dy)$$
for all $t \geq 0$ and $x \in \mathcal{O}$. Hence if τ is the Dirichlet boundary operator, then
$$\mathcal{T}_\tau(t)v(x) = -\int_{\partial\mathcal{O}} \frac{\partial G_\tau}{\partial n_y}(t,x,y) v(y) \sigma(dy), \qquad t \geq 0,\ x \in \mathcal{O}.$$

If τ is the Neumann boundary operator, then repeating the previous argument we obtain
$$\mathcal{T}_\tau(t)v(x) = \int_{\partial\mathcal{O}} G_\tau(t,x,y) v(y) \sigma(dy), \qquad t \geq 0,\ x \in \mathcal{O}.$$
□

9. Applications

Assume that the process $\xi(t)(x)$, $t \geq 0$, $x \in \partial\mathcal{O}$, is of one of the following two forms
$$\xi(t)(x) = \sum_k \lambda_k W_k(t) e_k(x), \qquad t \geq 0,\ x \in \mathcal{O}, \qquad (9.1)$$
or
$$\xi(t)(x) = \sum_k Z_k(t) e_k(x), \qquad t \geq 0,\ x \in \mathcal{O}, \qquad (9.2)$$
where (λ_k) is a sequence of real numbers, (e_k) is a sequence of measurable functions on $\partial\mathcal{O}$, (W_k) is a sequence of independent real-valued standard Wiener processes,

and (Z_k) is a sequence of uncorrelated real-valued pure jump Lévy processes, that is

$$Z_k(t) = a_k t + \int_0^t \int_{\{|z|\leq 1\}} z\widehat{\pi}_k(\mathrm{d}s,\mathrm{d}z) + \int_0^t \int_{\{|z|>1\}} \pi(\mathrm{d}s,\mathrm{d}z), \qquad t \geq 0,$$

where π_k are Poisson random measures each with the jump measure ν_k.

In the jump case, let us set

$$\lambda_{k,R} := a_k + \nu_k\{1 < |z| \leq R\}, \qquad \tilde{\lambda}_{k,R} := \lambda_{k,R} + \nu_k\{0 < |z| \leq R\}.$$

Then, for all k and R,

$$Z_k(t) = \lambda_{k,R} t + M_{k,R}(t) + \int_0^t \int_{\{|z|>R\}} \pi(\mathrm{d}s,\mathrm{d}z), \qquad t \geq 0,$$

where

$$M_{k,R}(t) := \int_0^t \int_{\{|z|\leq R\}} z\widehat{\pi}_k(\mathrm{d}s,\mathrm{d}z), \qquad t \geq 0.$$

Let

$$\tau_{k,R} := \inf\{t \geq 0 \colon |Z_k(t) - Z_k(t-)| \geq R\}.$$

Then $\tau_{k,R} \uparrow +\infty$ as $R \uparrow +\infty$. Moreover,

$$Z_k(t) = \lambda_{k,R} t + M_{k,R}(t) \quad \text{on } \{\tau_{k,R} \geq t\}.$$

Recall, see e.g. Ref. 25 [Lemma 8.22] or Ref. 26, that $M_{k,R}$ are square integrable martingales and that for any predictable process f and any $T > 0$,

$$\mathbb{E} \sup_{t \in [0,T]} \left| \int_0^t f(s)\mathrm{d}M_{k,R}(s) \right| \leq C \nu_k\{0 < |z| \leq R\} \int_0^T \mathbb{E}|f(s)|\mathrm{d}s,$$

and

$$\mathbb{E} \sup_{t \in [0,T]} \left| \int_0^t f(s)\mathrm{d}M_{k,R}(s) \right|^2 \leq C \int_{\{0<|z|\leq R\}} z^2 \nu_k(\mathrm{d}z) \int_0^T \mathbb{E}|f(s)|^2 \mathrm{d}s,$$

where C is a certain (independent of f, T, R and π_k) universal constant.

Recall that G_τ is the fundamental solution to the Cauchy problem for the wave equation with boundary operator τ. It turns out that the Dirichlet problem has a function valued solution if ξ is absolutely continuous. Therefore we present below the result on the Neumann problem. For specific examples see e.g. Refs. 7–9,19.

Theorem 9.1. *Assume that τ is the Neumann boundary operator. (i) If ξ is given by (9.1), then the solution u to (1.2) is a square integrable random field on $[0, +\infty) \times \mathcal{O}$ if and only if*

$$I(t,x) := \sum_k \lambda_k^2 \int_0^t \left| \int_{\partial \mathcal{O}} G_\tau(s,x,y) e_k(y) \sigma(\mathrm{d}y) \right|^2 \mathrm{d}s < \infty, \ t \geq 0, \ x \in \mathcal{O}.$$

Moreover,

$$u(t,x) = \sum_k \lambda_k \int_0^t \int_{\partial\mathcal{O}} G_\tau(t-s,x,y)e_k(y)d\sigma(y)dW_k(s), \quad t \geq 0, \ x \in \mathcal{O}$$

and $\mathbb{E}\,|u(t,x)|^2 \leq I(t,x)$ for all $t \geq 0$ and $x \in \mathcal{O}$.
(ii) If ξ is given by (9.2), then for all $R > 0$,

$$J_R(t,x) := \sum_k \tilde{\lambda}_{k,R} \int_0^t \left| \int_{\partial\mathcal{O}} G_\tau(s,x,y)e_k(y)\sigma(dy) \right| ds < \infty.$$

and the solution u to (1.2) is a random field if $\tau_R \to +\infty$ as $R \to +\infty$. Moreover,

$$u(t,x) = -\sum_k \int_0^t \int_{\partial\mathcal{O}} G_\tau(t-s,x,y)e_k(y)\sigma(dy)dZ_k(s), \quad t \geq 0, \ x \in \mathcal{O}$$

and $\mathbb{E}\,|u(t,x)|\chi_{\{t \leq \tau_R\}} \leq C J_R(t,x)$ for all $t > 0$, $x \in \mathcal{O}$, and $R > 0$.

Example 9.2. As an example of the wave equation with stochastic Dirichlet boundary condition, consider $d = 1$ and $\mathcal{O} = (0, +\infty)$. Then the fundamental solution G is given by

$$G(t,x,y) = \frac{1}{2}\left(\chi_{\{|x-y|<t\}} - \chi_{\{|x+y|<t\}}\right), \quad t \geq 0, \ x,y \geq 0.$$

Then

$$-\frac{\partial G}{\partial \mathbf{n}_y}(t,x,0) = \delta_t(x), \quad t,x \geq 0,$$

and, by Theorem 8.1, see also Example 6.1(iii), the solution to (1.2) with $u_0 = u_{0,1} = 0$, is given by

$$u(t,x) = \int_0^t \delta_{t-s}(x)d\xi(s), \quad t,x \geq 0.$$

In general, process $u = \big(u(t)\big)_{t\geq 0}$ takes values in a proper space of distributions. In fact, for a test function φ,

$$(u(t),\varphi) = \int_0^t \varphi(t-s)d\xi(s), \quad t \geq 0.$$

For similar results in the case of the transport equation see Examples 10.3 and 10.4.

10. The Transport equation

Let us describe an abstract framework in which we will study problem (1.3). Namely, in this case we put $\mathcal{H} = L^2(0, 2\pi)$ and $\mathcal{U} = \mathbb{R}$. We will identify $\mathcal{H}$ with the space of all locally square integrable 2π periodic functions $f\colon \mathbb{R} \mapsto \mathbb{R}$. Alternatively, we can identify $\mathcal{H}$ with the space $L^2(S^1)$, where S^1 is the standard unit circle equipped with the Haar measure (multiplied by 2π). In the space $\mathcal{H}$ we consider an operator $\mathcal{A}$ defined by

$$\mathrm{D}(\mathcal{A}) = H^{1,2}_{\mathrm{per}}(0, 2\pi), \qquad \mathcal{A}u = \frac{\mathrm{d}u}{\mathrm{d}x}, \tag{10.1}$$

where $H^{1,2}(0, 2\pi)$ is the Sobolev space of functions $u \in L^2(0, 2\pi)$ with the weak derivative $\frac{\mathrm{d}u}{\mathrm{d}x} \in L^2(0, 2\pi)$, and

$$H^{1,2}_{\mathrm{per}}(0, 2\pi) = \{u \in H^{1,2}(0, 2\pi) \colon u(0+) = u(2\pi-)\}.$$

It is easy to see that $\mathcal{A}$ generates a C_0-group $(U(t))_{t \in \mathbb{R}}$ in $\mathcal{H}$. In fact, this group is the standard translation group defined by

$$U(t)u(x) = u(t \dotplus x), \qquad t \in \mathbb{R},\ x \in (0, 2\pi), \tag{10.2}$$

where $\dotplus$ is the addition modulo 2π. Let τ be the boundary operator defined by

$$H^{1,2}(0, 2\pi) \ni u \mapsto \tau u = u(2\pi-) - u(0+) \in \mathbb{R}. \tag{10.3}$$

Note that by the Sobolev embedding theorem $H^{1,2}(0, 2\pi) \hookrightarrow C([0, 2\pi])$ and hence $u(2\pi-)$ and $u(0+)$ make sense for each $u \in H^{1,2}(0, 2\pi)$. Finally, we assume that $\xi = (\xi(t))_{t \geq 0}$ is an $\mathbb{R}$-valued càdlàg process defined on some complete filtered probability space $(\Omega, \mathcal{F}, (\mathcal{F}_t), \mathbb{P})$.

With all these notation we can now present an abstract form of problem (1.3), i.e.

$$\begin{cases} \dfrac{\partial u(t)}{\partial t} = \mathcal{A}u(t) & \text{for } t \in (0, \infty), \\ u(0, \cdot) = u_0 & \text{on } (0, 2\pi), \\ \tau[u(t)] = \dfrac{\mathrm{d}\xi(t)}{\mathrm{d}t} & \text{for } t \in (0, \infty). \end{cases} \tag{10.4}$$

The problem needs to be reformulated as, for example, on the one hand there is an expression $\mathcal{A}u(t)$ and on the other hand $\tau[u(t)]$ may be different from 0, and hence $u(t)$ may not belong to $\mathrm{D}(\mathcal{A})$. In order to give a proper definition of a mild solution to the above problem we will argue heuristically as in Section 5. For this we need to introduce a counterpart of the Dirichlet map D_τ from Section 4. To this aim let $z_0 \in H^{1,2}(0, 2\pi)$ be the unique solution of the following problem

$$\frac{\mathrm{d}}{\mathrm{d}x} z_0(x) = z_0(x),\ x \in (0, 2\pi), \qquad z_0(2\pi) - z_0(0) = 1.$$

Thus $z_0(x) = (\mathrm{e}^{2\pi} - 1)^{-1}\, \mathrm{e}^x$, $x \in (0, 2\pi)$.

Then we define a map $D_\tau \colon \mathbb{R} \ni \alpha \mapsto \alpha z_0 \in H^{1,2}(0, 2\pi)$. Note that the map D_τ satisfies the following

$$u \in H^{1,2}(0, 2\pi),\ \frac{\mathrm{d}u}{\mathrm{d}x} = u \text{ on } (0, 2\pi) \text{ and } \tau(u) = \alpha \iff D_\tau(\alpha) = u.$$

In other words, $\tau \circ D_\tau$ is the identity operator on $\mathbb{R}$.

Let u be a solution to (10.1). As in Section 5 we consider a new process y defined by the following formula

$$y(t) := u(t) - D_\tau \frac{d\xi}{dt}(t), \qquad t > 0. \tag{10.5}$$

Clearly $\tau y(t) = 0$ for $t > 0$ and $y(0) = u_0 - D_\tau \frac{d\xi}{dt}(0)$. Next we have

$$\begin{aligned}
\frac{\partial y}{\partial t} &= \frac{\partial u}{\partial t} - \frac{\partial}{\partial t}\left(D_\tau \frac{d\xi}{dt}(t)\right) \\
&= \mathcal{A}\left(y(t) + D_\tau \frac{d\xi}{dt}(t)\right) - \frac{\partial}{\partial t}\left(D_\tau \frac{d\xi}{dt}(t)\right) \\
&= \mathcal{A}y(t) + \left[D_\tau \frac{d\xi}{dt}(t) - \frac{\partial}{\partial t}\left(D_\tau \frac{d\xi}{dt}(t)\right)\right].
\end{aligned}$$

Continuing as in Section 5 we obtain

$$y(t) = U(t)y(0) + \int_0^t U(t-s)\left[D_\tau \frac{d\xi}{dr}(r) - \frac{\partial}{\partial r}\left(D_\tau \frac{d\xi}{dr}(r)\right)\right] dr.$$

On the other hand by integration by parts formula

$$-\int_0^t U(t-r) \frac{\partial}{\partial r}\left(D_\tau \frac{d\xi}{dr}(r)\right) dr = -D_\tau \frac{d\xi}{dt}(t)$$
$$+ U(t) D_\tau \frac{d\xi}{dt}(0) - \int_0^t \mathcal{A} U(t-r) D_\tau \frac{d\xi}{dr}(r) dr.$$

Hence, we infer that

$$y(t) = U(t)\left(y(0) + D_\tau \frac{d\xi}{dt}(0)\right) - D_\tau \frac{d\xi}{dt}(t)$$
$$+ \int_0^t (I - \mathcal{A}) U(t-r) \left[D_\tau \frac{d\xi}{dr}(r)\right] dr, \quad t \geq 0.$$

Therefore, in view of (10.5) we arrive at the following heuristic formula

$$u(t) = U(t)u_0 + \int_0^t (I - \mathcal{A}) U(t-r) \left[D_\tau \frac{d\xi}{dr}(r)\right] dr, \quad t \geq 0.$$

Since $(\mathcal{A}^*)'$ is an extension of $\mathcal{A}$, to obtain a meaningful version of the above equation one should write

$$u(t) = U(t)u_0 + \int_0^t (I - (\mathcal{A}^*)') U(t-r) \left[D_\tau \frac{d\xi}{dr}(r)\right] dr, \quad t \geq 0, \tag{10.6}$$

where U is the extension of the original group to $(D(\mathcal{A}^*))'$.

Summing up, we have shown (for more details see Ref. 4) that the transport problem can be written in the abstract form (1.1) with $E = D_\tau$ and $\lambda = 1$. It is known (see Ref. 4) that the condition (2.1) is satisfied.

It follows from Ref. 4 [Proposition 1.1, p.459] that the above integral defines a function belonging to $C([0, \infty); \mathcal{H})$ provided $\frac{d\xi}{dr} \in L^2_{\text{loc}}(0, \infty)$. However, we are interested in cases when this condition is no longer satisfied. Hence, if we perform integration by parts in the integral in (10.6) we obtain the following Anzatz for the solution to problem (10.1):

$$u(t) = U(t)u_0 + (I - (\mathcal{A}^*)') \left[D_\tau \xi(t) - U(t) D_\tau \xi(0) \right]$$
$$+ \int_0^t (I - \mathcal{A}^*)') \mathcal{A} U(t-r) \left[D_\tau \xi(r) \right] dr, \qquad t \geq 0. \tag{10.7}$$

Let us observe that formula (10.7) is a counterpart of formula (3.5) from Section 3. In some sense, this could be seen as a stochastic counterpart of a deterministic result from Ref. 4, see Proposition 1.1 on p. 459.

As far as the problem (10.1) is concerned, it remains to identify the space $(D(\mathcal{A}^*))'$ with an appropriate space of distributions. We have the following.

Proposition 10.3. *The space* $(D(\mathcal{A}^*))'$ *is equal to* $H^{-1,2}(0, 2\pi)$ *and the operator* $(\mathcal{A}^*)'$ *is equal to the weak derivative. In particular,*

$$((\mathcal{A}^*)'u, \varphi) = -\left(u, \frac{d\varphi}{dx}\right), \qquad u \subset \mathcal{H}, \varphi \in H^{1,2}(0, 2\pi).$$

Recall that $D_\tau(\alpha) = \alpha z_0$, where $z_0(x) = (e^{2\pi} - 1)^{-1} e^x$, $x \in (0, 2\pi)$. Thus, by the proposition above, for any $\varphi \in H^{1,2}(0, 2\pi)$,

$$((\mathcal{A}^*)'z_0, \varphi) = -\int_0^{2\pi} z_0(x) \frac{d\varphi}{dx}(x) dx$$
$$= -z_0(2\pi)\varphi(2\pi) + z_0(0)\varphi(0) + \int_0^{2\pi} z_0(x)\varphi(x) dx,$$

and hence

$$(I - (\mathcal{A}^*)') D_\tau[\alpha] = \left(\frac{e^{2\pi}}{e^{2\pi} - 1} \delta_{2\pi} - \frac{1}{e^{2\pi} - 1} \delta_0\right) \alpha. \tag{10.8}$$

Next not that, for any test function $\varphi \in H^{1,2}(0, 2\pi)$,

$$\left(\int_0^t U(t-s)\delta_{2\pi} d\xi(s), \varphi\right) = \int_0^t (\delta_{2\pi}, \varphi(t-s\dot{+}\cdot)) d\xi(s)$$
$$= \int_0^t \varphi(t - s\dot{+}2\pi) d\xi = \int_0^t \varphi(t - s\dot{+}0) d\xi$$
$$= \left(\int_0^t U(t-s)\delta_0 d\xi(s), \varphi\right).$$

Therefore, by (10.8),

$$\left(\int_0^t U(t-s)(I - (\mathcal{A}^*)') D_\tau d\xi(s), \varphi\right) = \int_0^t \varphi(t - s\dot{+}0) d\xi(s),$$

and in other words, we have the following result.

Proposition 10.4. *The solution u to problem (1.3) is an $H^{-1,2}(0,2\pi)$-valued process such that for any test function $\varphi \in H^{1,2}(0,2\pi)$,*

$$(u(t),\varphi) = (u_0,\varphi) + \int_0^t \varphi(t-s\dot{+}0)\mathrm{d}\xi(s), \qquad t \geq 0.$$

Example 10.3. Assume that ξ is a compound Poisson process defined on a probability space $(\Omega, \mathcal{F}, \mathbb{P})$ with the jump measure ν; that is

$$\xi(t) = \sum_{k=1}^{\Pi(t)} X_k, \qquad t \geq 0,$$

where Π is Poisson process with intensity $\nu(\mathbb{R})$ and X_k are independent random variable with the distribution $\nu/\nu(\mathbb{R})$. Let τ_k be the moments of jumps of Π. Then

$$\int_0^t \varphi(t-s\dot{+}0)\mathrm{d}\xi(s) = \sum_{\tau_k \leq t} \varphi(t-\tau_k\dot{+}0)X_k.$$

Since each τ_k has a absolutely continuous distribution (exponential), the formula can be extended to any $\varphi \in L^2(0,2\pi)$, and hence the solution is a cylindrical process in $L^2(0,2\pi)$; that is for each $t > 0$, $u(t)$ is a bounded linear operator from $L^2(0,2\pi)$ to $L^2(\Omega, \mathcal{F}, \mathbb{P})$.

Example 10.4. Let ξ be a Wiener process. Then again u is a distribution valued process and a cylindrical Gaussian random process in $L^2(0,2\pi)$.

References

1. E. Alòs and S. Bonaccorsi *Stability for stochastic partial differential equations with Dirichlet white-noise boundary conditions*, Infin. Dimens. Anal. Quantum Probab. Relat. Top. **5**, 465–481 (2002)
2. E. Alòs and S. Bonaccorsi, *Stochastic partial differential equations with Dirichlet white-noise boundary conditions*, Ann. Inst. H. Poincaré Probab. Statist. **38**, 125–154 (2002)
3. A. V. Balakrishnan, *Identification and stochastic control of a class of distributed systems with boundary noise*. CONTROL THEORY, NUMERICAL METHODS AND COMPUTER SYSTEMS MODELLING (Internat. Sympos., IRIA LABORIA, Rocquencourt, 1974), pp. 163–178, Lecture Notes in Econom. and Math. Systems, Vol. **107**, Springer, Berlin, 1975.
4. A. Bensoussan, G. Da Prato, M.C. Delfour, and S.K. Mitter, REPRESENTATION AND CONTROL OF INFINITE DIMENSIONAL SYSTEMS, Second edition. Systems & Control: Foundations & Applications. Birkhäuser Boston, Inc., Boston, MA, 2007.
5. Z. Brzeźniak, B. Goldys, G. Fabri, S. Peszat, and F. Russo, *Second order PDEs with Dirichlet white noise boundary conditions*, in preparation.
6. I. Chueshov and B. Schmalfuss, *Qualitative behavior of a class of stochastic parabolic PDEs with dynamical boundary conditions*, Discrete Contin. Dyn. Syst. **18**, 315–338 (2007)
7. R. Dalang and O. Lévêque, *Second order linear hyperbolic SPDE's driven by isotropic Gaussian noise on a sphere*, Ann. Probab. **32**, 1068–1099 (2004)

8. R. Dalang and O. Lévêque, *Second-order hyperbolic SPDE's driven by homogeneous Gaussian isotropic noise on a hyperplane*, Trans. Amer. Math. Soc. **358**, 2123–2159 (2006)
9. R. Dalang and O. Lévêque, *Second-order hyperbolic SPDE's driven by boundary noises*, Seminar on Stochastic Analysis, Random Fields and Applications IV, pp. 83–93, Progr. Probab., 58, Birkhuser, Basel, 2004.
10. G. Da Prato and J. Zabczyk, STOCHASTIC EQUATIONS IN INFINITE DIMENSIONS, Cambridge Univ. Press, Cambridge, 1992.
11. G. Da Prato and J. Zabczyk, *Evolution equations with white-noise boundary conditions*, Stochastics and Stochastics Rep. **42**, 167–182 (1993)
12. G. Da Prato and J. Zabczyk, ERGODICITY FOR INFINITE DIMENSIONAL SYSTEMS, Cambridge Univ. Press, Cambridge, 1996.
13. J. Duan and B. Schmalfuss, *The 3D quasigeostrophic fluid dynamics under random forcing on boundary*, Commun. Math. Sci. **1**, 133–151 (2003)
14. I. Lasiecka and R. Triggiani, *A cosine operator approach to modeling $L_2(0,T;L^2(\Gamma))$-boundary input hyperbolic equations*, Appl. Math. Optim. **7**, 35–93 (1981)
15. I. Lasiecka and R. Triggiani, *Regularity of hiperbolic equations under $L_2(0,T;L^2(\Gamma))$-Dirichlet boundary terms*, Appl. Math. Optim. **10**, 275–286 (1983)
16. I. Lasiecka and R. Triggiani, DIFFERENTIAL AND ALGEBRAIC RICCATI EQUATIONS WITH APPLICATIONS TO BOUNDARY/PIOINT CONTROL PROBLEMS, Lecture Notes in Control and Inform. Sci., vol. 165, Springer-Verlag, Berlin, Heidelberg, New York, 1991.
17. I. Lasiecka and R. Triggiani, CONTROL THEORY FOR PARTIAL DIFFERENTIAL EQUATIONS: CONTINUOUS AND APPROXIMATION THEORIES. II. ABSTRACT HYPERBOLIC-LIKE SYSTEMS OVER A FINITE TIME HORIZON, Encyclopedia of Mathematics and its Applications, 75. Cambridge Univ. Press, Cambridge, 2000.
18. I. Lasiecka, J. L. Lions, and R. Triggiani, *Non homogeneous boundary value problems for second order hyperbolic operators*, J. Math. Pures Appl. **65**, 149–192 (1986)
19. O. Lévêque, *Hyperbolic SPDE's driven by a boundary noise*, PhD Thesis 2452 (2001), EPF Lausanne.
20. J.L. Lions and E. Magenes, NON-HOMOGENEOUS BOUNDARY VALUE PROBLEMS AND APPLICATIONS I, Springer-Verlag, Berlin Heidenberg New York, 1972.
21. A. Lunardi, ANALYTIC SEMIGROUPS AND OPTIMAL REGULARITY IN PARABOLIC PROBLEMS, Birkhauser, 1995.
22. X. Mao and L. Markus, *Wave equations with stochastic boundary values*, J. Math. Anal. Appl. **177**, 315–341 (1993)
23. B. Maslowski, *Stability of semilinear equations with boundary and pointwise noise*, Ann. Scuola Norm. Sup. Pisa **22**, no. 1, 55–93 (1995)
24. A. Pazy, SEMIGROUPS OF LINEAR OPERATORS AND APPLICATIONS TO PARTIAL DIFFERENTIAL EQUATIONS, Springer, New York, 1983.
25. S. Peszat and J. Zabczyk, STOCHASTIC PARTIAL DIFFERENTIAL EQUATIONS DRIVEN BY LÉVY PROCESSES, Cambridge Univ. Press, Cambridge, 2007.
26. E. Saint Loubert Bié, *Étude d'une EDPS conduite par un bruit poissonien*, Probab. Theory Related Fields **111**, 287–321 (1998)
27. L. Schwartz, THÉORIE DES DISTRIBUTIONS I, II, Hermann & Cie., Paris, 1950, 1951.
28. R.B. Sowers, *Multidimensional reaction-diffusion equations with white noise boundary perturbations*, Ann. Probab. **22**, 2071–2121 (1994)

Chapter 2

Decoherent Information of Quantum Operations

Xuelian Cao, Nan Li[*] and Shunlong Luo[†]

School of Mathematics and Statistics, Huazhong University of Science and Technology, Wuhan 430074, China

Quantum operations (channels) are natural generalizations of transition matrices in stochastic analysis. A quantum operation usually causes decoherence of quantum states. Speaking in a broad sense, decoherence is the loss of some kind of correlations (in particular, entanglement). This is also recognized as the origin of the emergence of classicality. While this decoherence has been widely studied and characterized from various qualitative perspectives, there are relatively fewer quantitative characterizations. In this work, motivated by the notion of coherent information and consideration of correlating capability and transferring of correlations, we study an informational measure of decoherent capability of quantum operations in terms of quantum mutual information. The latter is usually regarded as a suitable measure of total correlations in a bipartite system. This measure possesses a variety of desirable properties required for quantitative characterizations of decoherent information, and is complementary to the coherent information introduced by Schumacher and Nielsen (Phys. Rev. A, **54**, 2629, 1996) in a loose sense. Apart from the significance in its own right, the decoherent information also provides an alternative and simple interpretation of, and sheds new light on, the coherent information. Several examples are worked out. A quantum Fano type inequality, an informational no-broadcasting result for bipartite correlations, and a continuity estimate for the decoherent information, are also established.

Contents

1 Introduction 24
2 Quantum mutual information and purification of mixed states 25
3 Decoherent information 26
4 No-broadcasting in terms of decoherent information 36
5 Continuity of the decoherent information 38
6 Discussion 40
References 41

---

[*]Academy of Mathematics and Systems Science, Chinese Academy of Sciences, Beijing 100190, China
[†]Academy of Mathematics and Systems Science, Chinese Academy of Sciences, Beijing 100190, China, luosl@amt.ac.cn

1. Introduction

Quantum states and quantum operations (channels) are fundamental objects in quantum theory,[20,22] and they are the natural quantum counterparts of probability densities and transition matrices in classical probability theory. The coupling between states, measurements and quantum operations is the basic starting point for generating probabilities of measurement outcomes and usually causes decoherence of quantum features. This decoherence is also the central phenomenon relating the quantum world to the classical realm.

Given a mixed state ρ, which is mathematically represented by a non-negative operator with unit trace on a Hilbert space, and a quantum operation $\mathcal{E}$, which is described by a trace-preserving completely positive linear mapping on quantum states,[20] in order to quantify the decoherence caused by the quantum operation, one naturally asks how the output state $\mathcal{E}(\rho)$ is different from the input state ρ, and furthermore, how the relations with their respective outside environments are changed.

We will distinguish two kinds of differences: The first is physical and the second is informational. As for the physical difference, there are many distance measures, such as the trace distance, the Hilbert-Schmidt distance, the Bures distance (a simple function of fidelity), etc.,[19,20] which quantify the formal difference between any two states. While these distance measures capture the formal difference of quantum states and play a significant role in the study of state disturbance, they are hardly directly useful in characterizing the intrinsic difference of *informational contents* of quantum states in the context of correlations, which lies in the heart of decoherence. To take an extreme example, any two unitarily equivalent states of the same quantum system have the same amount of informational contents and can be converted to each other by a unitary operation without loss of information, and thus should be considered informationally equivalent, but the above various distance measures fail to characterize this phenomenon.

As for the informational difference between quantum states, a seemingly necessary requirement for this kind of measure is that it should not distinguish any two states which can be converted to each other by quantum operations while preserving their correlations with other systems. In particular, the informational distance should be zero for any two states which are related by unitary operations.

In this work, we are concerned with informational difference and interested in how the operation $\mathcal{E}$ causes informational decoherence of a quantum state ρ in the context of correlations. The setup is as follows: From a fundamental point of view, any mixed state ρ can be interpreted as a marginal state of a pure state in a larger system (purification), that is, arising from the entanglement of the quantum system with an auxiliary system, and the informational content of ρ quantifies the correlating capability of the quantum system to the auxiliary system. Consider the task of sending the system state through a noisy channel in order to transmit the *initial correlations* (not just the initial system state). Since the auxiliary system is isolated and left undisturbed, the joint operation on the initial purified state is the tensor product of the system evolution $\mathcal{E}$ and the identity operation on the auxiliary system, and the final bipartite state will usually have less correlations. The purpose is to maintain correlations between the final system state and the auxiliary system as many as possible. In such a scenario, a fundamental and key quantity is the difference between the initial correlations and the final correlations. We interpret this quantity as a measure of decoherent information, with the correlations quantified by the quantum mutual information. It turns out that this measure is somewhat complementary to the celebrated coherent information introduced by Schumacher and Nielsen.[27] As a simple consequence, our approach simplifies their derivations and intuitions considerably, and furthermore, clarifies the true meaning of the coherent information as a hybrid quantity (which is however ambiguously interpreted in the literature). Moreover, the decoherent information, though rather trivially related to the coherent information from the mathematical viewpoint, is conceptually more intuitive, and can be more easily manipulated. Among other applications of the notion of the decoherent information, we evaluate the decoherent information for several important quantum operations, establish a quantum Fano type inequality, an informational no-broadcasting theorem, and a continuity estimate for the decoherent information.

The remaining part of this work is structured as follows. In Sect. 2, we review briefly the notions of quantum mutual information and purification, which will play crucial roles in our approach. In Sect. 3, working in the framework of Schumacher and Nielsen,[27] and Adami and Cerf,[1] we introduce a measure of decoherent information and enumerate its fundamental properties, which are essentially reformulations of the results in Refs. 1 and 27. However, the simple and intuitive derivations here do not depend on their results. We will emphasize the complementary relation between the decoherent information and the coherent information. Our main results in this section are the explicit evaluations of the decoherent information for several important quantum operations (channels). A quantum Fano type inequality is also established. As an interesting application of the decoherent information, we establish in Sect. 4 an informational no-broadcasting theorem. We provide a continuity estimate for the decoherent information in Sect. 5. Finally, we conclude with discussion in Sect. 6.

2. Quantum mutual information and purification of mixed states

For any bipartite quantum state ρ^{ab} of a composite quantum system $H^a \otimes H^b$ with marginal states (partial traces)

$$\rho^a = \text{tr}_b \rho^{ab}, \qquad \rho^b = \text{tr}_a \rho^{ab},$$

its quantum mutual information is defined as[1,30]

$$I(\rho^{ab}) := S(\rho^a) + S(\rho^b) - S(\rho^{ab}).$$

Here $S(\rho^a) := -\mathrm{tr}\rho^a \log\rho^a$ is the quantum entropy (von Neumann entropy) and the logarithm may be taken to any base larger than 1. Quantum mutual information is widely recognized and used as a natural measure of total correlations in a bipartite state, and there are a variety of mathematical, informational as well as physical arguments supporting this belief and usage.[1,10,16,29,30] In particular, see Ref. 16 for a concise and informative review.

A fundamental property of the quantum mutual information is the decreasing property under local quantum operations. More precisely, let $\mathcal{E}^a$ and $\mathcal{E}^b$ be quantum operations on the systems H^a and H^b, respectively, and put $\mathcal{E}^{ab} := \mathcal{E}^a \otimes \mathcal{E}^b$, then

$$I(\rho^{ab}) \geq I(\mathcal{E}^{ab}(\rho^{ab})).$$

This is a particular instance of the monotonicity of quantum relative entropy, which is an extremely important and fundamental property.[13,30]

In quantum mechanics, there are several interpretations of the notion of a density operator (mixed state). We will regard it as the marginal state of a pure state in a larger system. This has both the mathematical convenience and physical significance, and is the departure of our study of decoherence (of correlations) in this chapter. See Ref. 18 for a concise review of the status of density operators.

For any state ρ^b of a quantum system H^b (the unusual superscript here is for latter convenience), there is a pure state $|\Psi^{ab}\rangle$ of a composite system $H^a \otimes H^b$ such that

$$\rho^b = \mathrm{tr}_a |\Psi^{ab}\rangle\langle\Psi^{ab}|.$$

Here H^a is an auxiliary system. Moreover, ρ^b and $\rho^a = \mathrm{tr}_b |\Psi^{ab}\rangle\langle\Psi^{ab}|$ have the same quantum entropy. The most natural construction of a purification is as follows:[20] Let the spectral decomposition of ρ^b be

$$\rho^b = \sum_i \lambda_i |\psi_i^b\rangle\langle\psi_i^b|,$$

then

$$|\Psi^{ab}\rangle = \sum_i \sqrt{\lambda_i} |\psi_i^a\rangle \otimes |\psi_i^b\rangle$$

is a particular purification of ρ^b. Here $|\psi_i^a\rangle = |\psi_i^b\rangle$. Notice that purification is not unique, but this will have no consequence here because our approach will be independent of the purifications.

3. Decoherent information

Now consider a state ρ^b of a quantum system H^b and a quantum operation $\mathcal{E}^b$, thus with the input state ρ^b, we get the output state $\mathcal{E}^b(\rho^b)$. In order to quantify the informational decoherence caused by $\mathcal{E}^b$ on the state ρ^b, we compare the correlating capability of the original state ρ^b with an auxiliary system and that of the final state after the operation. For this purpose, we put the quantum system in the purified context, and consider the quantum state ρ^b as the marginal state of a larger system.[1,27] Thus let $|\Psi^{ab}\rangle$ be a pure state in a composite quantum system $H^a \otimes H^b$ which purifies ρ^b, that is, $\rho^b = \text{tr}_a |\Psi^{ab}\rangle\langle\Psi^{ab}|$ (here H^a is an auxiliary system). Let $\mathcal{I}^a$ be the identity operation acting on the auxiliary system H^a. After the joint operation $\mathcal{I}^a \otimes \mathcal{E}^b$ acting on the joint pure state $|\Psi^{ab}\rangle$, the output state is

$$\rho^{a'b'} := \mathcal{I}^a \otimes \mathcal{E}^b(|\Psi^{ab}\rangle\langle\Psi^{ab}|)$$

with marginal states

$$\rho^{b'} := \text{tr}_{a'} \rho^{a'b'} = \mathcal{E}^b(\rho^b), \qquad \rho^{a'} := \text{tr}_{b'} \rho^{a'b'}.$$

Before the action of the quantum operation $\mathcal{E}^b$, the correlating capability of ρ^b is quantified by the quantum mutual information $I(\rho^{ab})$ in the bipartite state

$$\rho^{ab} := |\Psi^{ab}\rangle\langle\Psi^{ab}|,$$

and after the action, the correlating capability of the output state $\rho^{b'} = \mathcal{E}^b(\rho^b)$ in this context is quantified by the quantum mutual information $I(\rho^{a'b'})$. Consequently, the difference

$$\mathcal{D}(\rho^b, \mathcal{E}^b) := I(\rho^{ab}) - I(\rho^{a'b'})$$

quantifies the loss of correlating capability of ρ^b due to the operation (decoherence map) $\mathcal{E}^b$, and thus serves as a natural measure of the decoherence. We will call $\mathcal{D}(\rho^b, \mathcal{E}^b)$ the *decoherent information* (as opposed to the coherent information), for its own sake and further for a reason that will be transparent after we establish the complementary nature between it and the coherent information of Schumacher and Nielsen[27] in the subsequent Eq. (3.6). The decoherent information is an intrinsic quantity, depending only on ρ^b and $\mathcal{E}^b$, and is independent of the purifications.

To gain an intuition about this notion, let us first consider some extreme examples.

Example 1. Let U be any unitary operator on H^b and consider the unitary operation $\mathcal{E}^b(\rho^b) = U\rho^b U^\dagger$. Because the evolution is unitary, there is not any decoherence here. Consequently, we expect that the decoherent information in this situation should be zero. This is indeed the case since

$$I(\rho^{a'b'}) = I(\mathbf{1}^a \otimes U \rho^{ab} \mathbf{1}^a \otimes U^\dagger) = 2S(\rho^a) = I(\rho^{ab}),$$

from which we obtain

$$\mathcal{D}(\rho^b, \mathcal{E}^b) = 0.$$

Here $\mathbf{1}^a$ is the identity operator on H^a.

Example 2. Let $\mathcal{E}^b$ be the completely depolarizing operation which transforms every state into the maximally mixed one:

$$\mathcal{E}^b(\sigma) = \frac{\mathbf{1}^{b'}}{d'}$$

(here $\mathbf{1}^{b'}$ is the identity operator on $H^{b'}$, and d' is the dimension of $H^{b'}$). Then since this is the complete decoherence and we expect the decoherent information equals to the correlating capability of the original state, that is, $2S(\rho^b)$ (total correlations of the original purified state). This is indeed the case because

$$\rho^{a'b'} = \rho^{a'} \otimes \frac{\mathbf{1}^{b'}}{d'}$$

and thus $I(\rho^{a'b'}) = 0$, from which we obtain

$$\mathcal{D}(\rho^b, \mathcal{E}^b) = I(\rho^{ab}) - I(\rho^{a'b'}) = 2S(\rho^b),$$

and we see that all correlations are lost.

Now, we list some of the main properties of the decoherent information.

(1) For any quantum state ρ^b, it holds that

$$0 \leq \mathcal{D}(\rho^b, \mathcal{E}^b) \leq 2S(\rho^b).$$

(2) $\mathcal{D}(\rho^b, \mathcal{E}^b) = 0$ if and only if $\mathcal{E}^b$ is invertible on ρ^b, that is, there exists a quantum operation $\mathcal{E}^{b'}$ such that $\mathcal{E}^{b'} \circ \mathcal{E}^b(\rho^b) = \rho^b$.

(3) $\mathcal{D}(\rho^b, \mathcal{E}^b) = 2S(\rho^b)$ if $\mathcal{E}^b$ is a completely depolarization operation.

(4) Let $\mathcal{E}^b$ and $\mathcal{E}^{b'}$ be two quantum operations acting on the H^b system and the $H^{b'}$ system, respectively, then

$$\mathcal{D}(\rho^b, \mathcal{E}^b) \leq \mathcal{D}(\rho^b, \mathcal{E}^{b'} \circ \mathcal{E}^b).$$

For item (1), the first inequality follows from the monotonicity of the quantum mutual information under local operations,[13,30] and the second inequality is trivial since

$$\mathcal{D}(\rho^b, \mathcal{E}^b) = I(\rho^{ab}) - I(\rho^{a'b'}) \leq I(\rho^{ab}) = 2S(\rho^b).$$

Item (2) is also a direct consequence of the monotonicity of the quantum mutual information and the equality condition as specified by Petz et al.[11,23] Item (3) follows from direct calculations, as shown in Example 2.

Item (4) also follows immediately from the monotonicity of quantum mutual information. This may be interpreted as a data processing inequality.

Following Schumacher and Nielsen,[27] let us recall the coherent information

$$\mathcal{C}(\rho^b, \mathcal{E}^b) := S(\rho^{b'}) - S(\rho^{a'b'}), \tag{3.1}$$

which is an intrinsic quantity, depending only on ρ^b and $\mathcal{E}^b$, and is actually the minus of the quantum conditional entropy $S(\rho^{a'b'}|\rho^{b'}) := S(\rho^{a'b'}) - S(\rho^{b'})$. The coherent information is often interpreted as a measure of degree of entanglement retained by the systems H^a and H^b,[27] a measure of quantumness of the correlations in the final joint state. However, these interpretations are ambiguous and cannot be taken too seriously for two reasons: First, it can be negative, and second, as we will see after we establish Eq. (3.6), the coherent information, as the difference of the entanglement entropy and the decoherent information, is actually a hybrid quantity. Nevertheless, the coherent information is a fundamental quantity and plays a significant role in the study of quantum error corrections and quantum channel capacities.[3,5,7,8,12,17,26–28]

We will also need the notion of entropy exchange,[26,27] which is defined as the von Neumann entropy

$$S_e(\rho^b, \mathcal{E}^b) := S(\rho^{a'b'}) \tag{3.2}$$

of the final state $\rho^{a'b'}$. This is an intrinsic quantity measuring the information exchanged between the system and its exterior world. To see this more clearly, let us introduce an environment system H^c with an initial pure state $|\psi^c\rangle$ in order to unitarily dilate the quantum operation $\mathcal{E}^b$ as

$$\mathcal{E}^b(\rho^b) = \text{tr}_c(V(\rho^b \otimes |\psi^c\rangle\langle\psi^c|)V^\dagger). \tag{3.3}$$

Here V is a unitary operator on the composite system $H^b \otimes H^c$ (system plus environment). Then as a whole combining the quantum system H^b, the auxiliary system H^a purifying the quantum state ρ^b, and the environment H^c dilating the quantum operation $\mathcal{E}^b$, we have a tripartite system $H^a \otimes H^b \otimes H^c$ and the initial tripartite pure state

$$|\Psi^{abc}\rangle := |\Psi^{ab}\rangle \otimes |\psi^c\rangle,$$

which is driven by the unitary operator $\mathbf{1}^a \otimes V$ to the final *pure* state

$$|\Psi^{a'b'c'}\rangle := \mathbf{1}^a \otimes V(|\Psi^{ab}\rangle \otimes |\psi^c\rangle).$$

See Figure 2.1 for a schematic illustration.

Because $\rho^{a'b'c'} := |\Psi^{a'b'c'}\rangle\langle\Psi^{a'b'c'}|$ is pure, for the marginal states, we have $S(\rho^{a'b'}) = S(\rho^{c'})$, and consequently,

$$S_e(\rho^b, \mathcal{E}^b) = S(\rho^{a'b'}) = S(\rho^{c'}) \tag{3.4}$$

is the entropy of the final state of the environment, which in turn can be interpreted as the net increase of the entropy of the environment since the initial entropy of the environment is zero (note that the initial state of the environment $|\psi^c\rangle$ is pure). In this context, we may also rewrite the coherent information as

$$\mathcal{C}(\rho^b, \mathcal{E}^b) = S(\rho^{b'}) - S(\rho^{c'}).$$

This equation corroborates in a loose sense the interpretation of the coherent information as a measure of retained entanglement since $S(\rho^{b'})$ is an upper bound of the entanglement of $H^{b'}$ with other system and $S(\rho^{c'})$ quantifies the net information flow to the environment.

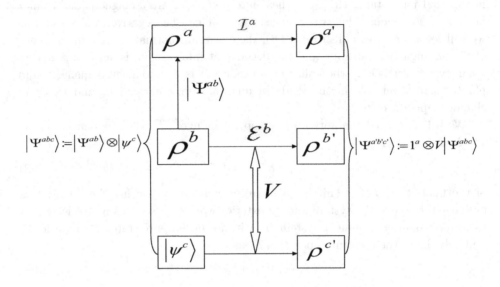

Figure 2.1. Tripartite purification of the quantum state ρ^b and the quantum operation $\mathcal{E}^b$. This is a combination of the purification of ρ^b and the unitarization of $\mathcal{E}^b$. Note that $\rho^{abc} := |\Psi^{abc}\rangle\langle\Psi^{abc}|$, $\rho^{ab} := |\Psi^{ab}\rangle\langle\Psi^{ab}|$, $\rho^{a'b'c'} := |\Psi^{a'b'c'}\rangle\langle\Psi^{a'b'c'}|$, $\rho^{a'b'} := (\mathcal{I}^a \otimes \mathcal{E}^b)(\rho^{ab})$, and various marginal states are obtained by taking partial traces, such as $\rho^{ab} = \text{tr}_c \rho^{abc}$, $\rho^a = \text{tr}_b \rho^{ab} = \text{tr}_{bc} \rho^{abc}$, $\rho^{a'} = \text{tr}_{b'} \rho^{a'b'} = \text{tr}_{b'c'} \rho^{a'b'c'}$, etc..

Combining Eqs. (3.1) and (3.2), we obtain

$$\mathcal{C}(\rho^b, \mathcal{E}^b) + S_e(\rho^b, \mathcal{E}^b) = S(\rho^{b'}). \tag{3.5}$$

Eq. (3.5) indicates that the coherent information and the entropy exchange are complementary to each other with respect to the *final* system state $\rho^{b'} = \mathcal{E}^b(\rho^b)$.

In contrast, it turns out that the decoherent information is complementary to the coherent information with respect to the *initial* system state ρ^b, thus indeed is a measure of lost coherent information. This is summarized in the following equality

$$\mathcal{C}(\rho^b, \mathcal{E}^b) + \mathcal{D}(\rho^b, \mathcal{E}^b) = S(\rho^b). \tag{3.6}$$

Since the von Neumann entropy $S(\rho^b)$ has a natural interpretation as the entanglement entropy of the initial purified state $|\Psi^{ab}\rangle$,[4,24] the above equation exhibits a decomposition of the entanglement quantity into two complementary parts: the decoherent information may be roughly interpreted as the lost, and the coherent information as the retained (which is in agreement with the original interpretation of Schumacher and Nielsen[27]). However, an irritating issue concerning the coherent information is that it may be negative, and thus the above interpretations of the coherent information are somewhat vague and ambiguous. In sharp contrast, the meaning of the decoherent information is clear: it is the loss of total correlations which may contain both classical and quantum parts.

To establish Eq. (3.6), note that $S(\rho^a) = S(\rho^b)$ (since $\rho^{ab} = |\Psi^{ab}\rangle\langle\Psi^{ab}|$ is pure) and $\rho^a = \rho^{a'}$ (since the evolution on the auxiliary system H^a is the identity), we have

$$\mathcal{D}(\rho^b, \mathcal{E}^b) = I(\rho^{ab}) - I(\rho^{a'b'})$$
$$= 2S(\rho^b) - (S(\rho^{a'}) + S(\rho^{b'}) - S(\rho^{a'b'}))$$
$$= S(\rho^b) - C(\rho^b, \mathcal{E}^b).$$

It is interesting to compare Eqs. (3.5) and (3.6). The former exhibits a decomposition of the final state entropy and the latter exhibits a decomposition of the initial state entropy. We also observe that the decoherent information and entropy exchange exhibit similar properties.

Consider the representation of the operation $\mathcal{E}^b$ in Eq. (3.3), we have a complementary quantum operation (channel)

$$\mathcal{E}^b_c(\rho^b) := \mathrm{tr}_b(V(\rho^b \otimes |\psi^c\rangle\langle\psi^c|)V^\dagger).$$

Due to the complementary nature of the two quantum operations $\mathcal{E}^b$ and $\mathcal{E}^b_c$, we expect that the decoherent information $\mathcal{D}(\rho^b, \mathcal{E}^b)$ and $\mathcal{D}(\rho^b, \mathcal{E}^b_c)$ should be complementary to each other in some sense. Indeed, we have the following identity

$$\mathcal{D}(\rho^b, \mathcal{E}^b) + \mathcal{D}(\rho^b, \mathcal{E}^b_c) = 2S(\rho^b), \tag{3.7}$$

which may be interpreted as an information conservation principle for the decoherent information of two complementary quantum operations.

To establish Eq. (3.7), note that $S(\rho^a) = S(\rho^b) = S(\rho^{a'})$, and since $\rho^{a'b'c'}$ is pure, we also have $S(\rho^{a'b'}) = S(\rho^{c'})$ and $S(\rho^{b'c'}) = S(\rho^{a'})$. Now by the definition of the decoherent information, we have

$$\mathcal{D}(\rho^b, \mathcal{E}^b) = I(\rho^{ab}) - I(\rho^{a'b'})$$
$$= 2S(\rho^b) - (S(\rho^{a'}) + S(\rho^{b'}) - S(\rho^{a'b'}))$$
$$= S(\rho^b) - S(\rho^{b'}) + S(\rho^{c'}),$$

and similarly,

$$\mathcal{D}(\rho^b, \mathcal{E}^b_c) = 2S(\rho^b) - I(\rho^{a'c'})$$
$$= 2S(\rho^b) - (S(\rho^{a'}) + S(\rho^{c'}) - S(\rho^{a'c'}))$$
$$= S(\rho^b) - S(\rho^{c'}) + S(\rho^{b'}).$$

Summing up the above equations yields the desired Eq. (3.7).

A particular interesting instance of Eq. (3.7) is that when $\mathcal{D}(\rho^b, \mathcal{E}^b) = 2S(\rho^b)$. In such a case, the H^b system is completely decoupled from the H^a system (the correlations between them are completely lost). However, we have $\mathcal{D}(\rho^b, \mathcal{E}^b_c) = 0$, which indicates that there is no decoherence between the H^a system and the environment H^c, or equivalently, the lost correlations between H^a and H^b reemerge as the correlations between $H^{a'}$ and the environment $H^{c'}$. In other words, the correlations, like incompressible fluid, are not destroyed, but only transferred to different systems.

Combining Eqs. (3.6) and (3.7), we have the following information conservation relation for the coherent information of two complementary quantum operations

$$\mathcal{C}(\rho^b, \mathcal{E}^b) + \mathcal{C}(\rho^b, \mathcal{E}^b_c) = 0.$$

We next relate the decoherent information to the entanglement fidelity, another important quantity.[26,27] Recall that the entanglement fidelity is defined as

$$F_e(\rho^b, \mathcal{E}^b) := \langle \Psi^{ab} | \rho^{a'b'} | \Psi^{ab} \rangle.$$

This quantity also characterizes how much entanglement is retained after the quantum operation. The quantum Fano inequality states that[20,26,27]

$$S_e(\rho^b, \mathcal{E}^b) \leq H(F_e) + (1 - F_e)\log(d^2 - 1). \quad (3.8)$$

Here $H(p) := -p\log p - (1-p)\log(1-p)$ is the binary Shannon entropy function, $F_e := F_e(\rho^b, \mathcal{E}^b)$ and d is the dimension of H^b. Now, we can formulate a quantum Fano type inequality for the decoherent information as

$$\mathcal{D}(\rho^b, \mathcal{E}^b) \leq S(\rho^b) - S(\rho^{b'}) + H(F_e) + (1 - F_e)\log(d^2 - 1) \quad (3.9)$$
$$\leq 2\Big(H(F_e) + (1 - F_e)\log(d^2 - 1)\Big). \quad (3.10)$$

In particular, if the quantum operation $\mathcal{E}^b$ is a von Neumann projective measurement, then

$$\mathcal{D}(\rho^b, \mathcal{E}^b) \leq H(F_e) + (1 - F_e)\log(d^2 - 1). \quad (3.11)$$

Inequality (3.9) follows from the combination of Eqs. (3.5) and (3.6) and inequality (3.8).

To prove inequality (3.10), note that by Eq. (3.4) and the quantum Fano inequality, we have

$$S(\rho^{c'}) \leq H(F_e) + (1 - F_e)\log(d^2 - 1).$$

Now by the fact that

$$S(\rho^b) = S(\rho^a) = S(\rho^{a'}) = S(\rho^{b'c'})$$

and the subadditivity of the quantum entropy $S(\rho^{b'c'}) \leq S(\rho^{b'}) + S(\rho^{c'})$, we obtain the desired result.

When $\mathcal{E}^b$ is a von Neumann projective measurement, we further have $S(\rho^b) \leq S(\rho^{b'})$, and the desired inequality (3.11) now follows from inequality (3.9).

By use of the decoherent information, we may recast the perfect error correction condition of Schumacher and Nielsen[27] in a more elegant way as

$$\mathcal{D}(\rho^b, \mathcal{E}^b) = 0.$$

Here perfect error correction means that there exists a further quantum operation $\mathcal{E}^{b'}$ which takes $\rho^{b'} := \mathcal{E}^b(\rho^b)$ to $\rho^{b''}$ such that the overall entanglement fidelity

$$F_e(\rho^b, \mathcal{E}^{b'} \circ \mathcal{E}^b) = 1.$$

We can give a simple proof of the above result without reliance on the original result of Schumacher and Nielsen,[27] whose proof is rather ingenious.

First, if $F_e(\rho^b, \mathcal{E}^{b'} \circ \mathcal{E}^b) = 1$, then $S(\rho^{b''}) = S(\rho^b)$, and by Eq. (3.10) (with $\mathcal{E}^{b'} \circ \mathcal{E}^b$ playing the role $\mathcal{E}^b$ there), we immediately obtain $\mathcal{D}(\rho^b, \mathcal{E}^{b'} \circ \mathcal{E}^b) = 0$, which implies that $\mathcal{D}(\rho^b, \mathcal{E}^b) = 0$ by the data processing inequality.

Conversely, if $\mathcal{D}(\rho^b, \mathcal{E}^b) = 0$, then

$$I(\rho^{ab}) = I(\rho^{a'b'}) = I(\mathcal{I}^a \otimes \mathcal{E}^b(\rho^{ab})).$$

Now by the equality condition of the monotonicity of relative entropy[11,23] (in particular, see Theorem 3 in Ref. [11]), there exists a quantum operation $\mathcal{E}^{b'}$ such that

$$\rho^{a''b''} := (\mathcal{I}^{a'} \otimes \mathcal{E}^{b'}) \circ (\mathcal{I}^a \otimes \mathcal{E}^b)(\rho^{ab}) = \rho^{ab},$$

which implies that $F_e(\rho^b, \mathcal{E}^{b'} \circ \mathcal{E}^b) = 1$.

We remark that by use of the decoherent information, it is natural to define the following quantity

$$\mathcal{D}(\mathcal{E}^b) := \mathcal{D}(1^b/d, \mathcal{E}^b)$$

as the decoherent information of the quantum operation $\mathcal{E}^b$. Here d is the dimension of H^b. The quantity $\mathcal{D}(\mathcal{E}^b)$ varies from zero for unitary operations to $2S(\rho^b)$ for the completely depolarizing channel. This notion should be useful in studying the decoupling capability of local quantum operations. Let us evaluate $\mathcal{D}(\mathcal{E}^b)$ for several widely used quantum operations (channels) on a qubit. In the following situations, we always have $\rho^a = \rho^b = \frac{1}{2}$ and we may take $|\Psi^{ab}\rangle = \frac{1}{\sqrt{2}}(|00\rangle + |11\rangle)$ such that

$$\rho^{ab} = |\Psi^{ab}\rangle\langle\Psi^{ab}| = \frac{1}{2}\begin{pmatrix} 1 & 0 & 0 & 1 \\ 0 & 0 & 0 & 0 \\ 0 & 0 & 0 & 0 \\ 1 & 0 & 0 & 1 \end{pmatrix}.$$

For any $p \in [0,1]$, recall that $H(p) = -p\log p - (1-p)\log(1-p)$ is the Shannon binary entropy function. The logarithm is taken to base 2 in the following calculations.

(1). Following Buscemi et al.,[6] a complete decohering channel acting on a qubit system is defined by the following quantum operation

$$\mathcal{E}^b(\sigma) = M \circ \sigma.$$

Here M is a correlation matrix, i.e., a non-negative definitive matrix with all diagonal elements being 1, and the product $\circ$ here denotes the Hadamard product (also called Schur product, entry-wise product of matrices). In particular, $\frac{1}{2}M$ may be regarded as a quantum state.

The most general 2×2 correlation matrix can be written as

$$M = \begin{pmatrix} 1 & \alpha \\ \bar{\alpha} & 1 \end{pmatrix}, \qquad |\alpha| < 1.$$

Now the final state

$$\rho^{a'b'} := \mathcal{I}^a \otimes \mathcal{E}^b(|\Psi^{ab}\rangle\langle\Psi^{ab}|) = \frac{1}{2}\begin{pmatrix} 1 & 0 & 0 & \alpha \\ 0 & 0 & 0 & 0 \\ 0 & 0 & 0 & 0 \\ \bar{\alpha} & 0 & 0 & 1 \end{pmatrix}.$$

The eigenvalues of this $\rho^{a'b'}$ are $\lambda_1 = \lambda_2 = 0, \lambda_3 = \frac{1-|\alpha|}{2}, \lambda_4 = \frac{1+|\alpha|}{2}$. Consequently,

$$\mathcal{D}(\mathcal{E}^b) = H\left(\frac{1-|\alpha|}{2}\right),$$

which turns out to be equal to the quantum entropy $S(\frac{1}{2}M)$.

(2). The Pauli channel is defined as

$$\mathcal{E}^b(\sigma) = p_0\sigma + p_1 X\sigma X + p_2 Y\sigma Y + p_3 Z\sigma Z,$$

where X, Y, Z are the Pauli spin matrices, and $\mathbf{p} = (p_0, p_1, p_2, p_3)$ is a probability distribution.

We have

$$\rho^{a'b'} := \mathcal{I}^a \otimes \mathcal{E}^b(|\Psi^{ab}\rangle\langle\Psi^{ab}|) = \frac{1}{2}\begin{pmatrix} p_0+p_3 & 0 & 0 & p_0-p_3 \\ 0 & p_1+p_2 & p_1-p_2 & 0 \\ 0 & p_1-p_2 & p_1+p_2 & 0 \\ p_0-p_3 & 0 & 0 & p_0+p_3 \end{pmatrix}.$$

The eigenvalues of $\rho^{a'b'}$ are p_0, p_1, p_2, p_3, consequently,

$$\mathcal{D}(\mathcal{E}^b) = -\sum_{i=0}^{3} p_i \log p_i.$$

The bit-flip channel and the depolarizing channel are special instances of the Pauli channel. In particular, the depolarizing channel

$$\mathcal{E}^b(\sigma) = p\sigma + (1-p)\frac{1}{2}\mathrm{tr}\sigma, \qquad -\frac{1}{3} \leq p \leq 1,$$

which is defined for any 2×2 matrix σ, can be viewed as a Pauli channel with

$$\mathbf{p} = \left(\frac{1+3p}{4}, \frac{1-p}{4}, \frac{1-p}{4}, \frac{1-p}{4}\right).$$

In this instance,

$$\mathcal{D}(\mathcal{E}^b) = 2 - \frac{1+3p}{4}\log(1+3p) - \frac{3-3p}{4}\log(1-p).$$

This will be compared with the next example.

(3). For the transpose depolarizing channel
$$\mathcal{E}^b(\sigma) = p\sigma^T + (1-p)\frac{1}{2}\mathrm{tr}\sigma, \qquad -1 \le p \le \frac{1}{3},$$
we have
$$\rho^{a'b'} = \mathcal{I}^a \otimes \mathcal{E}^b(|\Psi^{ab}\rangle\langle\Psi^{ab}|) = \frac{1}{2}\begin{pmatrix} \frac{1+p}{2} & 0 & 0 & 0 \\ 0 & \frac{1-p}{2} & p & 0 \\ 0 & p & \frac{1-p}{2} & 0 \\ 0 & 0 & 0 & \frac{1+p}{2} \end{pmatrix}.$$

The eigenvalues of $\rho^{a'b'}$ are
$$\lambda_1 = \lambda_2 = \lambda_3 = \frac{1+p}{4}, \quad \lambda_4 = \frac{1-3p}{4},$$
and therefore
$$\mathcal{D}(\mathcal{E}^b) = 2 - \frac{1-3p}{4}\log(1-3p) - \frac{3+3p}{4}\log(1+p).$$

(4). For the amplitude damping channel
$$\mathcal{E}^b(\sigma) = E_1 \sigma E_1^\dagger + E_2 \sigma E_2^\dagger$$
with $E_1 = \begin{pmatrix} 1 & 0 \\ 0 & \sqrt{1-p} \end{pmatrix}$, $E_2 = \begin{pmatrix} 0 & \sqrt{p} \\ 0 & 0 \end{pmatrix}$, $0 \le p \le 1$, we have
$$\rho^{a'b'} = \mathcal{I}^a \otimes \mathcal{E}^b(|\Psi^{ab}\rangle\langle\Psi^{ab}|) = \frac{1}{2}\begin{pmatrix} 1 & 0 & 0 & \sqrt{1-p} \\ 0 & 0 & 0 & 0 \\ 0 & 0 & p & 0 \\ \sqrt{1-p} & 0 & 0 & 1-p \end{pmatrix}.$$

with marginal states
$$\rho^{a'} = \frac{1}{2}\begin{pmatrix} 1 & 0 \\ 0 & 1 \end{pmatrix}, \quad \rho^{b'} = \frac{1}{2}\begin{pmatrix} 1+p & 0 \\ 0 & 1-p \end{pmatrix}.$$

The eigenvalues of $\rho^{a'b'}$ are $\lambda_1 = \lambda_2 = 0$, $\lambda_3 = \frac{p}{2}$, $\lambda_4 = 1 - \frac{p}{2}$. Consequently,
$$\mathcal{D}(\mathcal{E}^b) = 1 - H\left(\frac{1+p}{2}\right) + H\left(\frac{p}{2}\right).$$

(5). For the phase damping channel
$$\mathcal{E}^b(\sigma) = E_1 \sigma E_1^\dagger + E_2 \sigma E_2^\dagger$$
with $E_1 = \begin{pmatrix} 1 & 0 \\ 0 & \sqrt{1-p} \end{pmatrix}$, $E_2 = \begin{pmatrix} 0 & 0 \\ 0 & \sqrt{p} \end{pmatrix}$, $0 \le p \le 1$, we have
$$\rho^{a'b'} = \mathcal{I}^a \otimes \mathcal{E}^b(|\Psi^{ab}\rangle\langle\Psi^{ab}|) = \frac{1}{2}\begin{pmatrix} 1 & 0 & 0 & \sqrt{1-p} \\ 0 & 0 & 0 & 0 \\ 0 & 0 & 0 & 0 \\ \sqrt{1-p} & 0 & 0 & 1 \end{pmatrix}.$$

The eigenvalues of $\rho^{a'b'}$ are $\lambda_1 = \lambda_2 = 0$, $\lambda_3 = \frac{1-\sqrt{1-p}}{2}$, $\lambda_4 = \frac{1+\sqrt{1-p}}{2}$. Therefore,

$$\mathcal{D}(\mathcal{E}^b) = H\left(\frac{1-\sqrt{1-p}}{2}\right).$$

We compare the decoherent information between the depolarizing channel and the transpose depolarizing channel (see Figure 2.2), and that between the amplitude damping channel and the phase damping channel (see Figure 2.3). It is interesting to note that the amplitude damping channel is more decoherent than the phase damping channel with the same parameter p.

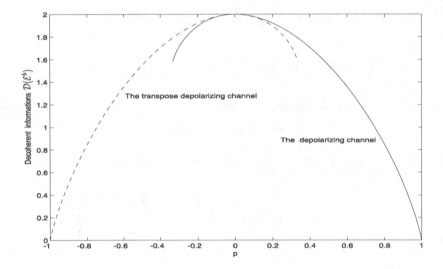

Figure 2.2. Graphs of the decoherent information $\mathcal{D}(\mathcal{E}^b)$ for the depolarizing channel and the transpose depolarizing channel versus the parameter p.

4. No-broadcasting in terms of decoherent information

In this section, we establish an informational no-broadcasting result for correlations. Just like no-cloning,[9,25,31] no-broadcasting is a fundamental characteristic of the quantum world and there are various characterizations.[2,14]

In the framework of the previous section, let $n \geq 2$ be any natural number and consider a quantum operation $\mathcal{E}^b$ which takes a mixed state ρ^b to an n-partite state $\rho^{b'} := \mathcal{E}^b(\rho^b)$ on the composite system $H^{b'_1} \otimes H^{b'_2} \otimes \cdots \otimes H^{b'_n}$. By taking partial traces over various subsystems of this composite state, we obtain n reduced states

$$\rho^{b'_k} := \mathrm{tr}_{\widehat{k}} \rho^{b'}, \qquad k = 1, 2, \cdots, n.$$

Here $\mathrm{tr}_{\widehat{k}}$ means taking partial trace with respect to all subsystems except the k-th

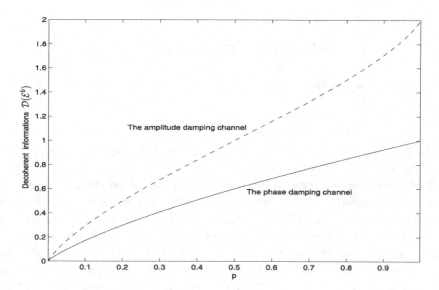

Figure 2.3. Graphs of the decoherent information $\mathcal{D}(\mathcal{E}^b)$ for the amplitude damping channel and the phase damping channel versus the parameter p.

one. Then we have n induced quantum operations
$$\mathcal{E}^{b_k} := \text{tr}_{\widehat{k}} \circ \mathcal{E}^b, \qquad k = 1, 2, \cdots, n.$$

Theorem 1. For $n \geq 2$, it holds that
$$\mathcal{D}(\rho^b, \mathcal{E}^{b_1}) + \mathcal{D}(\rho^b, \mathcal{E}^{b_2}) + \cdots + \mathcal{D}(\rho^b, \mathcal{E}^{b_n}) \geq nS(\rho^b). \tag{4.1}$$

To see this, by the fact $S(\rho^a) = S(\rho^{a'})$ and the definition of the decoherent information, we have
$$\mathcal{D}(\rho^b, \mathcal{E}^{b_1}) + \mathcal{D}(\rho^b, \mathcal{E}^{b_2}) + \cdots + \mathcal{D}(\rho^b, \mathcal{E}^{b_n})$$
$$= \sum_{k=1}^{n} \left(I(\rho^{ab}) - I(\rho^{a'b'_k}) \right)$$
$$= nI(\rho^{ab}) - \sum_{k=1}^{n} I(\rho^{a'b'_k})$$
$$= 2nS(\rho^a) - \sum_{k=1}^{n} \left(S(\rho^{a'}) - S(\rho^{a'b'_k}|\rho^{b'_k}) \right)$$
$$= nS(\rho^a) + \sum_{k=1}^{n} S(\rho^{a'b'_k}|\rho^{b'_k}),$$

where $S(\rho^{a'b'_k}|\rho^{b'_k}) := S(\rho^{a'b'_k}) - S(\rho^{b'_k})$ is the quantum conditional entropy (which may be negative). To prove inequality (4.1), it suffices to show that
$$\sum_{k=1}^{n} S(\rho^{a'b'_k}|\rho^{b'_k}) \geq 0.$$

But this follows from the fact that

$$2\sum_{k=1}^{n} S(\rho^{a'b'_k}|\rho^{b'_k}) = \sum_{k=1}^{n}\left(S(\rho^{a'b'_k}|\rho^{b'_k}) + S(\rho^{a'b'_{n+1-k}}|\rho^{b'_{n+1-k}})\right),$$

and each summand in the right hand side is nonnegative. The latter is due to the general inequality

$$S(\rho^{ab}|\rho^b) + S(\rho^{ac}|\rho^c) \geq 0,$$

which holds for any tripartite state ρ^{abc} and is actually equivalent to the strong subadditivity of the quantum entropy.

The above theorem puts an informational constraint on the ability of locally broadcasting the bipartite correlations between H^a and H^b to the n bipartite correlations between $H^{a'}$ and $H^{b'_k}$, $k = 1, 2, \cdots, n$. To gain an intuitive understanding of this, let $n = 2$ and consider the case $\mathcal{D}(\rho^b, \mathcal{E}^{b_1}) = 0$, which indicates that the correlations are perfectly transferred from $H^a \otimes H^b$ to $H^{a'} \otimes H^{b'_1}$. Now by the above theorem, we obtain $\mathcal{D}(\rho^b, \mathcal{E}^{b_2}) \geq 2S(\rho^b)$, but we clearly have $\mathcal{D}(\rho^b, \mathcal{E}^{b_2}) \leq 2S(\rho^b)$. Consequently, $\mathcal{D}(\rho^b, \mathcal{E}^{b_2}) = 2S(\rho^b)$, and this means that no correlations in $H^a \otimes H^b$ are transferred to $H^{a'} \otimes H^{b'_2}$.

5. Continuity of the decoherent information

We establish a continuity estimate for the decoherent information of quantum operation $\mathcal{D}(\mathcal{E}^b)$ in this section: If two quantum operations are close enough in some sense to be specified late, then their magnitudes of decoherent information are also close enough.

For any quantum operation

$$\mathcal{E}^b : \mathcal{S}(H^b) \to \mathcal{S}(H^{b'}),$$

its dual map

$$\mathcal{E}^{b*} : \mathcal{B}(H^{b'}) \to \mathcal{B}(H^b)$$

is defined via the duality relation

$$\mathrm{tr}\mathcal{E}^b(\rho)X = \mathrm{tr}\rho\mathcal{E}^{b*}(X), \qquad \rho \in \mathcal{S}(H^b),\ X \in \mathcal{B}(H^{b'}).$$

Here $\mathcal{B}(H^b)$ denotes the space of observables on the Hilbert space H^b. Let

$$\mathcal{E}_i^b : \mathcal{S}(H^b) \to \mathcal{S}(H^{b'}), \qquad i = 1, 2,$$

be two quantum operations, and put

$$\delta := \frac{1}{2}\|\mathcal{E}_1^{b*} - \mathcal{E}_2^{b*}\|_{\mathrm{cb}},$$

where the so-called norm of complete boundedness (cb-norm, for short) is defined as[21]

$$\|\mathcal{E}_1^{b*} - \mathcal{E}_2^{b*}\|_{\mathrm{cb}} := \sup_n \|\mathcal{I}_n^* \otimes (\mathcal{E}_1^{b*} - \mathcal{E}_2^{b*})\|_\infty. \tag{5.1}$$

Here $\mathcal{I}_n$ denotes the identity operation on the state space of an n-dimensional Hilbert space, and the norm $||\cdot||_\infty$ is the conventional operator norm for the dual maps of quantum operations considered as operators on the space of observables. In particular, from the duality relation between the operator norm $||\cdot||_\infty$ for any linear operator $\mathcal{K}$ on the space of observables $\mathcal{B}(H)$ and the trace norm $||\cdot||_1$ on $\mathcal{S}(H)$, we have

$$||\mathcal{K}||_\infty = \sup_{\rho \in \mathcal{S}(H)} ||\mathcal{K}^*(\rho)||_1. \tag{5.2}$$

Theorem 2. Let d be the dimension of H^b and $\delta := \frac{1}{2}||\mathcal{E}_1^{b*} - \mathcal{E}_2^{b*}||_{cb}$. If $\delta \in (0, 1 - \frac{1}{d})$, then

$$|\mathcal{D}(\mathcal{E}_1^b, \rho^b) - \mathcal{D}(\mathcal{E}_2^b, \rho^b)| \leq 2H(\delta) + \delta \log(d-1)(d^2-1). \tag{5.3}$$

Here $H(x) := -x\log x - (1-x)\log(1-x)$ is the binary Shannon entropy function.

To establish the above result, let $\rho^{ab} = |\Psi^{ab}\rangle\langle\Psi^{ab}|$ with $|\Psi^{ab}\rangle \in H^a \otimes H^b$ a purification of $\rho^b = \text{tr}_a \rho^{ab}$ (here $H^a = H^b$). Let $\rho^{ab_i} = \mathcal{I}^a \otimes \mathcal{E}_i^b(\rho^{ab})$, $i = 1, 2$. By the definition, we have

$$\mathcal{D}(\mathcal{E}_1^b, \rho^b) = S(\rho^b) - S(\rho^{b_1}) + S(\rho^{ab_1}),$$
$$\mathcal{D}(\mathcal{E}_2^b, \rho^b) = S(\rho^b) - S(\rho^{b_2}) + S(\rho^{ab_2}),$$

from which we obtain

$$|\mathcal{D}(\mathcal{E}_1^b, \rho^b) - \mathcal{D}(\mathcal{E}_2^b, \rho^b)| = |S(\rho^{b_2}) - S(\rho^{b_1}) + S(\rho^{ab_1}) - S(\rho^{ab_2})|$$
$$\leq |S(\rho^{b_1}) - S(\rho^{b_2})| + |S(\rho^{ab_1}) - S(\rho^{ab_2})|.$$

Let $\gamma_1 = \frac{1}{2}\text{tr}|\rho^{b_1} - \rho^{b_2}|$, $\gamma_2 = \frac{1}{2}\text{tr}|\rho^{ab_1} - \rho^{ab_2}|$. Applying the result in Zhang,[32] which states that

$$|S(\rho) - S(\sigma)| \leq H(\gamma) + \gamma \log(m-1),$$

where $\gamma = \frac{1}{2}\text{tr}|\rho - \sigma|$, and m is the dimension of the system, we have

$$|S(\rho^{b_1}) - S(\rho^{b_2})| \leq H(\gamma_1) + \gamma_1 \log(d-1),$$

and

$$|S(\rho^{ab_1}) - S(\rho^{ab_2})| \leq H(\gamma_2) + \gamma_2 \log(d^2-1).$$

Let

$$f(\gamma_1) = H(\gamma_1) + \gamma_1 \log(d-1), \quad g(\gamma_2) = H(\gamma_2) + \gamma_2 \log(d^2-1),$$

then

$$|\mathcal{D}(\mathcal{E}_1^b, \rho^b) - \mathcal{D}(\mathcal{E}_2^b, \rho^b)| \leq f(\gamma_1) + g(\gamma_2).$$

Note that $f(\gamma_1)$ is monotone increasing when $\gamma_1 \in (0, 1 - \frac{1}{d})$, and from the monotonicity of the trace distance, we have $\gamma_1 \leq \gamma_2$. Consequently,

$$f(\gamma_1) \leq f(\gamma_2),$$

and when $\gamma_2 \in (0, 1-\frac{1}{d})$, we have
$$|\mathcal{D}(\mathcal{E}_1^b, \rho^b) - \mathcal{D}(\mathcal{E}_2^b, \rho^b)| \leq f(\gamma_2) + g(\gamma_2)$$
$$= 2H(\gamma_2) + \gamma_2 \log(d-1)(d^2-1),$$

From
$$\gamma_2 = \frac{1}{2}\mathrm{tr}|\rho^{ab_1} - \rho^{ab_2}|$$
$$= \frac{1}{2}\|\rho^{ab_1} - \rho^{ab_2}\|_1$$
$$= \frac{1}{2}\|\mathcal{I}^a \otimes \mathcal{E}_1^b(\rho^{ab}) - \mathcal{I}^a \otimes \mathcal{E}_2^b(\rho^{ab})\|_1$$
$$= \frac{1}{2}\|(\mathcal{I}^a \otimes \mathcal{E}_1^b - \mathcal{I}^a \otimes \mathcal{E}_2^b)(\rho^{ab})\|_1$$
$$= \frac{1}{2}\|\mathcal{I}^a \otimes (\mathcal{E}_1^b - \mathcal{E}_2^b)(\rho^{ab})\|_1$$
$$\leq \frac{1}{2}\|\mathcal{I}^{a*} \otimes (\mathcal{E}_1^{b*} - \mathcal{E}_2^{b*})\|_\infty \quad \text{(by Eq. (5.2))}$$
$$\leq \frac{1}{2}\|\mathcal{E}_1^{b*} - \mathcal{E}_2^{b*}\|_{\mathrm{cb}} \quad \text{(by Eq. (5.1))},$$

we conclude that $\gamma_2 \leq \delta$. Let
$$h(\gamma_2) = 2H(\gamma_2) + \gamma_2 \log(d-1)(d^2-1).$$
Since $h(\gamma_2)$ is increasing in the interval $(0, 1 - \frac{1}{\sqrt{(d-1)(d^2-1)}+1})$, and $1 - \frac{1}{d} \leq 1 - \frac{1}{\sqrt{(d-1)(d^2-1)}+1}$, we have $h(\gamma_2) \leq h(\delta)$ when $\delta \in (0, 1-\frac{1}{d})$. Thus when $\delta \in (0, 1-\frac{1}{d})$, the desired inequality (5.3) follows.

6. Discussion

By exploiting the difference between the quantum mutual information before and after a quantum operation, we have introduced the notion of decoherent information, which quantifies how much correlation information is lost, and which in turn characterizes certain aspect of decoherence caused by a quantum operation. The decoherent information is somewhat complementary to the coherent information with respect to the initial state. The informational meaning and properties of the former are more transparent and can be more simply derived than the latter from the monotonicity of quantum mutual information. By use of the decoherent information, we have not only provided an alternative and intrinsic interpretation of the coherent information, but have also gained some new insight. We have also established a quantum Fano type inequality for the decoherent information, an informational no-broadcasting result and a continuity estimate for the decoherent information. It is hoped that the notion of decoherent information may be useful in studying decoherence phenomena and quantum channel capacities.

Acknowledgement

This work was supported by NSFC, Grant No. 10771208, and by the science Fund for Creative Research Groups, Grant No. 10721101.

References

1. C. Adami and N. J. Cerf, von Neumann capacity of noisy quantum channels, Phys. Rev. A **56**, 3470 (1997).
2. H. Barnum, C. M. Caves, C. A. Fuchs, R. Jozsa, and B. Schumacher, Noncommuting mixed states cannot be broadcast, Phys. Rev. Lett. **76**, 2818 (1996).
3. H. Barnum, M. A. Nielsen, and B. Schumacher, Information transmission through a noisy quantum channel, Phys. Rev. A **57**, 4153 (1998).
4. C. H. Bennett, H. J. Bernstein, S. Popescu, and B. Schumacher, Concentrating partial entanglement by local operations, Phys. Rev. A **53**, 2046 (1996).
5. C. H. Bennett, P. W. Shor, J. A. Smolin, and A. V. Thapliyal, Entanglement-assisted classical capacity of noisy quantum channels, Phys. Rev. Lett. **83**, 3081 (1999); Entanglement-assisted capacity of a quantum channel and the reverse Shannon theorem, IEEE Trans. Inform. Theory **48**, 2637 (2002).
6. F. Buscemi, G. Chiribella, and G. M. D'Ariano, Inverting quantum decoherence by classical feedback from the environment, Phys. Rev. Lett. **95**, 090501 (2005).
7. I. Devetak and P. W. Shor, The capacity of a quantum channel for simultaneous transmission of classical and quantum information, Commun. Math. Phys. **256**, 287 (2005).
8. I. Devetak, The private classical capacity and quantum capacity of a quantum channel, IEEE Trans. Inform. Theory **51**, 44 (2005).
9. D. Dieks, Communication by EPR devices, Phys. Lett. A **92**, 271 (1982).
10. B. Groisman, S. Popescu, and A. Winter, Quantum, classical, and total amount of correlations in a quantum state, Phys. Rev. A **72**, 032317 (2005).
11. P. Hayden, R. Jozsa, D. Petz, and A. Winter, Structure of states which satisfy strong subadditivity of quantum entropy with equality, Commun. Math. Phys. **246**, 359 (2004).
12. A. S. Holevo, On entanglement-assisted classical capacity, J. Math. Phys. **43**, 4326 (2002).
13. B. Ibinson and A. Winter, All inequalities for the relative entropy, Commun. Math. Phys. **269**, 223 (2007).
14. A. Kalev and I. Hen, No-broadcasting theorem and its classical counterpart, Phys. Rev. Lett. **100**, 210502 (2008).
15. D. Kretschmann, D. Schlingermann, and R. F. Werner, The information-disturbance tradeoff and the continuity of Stinespring's representation, IEEE Trans. Inform. Theory **54**, 1708 (2008).
16. N. Li and S. Luo, Total versus quantum correlations in quantum states, Phys. Rev. A **76**, 032327 (2007).
17. S. Lloyd, Capacity of the noisy quantum channel, Phys. Rev. A **55**, 1613 (1997).
18. S. Luo, Quantum versus classical uncertainty, Theore. Math. Phys. **143**, 681 (2005).
19. S. Luo and Q. Zhang, Informational distance on quantum-state space, Phys. Rev. A **69**, 032106 (2004).
20. M. A. Nielsen and I. L. Chuang, Quantum Computation and Quantum Information (Cambridge University Press, Cambridge, UK, 2000).

21. V. I. Paulsen, Completely Bounded Maps and Operator Algebras (Cambridge University Press, Cambridge, 2002).
22. A. Peres, Quantum Theory: Concepts and Methods (Kluwer, Dordrecht, 1993).
23. D. Petz, Monotonicity of quantum relative entropy revisited, Rev. Math. Phys. **15**, 79 (2003).
24. S. Popescu and D. Rohrlich, Thermodynamics and the measure of entanglement, Phys. Rev. A **56**, R3319 (1997).
25. V. Scarani, S. Iblisdir, and N. Gisin, Quantum cloning, Rev. Mod. Phys. **77**, 1225 (2005).
26. B. Schumacher, Sending entanglement through noisy quantum channels, Phys. Rev. A **54**, 2614 (1996).
27. B. Schumacher and M. A. Nielsen, Quantum data processing and error correction, Phys. Rev. A **54**, 2629 (1996).
28. B. Schumacher and M. D. Westmoreland, Quantum privacy and quantum coherence, Phys. Rev. Lett. **80**, 5695 (1998).
29. B. Schumacher and M. D. Westmoreland, Quantum mutual information and the one-time pad, Phys. Rev. A **74**, 042305 (2006).
30. V. Vedral, The role of relative entropy in quantum information theory, Rev. Mod. Phys. **74**, 197 (2002).
31. W. K. Wootters and W. H. Zurek, A single quantum cannot be cloned, Nature **299**, 802 (1982).
32. Z. Zhang, Uniform estimates on the Tsallis entropies, Lett. Math. Phys. **80**, 171-181 (2007).

Chapter 3

Stabilization of Evolution Equations by Noise

Tomás Caraballo[1] and Peter E. Kloeden[2]

[1] Dpto. Ecuaciones Diferenciales y Análisis Numérico,
Universidad de Sevilla, 41080–Sevilla, Spain,
caraball@us.es

[2] Institut für Mathematik, Goethe Universität,
D-60054 Frankfurt am Main, Germany,
kloeden@math.uni-frankfurt.de

Some recent results on stabilization by noise in systems modeled by evolution equations is reviewed, mostly for partial differential equations but also for delay differential equations and systems without uniqueness.

Keywords: Stochastic partial differential equations. Itô noise, Stratonovich noise, exponential stochastic stability, stabilization, destabilization.

2000 AMS Subject Classification: 35R10 35B40 47H20 58F39 73K70

Contents

1	Introduction	44
2	Linear PDEs	46
	2.1 Persistence of stability and stabilization by Itô noise	48
	2.2 Destabilization by Itô noise	49
3	Linear PDEs without fully commuting noise	50
	3.1 Stabilization by simple multiplicative Itô noise	50
	3.2 Stabilization by Stratonovich noise	51
4	Nonlinear PDEs	53
	4.1 Stabilization by Itô noise	54
	4.2 Stabilization by Stratonovich noise	59
5	Other types of evolution equations and models	60
	5.1 Delay differential equations	60
	5.2 Stabilization of evolution inclusions and PDEs without uniqueness	60
	5.3 Stabilization of stationary solutions of a stochastic PDE	63
	5.4 Other types of problems	63
References		64

1. Introduction

Stabilization by noise has a long history dating back, in engineering practice, to the early days of the industrial revolution. Attempts to understand the phenomenon mathematically in terms of stochastic differential equations (SDEs) began in the 1960s, at first mainly for finite dimensional systems and in the past decade for infinite dimensional systems, that is for evolution equations, by which is meant mainly partial differential equations (PDEs), but also includes delay differential equations (DDEs).

The mechanism behind stabilization is intuitively quiet simple. Consider a linear system for which the zero steady state solution is unstable, but not unstable in all directions. The idea is to apply noise in an appropriate way to drive the dynamics in the unstable directions into the stable directions. Stabilization thus requires the system to be sufficiently strongly stable in certain directions to overcome the effects of the unstable directions. This restricts, in general, the classes of systems which can be stabilized.

Another important consideration is whether the SDEs are interpreted in the Itô and Stratonovich senses. This is essentially a modelling issue, but has significant consequences. For example, the scalar SDE

$$dX_t = aX_t\,dt + bX_t\,dW_t,$$

where $a > 0$ and W_t is a standard Wiener process, has the Itô and Stratonovich solutions

$$X_t = e^{(a-\frac{1}{2}b^2)t + bW_t} X_0 \quad \text{and} \quad X_t = e^{at + bW_t} X_0.$$

It follows by the properties of the Wiener process (see Arnold[3] or Kloeden & Platen[36]) that the zero solution is pathwise exponentially stable for the Itô SDE for b^2 large enough, while the same is not true for the Stratonovich SDE. This seems to imply that Itô noise has a more profound stabilizing effect than Stratonovich noise. However, this argument is somewhat misleading. Indeed, the above system has no stable directions and the interplay between stable directions and Stratonovich noise can lead to stabilization in higher dimensional systems, as will be seen in some examples below. See Caraballo[10] for a detailed discussion on the effects of Itô and Stratonovich formalisms on stabilization.

There is an extensive literature about such problems for finite-dimensional systems (see for example Arnold,[5] Arnold et al.,[7] Arnold & Kloeden,[8] Mao,[43] Scheutzow[46]). Many results on the stabilization and destabilization produced by both Itô and Stratonovich noise have been obtained, and these have also been applied to construct feedback stabilizers, which are an important tool in control problems. Although in each particular situation one or other choice of the noise may be more appropriate, stabilization by Stratonovich noise might be considered more significant and has a non-trivial literature with both mathematical and engineering contributions (see Arnold,[5] Arnold et al.[7] and the references therein). Since such noise behaves like a periodic zero-mean feedback control, its stabilizing effect is unexpected and very intriguing. In the finite-dimensional case, Arnold and his collaborators proved that the linear differential system

$$\frac{dx}{dt} = Ax, \qquad (1.1)$$

can be stabilized by the addition of a collection of multiplicative noisy terms,

$$dX_t = AX_t\,dt + \sum_{i=1}^{d} B_i X_t \circ dW_t^i, \qquad (1.2)$$

where the W_t^i are mutually independent Wiener processes and the B_i are suitable skew-symmetric matrices, if and only if

$$\operatorname{tr} A < 0. \qquad (1.3)$$

Since the trace of the matrix equals the sum of its eigenvalues, this indicates that the system must have stable directions which are sufficiently strongly attracting to overcome the effects of the unstable ones. (The stabilization and destabilization of nonlinear ordinary differential equations has been considered by Appleby et al.[2] and the papers cited therein).

The corresponding problem for linear partial differential equations remained open for a long time (and in fact is still open in the most general case) because it needs a version of the renowned Oseledec Multiplicative Ergodic Theorem for infinite-dimensional spaces, which has not yet been established as far as we know. However, Caraballo & Robinson[26] were able to circumvent this difficult by stabilizing the linear PDE with a *finite* sum of Stratonovich terms as in (1.2), which allowed them to project onto a finite-dimensional subspace and thus use finite-dimensional arguments. From an engineering perspective finite-dimensional noise is not necessarily a restriction and, indeed, is possibly more realistic.

To the best of our knowledge, the problem of stabilization of *nonlinear* PDEs by Stratonovich noise is still an unsolved problem in general, although it has been shown to be possible in a number of interesting applications by using Itô's noise (see Caraballo & Langa,[11] Caraballo et al.,[14,20,23,24] Kwiecinska,[39] and Leha et al.[41] amongst many others).

Our aim in this article is to present some of the results that have been obtained and to indicate the basic techniques used. To be more precise, we consider a linear evolution equation on a separable Hilbert space H given by

$$\frac{du}{dt} = Au, \qquad (1.4)$$

where $A : D(A) \subset H \mapsto H$ is a linear operator which has a sequence of eigenvalues λ_j with associated eigenfunctions e_j. We assume that these eigenfunctions form an orthonormal basis of H and that the eigenvalues λ_j are bounded above (but not necessarily below), so that they can be ordered in the form $\lambda_1 \geq \lambda_2 \geq \ldots$. In addition, we denote by $|\cdot|$ the norm in H and by $(\cdot,\cdot)$ its associated scalar product.

We will consider the stochastically perturbed evolution equation

$$dU_t = AU_t\, dt + \sum_{i=1}^{d} B_i U_t \circ dW_t^i, \qquad (1.5)$$

where the $B_i : D(B_i) \subset H \mapsto H$ are linear operators and the W_t^i are mutually independent Wiener process on the same probability space $(\Omega, \mathcal{F}, \mathbb{P})$.

We will report in the following sections the differences on the behaviour of problems (1.4) and (1.5) In fact, most of the analysis carried out in this field is concerned only with the stabilization of the trivial solution. Nevertheless, going deeper into the investigation of the nonlinear models, we can find in the stochastic models some special solutions called *stationary* but which are not stationary in the deterministic sense, i.e., steady state solutions. These solutions sometimes become random attractors for some systems, so their existence and properties are very important. Some preliminary results have been obtained in Caraballo et al.[16,17]

2. Linear PDEs

We first consider some properties of the solutions of a linear stochastic PDE and a result on the exponential stability of its zero solution. This will allow us to point out the different effects that the interpretation of the noise may produce in the final result as well as to characterize the stabilization of linear PDEs by Stratonovich noise.

To start we consider the Itô formulation of the linear SPDE. We can thus apply a result due to Da Prato & Zabczyk[31] which ensures the equivalence of the stochastic PDE to a pathwise nonautonomous deterministic PDE, i.e. a random PDE. Then, we transform our Stratonovich model to an equivalent Itô model and apply this result. (This equivalence has been proved by Kunita[37] for suitable partial differential operators and we implicitly assume that we are considering this case).

Consider the Cauchy problem for the linear Itô SPDE,

$$dU_t = AU_t\, dt + \sum_{k=1}^{d} B_k U_t\, dW_t^k, \qquad U_0 = u_0 \in H, \qquad (2.1)$$

where $A : D(A) \subset H \to H$ and the $B_k : D(B_k) \subset H \to H$, $k = 1, \cdots, d$, are generators of C_0-semigroups $S_A(t)$ and S_k, respectively, and the $W_t^1, \cdots, W_t^d$ are independent real Wiener processes on the same probability space $(\Omega, \mathcal{F}, \mathbb{P})$.

We also need the following assumptions:

Assumption 1 (1ex). (A1) , *The operators $B_1, \cdots, B_d$ generate mutually commuting C_0-groups S_k.*
(A2) $D(B_k^2) \supset D(A)$ for $k = 1, \cdots, d$ and $\bigcap_{k=1}^{d} D((B_k^*)^2)$ is dense in H, where B_k^* denotes the adjoint operator of B_k.

(A3) $C = A - \frac{1}{2}\sum_{k=1}^{d} B_k^2$ generates a C_0-semigroup S_C.

Given a realization of the Wiener processes $W_t^k(\omega)$ for fixed $\omega \in \Omega$ we define

$$T_\omega(t) = \prod_{k=1}^{d} S_k(W_t^k(\omega)) \quad \text{and} \quad v(t) = T_\omega^{-1}(t)\,u(t), \quad t \geq 0, \tag{2.2}$$

and we consider the auxiliary deterministic system

$$\frac{dv}{dt}(t) = T_\omega^{-1}(t)\,C\,T_\omega(t)\,v(t), \qquad v(0) = u_0, \tag{2.3}$$

which is a deterministic Cauchy problem depending on the parameter ω. The following result, along with the definition of a strong solution, can be found in Da Prato & Zabczyk:[31]

Proposition 2.1. *Suppose that Assumptions (A1)–(A3) are satisfied. Then, if U_t is a strong solution of (2.1), the process $v(t,\omega)$ defined by (2.2) satisfies (2.3). Conversely, if v is a predictable process whose trajectories are continuously differentiable and satisfy (2.3), $\mathbb{P}$-a.s., then the process $U_t(\omega) := T_\omega(t)v(t,\omega)$ takes values in $D(C)$, $\mathbb{P}$-a.s., and for almost all t, and is a strong solution to (2.1).*

A sufficient condition ensuring the solvability of (2.3) which is useful in applications can also be found in Da Prato & Zabczyk[31], pp. 177–179.

Now consider the Stratonovich version of the problem, i.e,

$$dU_t = AU_t\,dt + \sum_{k=1}^{d} B_k U_t \circ dW_t^k, \qquad U_0 = u_0 \in H. \tag{2.4}$$

To obtain existence of its solutions we consider its equivalent Itô version

$$dU_t = \left((A + \frac{1}{2}\sum_{k=1}^{d} B_k^2\right) U_t\,dt + \sum_{k=1}^{d} B_k U_t\,dW_t^k, \qquad U_0 = u_0 \in H. \tag{2.5}$$

Instead of (A3), we assume

Assumption 2.

(A3') $C = A + \frac{1}{2}\sum_{k=1}^{d} B_k^2$ generates a C_0-semigroup S_C,

as well as (A1) and (A2), then, thanks to Proposition 2.1, problem (2.4) can be rewritten equivalently as

$$\frac{dv}{dt}(t) = T_\omega^{-1}(t)AT_\omega(t)\,v(t), \quad v(0) = u_0, \tag{2.6}$$

The following result of Caraballo & Robinson[26] (see also Caraballo & Langa[11]) characterizes the asymptotic stability of the Stratonovich model (2.4) under the assumption that all the operators involved in the equation mutually commute. We call this the *fully commuting* case and show how, under the same assumptions, the Itô equation (2.1) may exhibit very different asymptotic behavior.

Theorem 2.1. *In addition to Assumptions (A1)–(A2) and (A3'), suppose that A commutes with each $S_k(t)$. Then, the strongly continuous semigroup $S_A(t)$ generated by A is exponentially stable, i.e., there exist M_0, $\gamma > 0$ such that*

$$|S_A(t)| \leq M_0 e^{-\gamma t} \text{ for all } t > 0,$$

if and only if there exist α, $C > 0$ and $\Omega_0 \subset \Omega$ with $\mathbb{P}(\Omega_0) = 0$ such that for any $\omega \notin \Omega_0$ there exists $T(\omega) > 0$ such that the solution U_t of (2.4) satisfies

$$|U_t| \leq C|u_0|e^{-\alpha t} \text{ for } t \geq T(\omega).$$

In the fully commuting case, the stability properties of the deterministic problem (1.4) and the stochastic (2.4) are thus equivalent, so we can ensure the adequacy of the deterministic model to the stochastic real phenomenon.

However, if we interpret the noise in the sense of Itô, then we may have very different results. For example, it may happen that (1.4) is stable and (2.1) remains stable (persistence of stability from the deterministic to the stochastic model), or that (1.4) is unstable and (2.1) becomes stable (stabilization produced by the noise), or that (1.4) is stable and (2.1) becomes unstable (destabilization). We will illustrate these scenarios in the following subsections.

2.1. Persistence of stability and stabilization by Itô noise

Let $\mathcal{O}$ be a bounded domain in $\mathbb{R}^d$ ($d \leq 3$) with C^∞-boundary, and consider the noisy reaction-diffusion equation

$$\begin{cases} \dfrac{\partial u(t,x)}{\partial t} = \Delta u(t,x) + \alpha u(t,x) + \gamma u(t,x)\dot{W}_t, \\ u(t,x) = 0, \quad t > 0, \quad x \in \partial\mathcal{O}, \\ u(0,x) = u_0(x), \quad x \in \mathcal{O}, \end{cases} \quad (2.7)$$

where Δ denotes the Laplacian operator and W_t is a scalar Wiener process and $\dot{W}_t$ the corresponding Gaussian white noise.

To set this problem in our framework as an Itô equation, we take $H = L^2(\mathcal{O})$, $A = \Delta + \alpha I$ and $B = \gamma I$. Then $D(A) = H_0^1(\mathcal{O}) \cap H^2(\mathcal{O})$. Let $\lambda_1 > 0$ denote the first eigenvalue of $-\Delta$. Then, as a consequence of Theorem 3.1 in Subsection 3.1 (see also Kwiecinska[38]), it is easy to check that the null solution of (2.7) is exponentially stable, $\mathbb{P}$-a.s., if the parameters in the equation satisfy

$$2(\alpha - \lambda_1) - \gamma^2 < 0.$$

Let us now discuss what this condition means.

First, notice that when $\alpha < \lambda_1$, the deterministic equation (i.e. Eq. (2.7) with $\gamma = 0$) is exponentially stable. Then, for any $\gamma \in \mathbb{R}$ (in other words, no matter how large or small the intensity of the noise might be), the stochastic equation (2.7) remains exponentially stable, $\mathbb{P}$-a.s. Consequently, the stability persists in the presence of noise.

When $\alpha > \lambda_1$, however, the deterministic equation is not stable (see, for instance, Example 2.2 below for a more detailed analysis in a case of one spatial dimension). But, if we choose γ large enough so that $2(\alpha - \lambda_1) - \gamma^2 < 0$, then the stochastic equation becomes exponentially stable, $\mathbb{P}$-a.s. Consequently, noise with large intensity stabilizes the system.

2.2. Destabilization by Itô noise

Consider again the deterministic heat equation, now in one spatial dimension:

$$\begin{cases} \dfrac{\partial u(t,x)}{\partial t} = \dfrac{\partial^2 u(t,x)}{\partial x^2} + \alpha u(t,x), & t > 0, \quad 0 < x < \pi, \\ u(t,0) = u(t,\pi) = 0, & t > 0, \\ u(0,x) = u_0(x), & x \in [0,\pi]. \end{cases} \quad (2.8)$$

Set $H = L^2([0,\pi])$ and $A = \frac{\partial^2}{\partial x^2} + \alpha$, so $D(A) = H_0^1([0,\pi]) \cap H^2([0,\pi])$.
This system has the explicit solution

$$u(t,x) = \sum_{n=1}^{\infty} a_n e^{-(n^2 - \alpha)t} \sin nx$$

with initial value $u_0(x) = \sum_{n=1}^{\infty} a_n \sin nx$, and the zero solution is exponential stable if and only if $\alpha < n^2$ for all $n \in \mathbb{N}$, ie., if and only if $\alpha < 1$.

Consider now the problem

$$dT_t = A U_t\, t + B U_t\, dW_t, \quad (2.9)$$

where B is defined by

$$Bu(x) = \delta \dfrac{\partial u(x)}{\partial x}$$

for any $u \in H_0^1([0,\pi])$ and some $\delta \in \mathbb{R}$.

If we choose δ such that

$$1 - \alpha \leq \dfrac{\delta^2}{2} < 1,$$

then the stochastic problem becomes unstable. Indeed, denoting by $C = A - \frac{1}{2}B^2$, the stability of problem (2.9) is equivalent to the stability of

$$dU_t = C U_t\, dt + B U_t \circ dW(t), \quad U_0 = u_0. \quad (2.10)$$

But, due to the commutativity property of these operators, Theorem 2.1 ensures that the stability of (2.10) is equivalent to the stability of the deterministic problem

$$\begin{cases} \dfrac{\partial u(t,x)}{\partial t} = \left(1 - \dfrac{\delta^2}{2}\right) \dfrac{\partial^2 u(t,x)}{\partial x^2} + \alpha u(t,x), \\ u(t,0) = u(t,\pi) = 0, \ t > 0, \\ u(0,x) = u_0(x), \ x \in [0,\pi]. \end{cases}$$

The zero solution of this system is exponentially stable if and only if

$$\alpha < 1 - \frac{\delta^2}{2}.$$

Since our constants satisfy the opposite inequality, we see that the noise destabilizes the deterministic exponentially stable system.

In conclusion, it is clear in this fully commuting case that we should be very careful how we interpret the noise since the behavior of the resulting stochastic models may then be completely different. More precisely, the Stratonovich noise does not modify the stability properties of the deterministic model, while Itô noise can produce very different effects.

3. Linear PDEs without fully commuting noise

Under fully commuting assumptions in the previous subsection, the zero solution of the deterministic problem is exponentially stable if and only if the stochastically perturbed equation (Stratonovich sense) has the same property. However, an immediate question arises. When no commutativity holds between A and some B_k, then there are some sufficient conditions in Caraballo & Robinson,[26] which ensure the persistence of exponential stability of the deterministic system in the stochastic model.

3.1. *Stabilization by simple multiplicative Itô noise*

A simple multiplicative noise in the Itô sense will stabilize the deterministic linear partial differential equation (1.4) in many cases, so we do not need to worry too much about looking for a very complicate expression of the noise. A term like

$$\sigma u \dot{W}(t)$$

can produce that effect. This stabilization can be produced for more general terms and, in some cases, we can even determine the decay rate of the solutions (exponential, sub- or super-exponential), see Caraballo et al.[14]

The following result is a particular situation of a much more general nonlinear theorem (see Section 4 for more details). First, recall that a linear operator A generates a strongly continuous semigroup $S_A(t)$ satisfying

$$|S_A(t)| \leq e^{\alpha t}, \tag{3.1}$$

for some $\alpha \in \mathbb{R}$, if and only if $(Au, u) \leq \alpha |u|^2$ for all $u \in D(A)$.

Theorem 3.1. *Assume that A generates a strongly continuous semigroup $S_A(t)$ satisfying (3.1) for some $\alpha \in \mathbb{R}$, and that $B : D(B) \subset H \mapsto H$ is a linear (bounded or unbounded) operator with $D(A) \subset D(B)$. Suppose also that the two following hypotheses hold:*

i) There exists $\beta \in \mathbb{R}$ such that

$$(Au, u) + \frac{1}{2}|Bu|^2 \leq \beta|u|^2, \quad \forall u \in D(A) \tag{3.2}$$

(which is immediately fulfilled for $\beta = \alpha + \frac{1}{2}\|B\|^2$, if B is bounded).

ii) There exist $b, \tilde{b} \in \mathbb{R}$ with $0 \leq b \leq \tilde{b}$ such that

$$b|u|^2 \leq (u, Bu) \leq \tilde{b}|u|^2, \quad \forall u \in D(B). \tag{3.3}$$

Then, for every $u_0 \in D(A), u_0 \neq 0$, the solution $U_t := u(t, \omega; 0, u_0)$ to the problem

$$dU_t = AU_t\, dt + BU_t\, dW_t, \quad U_0 = u_0 \in H,$$

satisfies

$$\limsup_{t \to +\infty} \frac{1}{t} \log |U_t|^2 \leq -(b^2 - \beta), \quad \mathbb{P}-a.s.$$

In the particular case that B is defined by $Bu = bu$ with $b \in \mathbb{R}$, then $\beta = \alpha + \frac{1}{2}b^2$ and, hence, $b^2 - \beta = \frac{1}{2}b^2 - \alpha$, which is positive when b^2 is large enough. Thus the zero solution will be stabilized by Itô noise of the above type with sufficiently large intensity.

3.2. Stabilization by Stratonovich noise

To obtain the same effect using Stratonovich noise turns out, however, to be a completely different and much more difficult problem as mentioned in the Introduction. But, surprisingly, a very simple trick (discovered long time after the results in the finite-dimensional case were obtained) allows one to prove that the negative trace assumption (1.3) is a necessary and sufficient condition for the stabilization of a linear PDE by a suitable Stratonovich noise (see Caraballo & Robinson[26] for a detailed exposition on this problem). Instead of presenting this stabilization result here, we will motivate the problem with an example, which provides the basic idea of the proof of the theorem.

Consider the following one-dimensional heat equation

$$\begin{cases} \dfrac{\partial u(t,x)}{\partial t} = \dfrac{\partial^2 u(t,x)}{\partial x^2} + 2u(t,x), \quad t > 0, \quad 0 < x < \pi, \\ u(t,0) = u(t,\pi) = 0, \quad t > 0, \\ u(0,x) = u_0(x), \quad x \in [0,\pi]. \end{cases} \tag{3.4}$$

This problem can be formulated in our framework by setting $H = L^2([0,\pi])$ and $A = \frac{\partial^2}{\partial x^2} + 2I$, so $D(A) = H_0^1([0,\pi]) \cap H^2([0,\pi])$. Recall from Section 2.2 that this problem has the explicit solution

$$u(t,x) = \sum_{n=1}^{\infty} a_n e^{-(n^2-2)t} \sin nx$$

for the initial value $u_0(x) = \sum_{n=1}^{\infty} a_n \sin nx$. Hence, it is clear that the zero solution of the problem (3.4) is not stable. However, we will see that by an appropriate choice of operators $B_k : H \to H$, $k = 1, \cdots, d$, the zero solution of the system

$$dU_t = AU_t\, dt + \sum_{k=1}^{d} B_k U_t \circ dW_t^k, \qquad (3.5)$$

is exponentially stable with probability one. Note that the operators B_k cannot commute with A here.

The operator A has eigenvalues $\lambda_n = 2 - n^2$, $n \geq 1$, with associated eigenfunctions $e_n = \sqrt{\frac{2}{\pi}} \sin nx$, which form an orthonormal basis of the Hilbert space H. Hence any $u \in H$ can be represented in the form

$$u = \sum_{k \geq 1} (u, e_k) e_k = \sum_{k \geq 1} u_k e_k.$$

We define $B : H \mapsto H$ via $Be_1 := -\sigma e_2$, $Be_2 := \sigma e_1$ and $Be_n := 0$ for any $n \geq 3$. This is a linear operator, which does not commute with A. Using the Fourier representation for the solution $u(t)$ to (3.5), this problem can be re-written as

$$\begin{cases} \sum_{k \geq 1} du_k(t) e_k = \sum_{k \geq 1} \lambda_k u_k(t) e_k\, dt + [\sigma u_2(t) e_1 - \sigma u_1(t) e_2] \circ dW(t) \\ u(0) = u_0 = \sum_{k \geq 1} u_{0,k} e_k. \end{cases} \qquad (3.6)$$

Identifying the coefficients gives two coupled problems, the first being a 2-dimensional stochastic ordinary differential system and the second an infinite-dimensional system which is exponentially stable (since $\lambda_n < 0$ for all $n \geq 3$), namely

$$\begin{cases} d\begin{pmatrix} u_1(t) \\ u_2(t) \end{pmatrix} = \begin{pmatrix} \lambda_1 & 0 \\ 0 & \lambda_2 \end{pmatrix} \begin{pmatrix} u_1(t) \\ u_2(t) \end{pmatrix} dt + \begin{pmatrix} 0 & \sigma \\ -\sigma & 0 \end{pmatrix} \begin{pmatrix} u_1(t) \\ u_2(t) \end{pmatrix} \circ dW_t \\ u_1(0) = u_{0,1}, \qquad u_2(0) = u_{0,2} \end{cases} \qquad (3.7)$$

and

$$\sum_{k \geq 3} du_k(t) e_k = \sum_{k \geq 1} \lambda_k u_k(t) e_k\, dt, \qquad \sum_{k \geq 3} u_k(0) e_k = \sum_{k \geq 3} u_{0,k} e_k. \qquad (3.8)$$

The matrix

$$\begin{pmatrix} 0 & 1 \\ -1 & 0 \end{pmatrix}$$

is a basis for the linear space of skew symmetric 2×2 matrices, so results in Arnold et al.[7] show that the leading Lyapunov exponent of solutions to (3.7) tends to $\frac{1}{2}(\lambda_1 + \lambda_2) = -1/2$ as the intensity parameter σ grows to $+\infty$. Moreover, the leading Lyapunov exponent for the solutions to (3.8) is $\lambda_3 = -7$, so we can ensure that the top Lyapunov exponent for the solutions of (3.6) is negative.

The main idea for the stabilization of evolution equations is thus to decompose the problem into two new problems: a finite-dimensional one which can be stabilized by using previously available methods from the finite dimensional framework, and an infinite-dimensional system which is already exponentially stable. This idea can be extended in a general way to enable the stabilization of a wide class of deterministic PDEs, which appear very frequently in applications.

Consider again the deterministic infinite-dimensional linear system (1.4). The main stabilization result of this article is the following one from Caraballo & Robinson[26].

Theorem 3.2. *Assume that the trace of A is negative, i.e.,*

$$\operatorname{tr} A := \sum_{j=1}^{\infty} \lambda_j < 0. \tag{3.9}$$

Then, there exist linear operators $B_k : H \mapsto H$, $k = 1, \cdots, d$, such that the zero solution of

$$dU_t = AU_t\, dt + \sum_{j=1}^{d} B_k U_t \circ dW_t^k, \tag{3.10}$$

is exponentially stable, $\mathbb{P}$-a.s. The operators B_k are such that for some $N > 0$, the $N \times N$ matrices $D_1, \cdots, D_k$ defined by

$$D_k = \begin{pmatrix} (B_k e_1, e_1) & (B_k e_2, e_1) & \cdots & (B_k e_N, e_1) \\ (B_k e_1, e_2) & (B_k e_2, e_2) & \cdots & (B_k e_N, e_2) \\ \vdots & \vdots & \ddots & \vdots \\ (B_k e_1, e_N) & (B_k e_2, e_N) & \cdots & (B_k e_N, e_N) \end{pmatrix}$$

are skew-symmetric.

Conversely, if there exist linear operators $B_k : H \mapsto H$, $k = 1, \cdots, d$, with the above properties, for which the zero solution of (3.10) is exponentially stable with probability one, then the trace of A is negative.

4. Nonlinear PDEs

The objective of this section is twofold. First, we will show that there is a well developed theory concerning the stabilization of nonlinear PDEs by Itô noise with applications to several interesting examples. On the other hand, since not much is known about the same topic involving Stratonovich noise, we will analyze a particular example (which can, in a sense, be considered as canonical) in which the previous Theorem 3.2 combined with some order preserving properties allow one to establish stabilization for the Chafee-Infante equation by Stratonovich noise. A more complete study for more general nonlinear equations is a topic for further research.

4.1. Stabilization by Itô noise

Let H be a real separable Hilbert space and let V a real reflexive and separable Banach space such that $V \hookrightarrow H \equiv H' \hookrightarrow V'$, where the injections are continuous and dense, and both V and V' are uniformly convex. Further, denote by $\|\cdot\|$, $|\cdot|$ and $\|\cdot\|_*$ the norms in V, H and V', respectively; by $\langle \cdot, \cdot \rangle$ the duality product between V, V', and by $(\cdot, \cdot)$ the scalar product in H. Finally, let a_1 be the constant of the injection $V \hookrightarrow H$, i.e,

$$a_1 |u|^2 \leq \|u\|^2 \text{ for all } u \in V.$$

Now consider the Cauchy problem

$$\frac{du}{dt} = F(t, u), \qquad u(0) = u_0 \in H, \tag{4.1}$$

where $F(t, \cdot) : V \mapsto V'$, $t \in \mathbb{R}_+$, is a family of (nonlinear) operators satisfying $F(t, 0) = 0$ and the following hypothesis:

Assumption 3. *There exist a continuous function $\nu(\cdot)$ and a real number $\nu_0 \in \mathbb{R}$ such that*

$$2 \langle u, F(t, u) \rangle \leq \nu(t) |u|^2, \text{ for all } u \in V, \tag{4.2}$$

where

$$\limsup_{t \to \infty} \frac{1}{t} \int_0^t \nu(s) \, ds \leq \nu_0. \tag{4.3}$$

Moreover, assume that for each $u_0 \in H$, there exists a unique strong solution $u(t) := u(t; u_0)$ to (4.1) with $u(t; u_0) \in L^2(0, T; V) \cap C^0([0, T]; H)$. Observe that, when $F(t, \cdot)$ satisfies a coercivity condition of the type

$$2 \langle u, F(t, u) \rangle \leq -\varepsilon \|u\|^p + \alpha |u|^2, \quad \forall u \in V,$$

for certain parameters $\varepsilon > 0$, $\alpha \in \mathbb{R}, p > 1$, and a monotonicity hypothesis, there exists a unique strong solution $u = u(t; u_0)$ to (4.1) in $L^p(0, T; V) \cap C^0([0, T]; H)$, see Lions.[42] This coercivity assumption obviously implies (4.2).

We will see that (4.1) can be stabilized by using a stochastic perturbation of the kind $g(t, u(t)) \, dW_t$, where W_t is (for simplicity) a standard real Wiener process defined on a certain complete probability space $(\Omega, \mathcal{F}, \mathbb{P})$ with filtration $(\mathcal{F}_t)_{t \geq 0}$, and $g(t, \cdot) : H \to H$ satisfies $g(t, 0) = 0$ and the following condition

Assumption 4.

$$|g(t, u) - g(t, v)|^2 \leq \lambda(t) |u - v|^2 \quad \forall t \in \mathbb{R}_+, \forall u, v \in H, \tag{4.4}$$

where $\lambda(\cdot)$ is a nonnegative continuous function such that

$$\limsup_{t \to \infty} \frac{1}{t} \int_0^t \lambda(s) \, ds \leq \lambda_0 \in \mathbb{R}_+. \tag{4.5}$$

We suppose that for each $u_0 \in H$ the stochastically perturbed problem
$$dU_t = F(t, U_t)\, dt + g(t, U_t)\, dW_t, \qquad u(0) = u_0 \in H, \qquad (4.6)$$
has a unique strong solution to (4.6) in $I^p(0, T; V) \cap L^2(\Omega; C^0([0, T]; H))$ for all $T > 0$ and a certain $p > 1$, where $I^p(0, T; V)$ denotes the space of all V-valued measurable processes U_t satisfying
$$\mathbb{E}\int_0^T \|U_t\|^p\, dt < +\infty$$
(see for instance Pardoux[45] for conditions under which there exists a unique solution for each $u_0 \in L^2(\Omega, \mathcal{F}_0, \mathbb{P}; H)$).

Finally, we assume the existence of a Lyapunov-like functional $W : \mathbb{R}_+ \times H \to \mathbb{R}_+$ which is a $C^{1,2}$-positive functional such that $W'_u(t, u) \in V$ for any $u \in V$ and $t \in \mathbb{R}_+$ and we define operators L and Q as
$$LW(t, u) = W'_t(t, u) + <W'_u(t, u), F(t, u)> + \frac{1}{2}\left(W''_{u,u}(t, u)g(t, u), g(t, u)\right)$$
and
$$QW(t, u) = \left(W'_u(t, u), g(t, u)\right)^2$$
for each $u \in V$ and $t \in \mathbb{R}_+$.

The following theorem is from Caraballo et al.[23]

Theorem 4.1. *Assume that the solution of (4.6) satisfies that $|u(t)| \neq 0$ for all $t \geq 0$, $\mathbb{P}$-a.s., provided $|u_0| \neq 0$, $\mathbb{P}$-a.s. Let $V \in C^2(H; \mathbb{R}_+)$ and let ψ_1 and ψ_2 be two real-valued continuous functions on $\mathbb{R}_+$ with $\psi_2 \geq 0$. Assume that there exist $p > 0$, $\gamma \geq 0$ and $\theta \in \mathbb{R}$ such that*
(a). $|u|^p \leq W(u), \qquad \forall u \in V;$
(b). $LW(t, u) \leq \psi_1(t)W(u), \qquad \forall u \in V,\ t \in \mathbb{R}_+;$
(c). $QW(t, u) \geq \psi_2(t)W^2(u), \qquad \forall u \in V,\ t \in \mathbb{R}_+;$
(d). $\limsup\limits_{t \to \infty} \dfrac{\int_0^t \psi_1(s)\,ds}{t} \leq \theta, \qquad \liminf\limits_{t \to \infty} \dfrac{\int_0^t \psi_2(s)\,ds}{t} \geq 2\gamma.$

Then, the unique strong solution U_t of (4.6) satisfies
$$\limsup_{t \to \infty} \frac{\log |U_t|}{t} \leq -\frac{\gamma - \theta}{p}, \qquad \mathbb{P} - a.s.,$$
whenever $U_0 = u_0 \in H$ is an $\mathcal{F}_0$-measurable random vector such that $|u_0| \neq 0$ a.s. In particular, if $\gamma > \theta$, then the solution is $\mathbb{P}$-a.s. exponentially stable.

A direct application of Theorem 4.1 with the function $W(t, u) = |u|^2$ gives the following result:

Theorem 4.2. *Assume that the solution of (4.6) satisfies $|u(t, u_0)| \neq 0$ for all $t \geq 0$, $\mathbb{P}$-a.s., provided $|u_0| \neq 0$, $\mathbb{P}$-a.s. In addition to hypotheses (4.2) – (4.5), assume that*
$$(g(t, u), u)^2 \geq \rho(t)|u|^4, \qquad \forall u \in H, \qquad (4.7)$$

where $\rho(\cdot)$ is a nonnegative continuous function such that

$$\liminf_{t\to\infty} \frac{1}{t} \int_0^t \rho(s)\,ds \geq \rho_0, \quad \rho_0 \in \mathbb{R}_+. \qquad (4.8)$$

Then, the solution $U_t = u(t, u_0)$ of (4.6) satisfies

$$\limsup_{t\to\infty} \frac{1}{t} \log |U_t|^2 \leq -(2\rho_0 - \nu_0 - \lambda_0), \quad \mathbb{P}-a.s. \qquad (4.9)$$

for any such $u_0 \in H$.

In particular, if $2\rho_0 > \nu_0 + \lambda_0$, then the zero solution of equation (4.6) is $\mathbb{P}$-a.s. exponentially stable.

4.1.1. A general nonlinear example

We are now going to apply Theorem 4.2 to analyze the pathwise stability of a nonlinear stochastic partial differential equation, which has been investigated by Pardoux[45] and Caraballo & Liu[22] amongst others, namely

$$dU_t = A(t, U_t)\,dt + B(t, U_t)\,dW_t, \quad u(0) = u_0 \in H, \qquad (4.10)$$

where $A(t, \cdot) : V \mapsto V'$ is a family of nonlinear operators defined for almost every t satisfying $A(t, 0) = 0$ for $t \in \mathbb{R}_+$, and $B(t, \cdot) : V \mapsto H$ satisfies
(b.1) $B(t, 0) = 0$;
(b.2) There exists $k > 0$ such that

$$|B(t, y) - B(t, x)| \leq k\|y - x\|, \quad \forall x, y \in V, \ t\text{-a.e.}$$

The following result is proved in Caraballo & Liu.[22]

Theorem 4.3. *Assume, in addition to (b.1)–(b.2), the following coercivity condition: there exist $\alpha > 0$, $p > 1$ and $\lambda \in \mathbb{R}$ such that*

$$2\langle x, A(t, x)\rangle + |B(t, x)|^2 \leq -\alpha\|x\|^p + \lambda|x|^2. \qquad (4.11)$$

for almost all $t \in \mathbb{R}_+$ and $x \in V$.

Then, there exists $r > 0$ such that

$$\mathbb{E}|U_t|^2 \leq \mathbb{E}|u_0|^2 e^{-rt}, \quad \forall t \geq 0,$$

where $U_t = u(t; u_0)$, if at least one of the following conditions hold: (i) $\lambda < 0$ or (ii) $\lambda\beta^2 - \alpha < 0$ and $p = 2$.

Furthermore, under the same assumptions, the solution is $\mathbb{P}$-a.s. exponentially stable, i.e., there exist positive constants ξ, η and a subset $\Omega_0 \subset \Omega$ with $\mathbb{P}(\Omega_0) = 0$ such that, for each $\omega \notin \Omega_0$, there exists a positive random number $T(\omega)$ satisfying

$$|U_t(\omega)|^2 \leq \eta|u_0|^2 e^{-\xi t}, \quad \forall t \geq T(\omega).$$

Observe that, in many applications, conditions (i) and (ii) mean that the term containing B must be small enough with respect to A. For example, consider the Sobolev spaces $V = W_0^{1,p}(\mathcal{O})$ and $H = L^2(\mathcal{O})$ with their usual inner products for some $2 \leq p < +\infty$, where $\mathcal{O}$ is an open bounded subset in $\mathbb{R}^d$ with regular boundary. Suppose that the operator $A : V \mapsto V'$ is defined by

$$\langle v, Au \rangle = -\sum_{i=1}^{N} \int_{\mathcal{O}} \left|\frac{\partial u(x)}{\partial x_i}\right|^{p-2} \frac{\partial u(x)}{\partial x_i} \frac{\partial v(x)}{\partial x_i} \, dx + \int_{\mathcal{O}} au(x)v(x) \, dx, \quad \forall u, v \in V,$$

where $a \in \mathbb{R}$, and consider B of the form $B(t,u) \equiv bu$, where $b \in \mathbb{R}$. Finally, let W_t be a standard real Wiener process. Then,

$$2\langle x, A(t,x) \rangle + |B(t,x)|^2 = -2\|x\|^p + 2a|x|^2 + b^2|x|^2, \quad \forall x \in V, \tag{4.12}$$

so (4.11) holds with equality for

$$\alpha = 2 \quad \text{and} \quad \lambda = 2a + b^2.$$

Condition (i) requires $2a + b^2 < 0$, so $a < 0$ and $b^2 < -2a$. On the other hand, condition (ii) holds whenever $(2a + b^2)a_1^{-1} - 2 < 0$, that is, when $b^2 < 2a_1 - 2a$. Thus, Theorem 4.3 guarantees the exponential stability of paths, $\mathbb{P}$-a.s., only for these values of a and b, which means that the zero solution of the deterministic system

$$\frac{du}{dt} = A(t, u(t)) \tag{4.13}$$

is exponentially stable and the random perturbation is small enough. However, Theorem 4.2 ensures exponential stability for sufficiently large perturbations even though the deterministic system is unstable. In this case, it is not difficult to see that

$$2\langle x, A(t,x) \rangle = -2\|x\|^p + 2a|x|^2 \leq \begin{cases} 2a|x|^2 & \text{if } p > 2, \\ (2a - 2a_1)|x|^2 & \text{if } p = 2, \end{cases} \tag{4.14}$$

so

$$\lambda_0 = \rho_0 = b^2, \quad \nu(t) = \nu_0 = \begin{cases} 2a, & \text{if } p > 2, \\ 2a - 2a_1, & \text{if } p = 2. \end{cases} \tag{4.15}$$

Thus, Theorem 4.2 yields

$$\limsup_{t \to \infty} \frac{1}{t} \log |U_t|^2 \leq \begin{cases} -(b^2 - 2a) & \text{if } p > 2, \\ -(b^2 - 2a + 2a_1) & \text{if } p = 2. \end{cases}$$

Consequently, we have pathwise exponential stability, $\mathbb{P}$-a.s., if

$$b^2 > \begin{cases} 2a & \text{if } p > 2, \\ 2a - 2a_1 & \text{if } p = 2. \end{cases}$$

In general, we have the following result:

Theorem 4.4. *Assume* (b.1), (b.2), (4.11) *and that there exists a nonnegative continuous function* $b = b(t)$ *such that*

$$(B(t,x), x)^2 \geq b(t)|x|^4 \quad \forall x \in V, \tag{4.16}$$

with

$$\liminf_{t \to +\infty} \frac{1}{t} \int_0^t b(s)\, ds \geq b_0 \in \mathbb{R}_+. \tag{4.17}$$

Then, $\mathbb{P}$*-a.s.,*

$$\limsup_{t \to +\infty} \frac{1}{t} \log |U_t|^2 \leq \begin{cases} -(2b_0 - \lambda) & \text{if } p > 1, \\ -(2b_0 - \lambda + \alpha a_1) & \text{if } p = 2, \end{cases} \tag{4.18}$$

for any solution U_t *with* $U_0 = u_0 \in L^2(\Omega, \mathcal{F}_0, \mathbb{P}; H)$ *such that* $|u_0| \neq 0$, $\mathbb{P}$*-a.s.*

Observe that if $\lambda < 0$, then (4.10) is pathwise exponentially stable, $\mathbb{P}$-a.s., for all $p > 1$ and all $b_0 \in \mathbb{R}_+$. However, when $\lambda > 0$, then (4.10) is stable if $2b_0 > \lambda$ (for $p \neq 2$) or $2b_0 > \lambda - \alpha a_1$ (for $p = 2$). Taking into account our previous theorems, we can summarize the analysis for the preceding example:

Case 1: The nonlinear problem, i.e., $p > 2$. In this case the problem is exponentially stable for all $b \in \mathbb{R}$ when $a \leq 0$. However, if $a > 0$ Theorem 4.3 gives stability provided $b^2 > 2a$. We do not know what happens here if $a > 0$ and $b^2 \leq 2a$.

Case 2: The linear problem, i.e., $p = 2$. As in the preceding case, when $a \leq 0$ the system is $\mathbb{P}$-a.s. exponentially stable for all $b \in \mathbb{R}$. But, if $a > 0$, then we need to check (ii), which requires $b^2 < 2a_1 - 2a$, or we have $b^2 > 2a - 2a_1$. Hence, if $a \leq a_1$, then $\mathbb{P}$-a.s. exponential stability follows for all $b \in \mathbb{R}$. However, when $a > a_1$, we can only ensure stability for $b^2 > 2a - 2a_1$ and we do not know what happens for $b^2 \leq 2a - 2a_1$.

Remark 4.1. Theorem 4.3 is a particular case of a more general result which ensures stabilization with general decay rate (super- or sub-exponential). These results can be used to construct stabilizers of PDEs, see Caraballo et al.[14] for more details on these topics.

In conclusion, our results guarantee exponential stability for a wide range of values of a and b. Of course, this means that, given the deterministic system (4.13), the perturbed system becomes exponentially stable when the parameter of the noise is of the type $bx(t)\, dW_t$ and satisfies the conditions above. Otherwise, when we do not know whether the system is stable or not, what can we say? Is it possible to add another stochastic term in order to stabilize the stochastic PDE? The answer is positive and some results on this direction can be found in Caraballo et al.[23]

4.2. Stabilization by Stratonovich noise

There are no general results that we know of concerning the stabilization of nonlinear PDEs by Stratonovich noise. The problem appears to be both difficult and challenging. The only work in this direction involves a canonical model whose dynamics is very well known in the deterministic case. This is the Chafee-Infante equation

$$\frac{\partial u}{\partial t} = \Delta u + \beta u - u^3, \qquad x \in \mathcal{O}, \quad u|_{\partial \mathcal{O}} = 0, \tag{4.19}$$

where $\mathcal{O}$ is a smooth bounded domain in $\mathbb{R}^n$.

It was shown in Caraballo et al.[13] that the *nonlinear* equation (4.19) can be stabilized by adding a collection of noisy terms similar to the linear case in Section 2, leading to the system

$$dU_t = \left[\Delta U_t + \beta U_t - U_t^3\right] dt + \sum_{k=1}^{d} B_k U_t \circ dW_t^k. \tag{4.20}$$

Essentially, it is shown that solutions of (4.20) can be bounded using appropriate *positive* solutions of the linear equation

$$dU_t = [\Delta U_t + \beta U_t] dt + \sum_{k=1}^{d} B_k u \circ dW_t^k. \tag{4.21}$$

Since (4.21) can be stabilized via a suitable choice of B_k, so can (4.20). The proof makes essential and repeated use of the order-preserving properties of (4.20).

To set this problem in a suitable context, we choose $H = L^2(\mathcal{O})$, denote by $-\mathbb{A}$ the linear operator in H associated to the Laplacian, and then take $A = \mathbb{A} + \beta I$, which clearly satisfies the conditions of Theorem 3.2. Finally, we let N be the smallest integer such that $\sum_{j=1}^{N}(\beta - \lambda_j) < 0$. It follows that there exist linear operators $B_k : H \to H$ such that the zero solution of

$$dU_t = [-\mathbb{A} U_t + \beta U_t] dt + \sum_{k=1}^{d} B_k U_t \circ dW_t^k \tag{4.22}$$

is exponentially stable, $\mathbb{P}$-a.s.

This fact can be used to deduce the stabilization of the nonlinear equation via the addition of the same noisy terms. The proof of the next result can be found in Caraballo et al.:[13]

Theorem 4.5. *There exist bounded linear operators* $B_k : H \mapsto H$ *and independent real Wiener processes* W^k, $k = 1, \ldots d$, *such that the zero solution of*

$$dU_t = \left[-\mathbb{A} u + \beta U_t - U_t^3\right] dt + \sum_{k=1}^{d} B_k U_t \circ dW_t^k \tag{4.23}$$

is exponentially stable, $\mathbb{P}$-*a.s.*

This simple, but illustrative example may help to solve the stabilization problem for more general nonlinear PDEs that arise in applications. To the best of our knowledge, this is still an open problem.

5. Other types of evolution equations and models

In this final section we briefly comment on some other types of evolution models whose stabilization by noise have been or are currently under investigation.

5.1. *Delay differential equations*

Some models require the inclusion of delay terms in the equations. The problem of stabilization of delay (ordinary or partial) differential equations is also an important task. In the finite dimensional context, there are only a few results on the stabilization by the Itô noise when the delay is small enough (see Appleby & Mao[1]). So far nothing is known for systems with arbitrary delay (finite or infinite), either with Itô or Stratonovich noise.

5.2. *Stabilization of evolution inclusions and PDEs without uniqueness*

It often happens in applications that a model is better described if we use a set-valued differential equation, i.e., a differential inclusion. In addition, there are other situations in which we cannot ensure uniqueness of solutions for certain evolution equations. These two cases yield to the framework of set-valued dynamical systems. The stabilization analysis carried out in the single-valued case is also important and interesting in both of these situation. For the sake of brevity, we will only exhibit some of the few results which are known on this field, so far (see Caraballo et al.[21] for more details).

Let V be a separable and reflexive Banach space (with norm $||\cdot||$ and inner product $((\cdot,\cdot))$), and consider a Hilbert space H (with norm $|\cdot|$ and inner product $(\cdot,\cdot)$). If we identify H with its dual space, then we can identify H with a subspace of V', so that we have $V \hookrightarrow H \hookrightarrow V'$, where the previous inclusions are continuous and dense. We will denote by $||\cdot||_*$ the norm in V' and by $\langle\cdot,\cdot\rangle$ the duality product between V and V'.

Let us first consider the following stochastic evolution inclusion in the Itô sense:

$$\begin{cases} \dfrac{du(t)}{dt} \in Au(t) + F(u(t)) + \sum_{i=1}^{d} B_i u(t) \dfrac{d}{dt}W_t^i, & 0 \le t < +\infty, \\ u(0) = u_0 \in H, \end{cases} \quad (5.1)$$

where $W_t^1, W_t^2, \cdots, W_t^d$ are mutually independent standard Wiener processes over the same filtered probability space $(\Omega, \mathcal{F}, \{\mathcal{F}_t\}_{t\ge 0}, \mathbb{P})$, $B_i : H \to H$ is a linear operator for $i = 1,\ldots, d$, $A : V \to V'$ is a linear A operator which is the infinitesimal generator of a strongly continuous semigroup (i.e., of class C_0) denoted by $S(t)$. As we are interested in analyzing the behaviour of the variational solutions of (5.1) (see below for the definition), we need to assume some additional hypotheses ensuring their existence. To be more precise, we need the following assumptions:

Assumption 5.

(A4) Coercivity: There exist $\alpha > 0, \lambda \in \mathbb{R}$ such that

$$-2\langle Au, u\rangle + \lambda |u|^2 \geq \alpha ||u||^p, \quad \text{for all } u \in V, \tag{5.2}$$

where $p > 1$ is fixed.

Assumption 6.

(A5) Boundedness: There exists $\beta > 0$ such that

$$||Au||_* \leq \beta ||u||^{p-1}, \quad \text{for all } u \in V. \tag{5.3}$$

Notice that, in the case $p = 2$, condition (5.2) implies that the operator A is the generator of a strongly continuous semigroup (see Dautray & Lions [32, page 388]).

Assumption 7. *The set-valued term $F : H \to 2^H$ satisfies:*

(F1) F has closed, bounded, convex, non-empty values.
(F2) There exists $C > 0$ such that

$$dist_H(F(u), F(v)) \leq C|u - v|, \forall u, v \in H,$$

where $dist_H(\cdot, \cdot)$ denotes the Hausdorff distance between bounded sets.
(F3) $F(0) = 0$.

Under the preceding assumptions (in fact without assuming (5.2), (5.3) and (F3)), Theorem 2.1 in Da Prato & Frankowska[30] ensures the existence of at least one solution $u(\cdot)$ of (5.1) for any random variable $u_0 \in L^p(\Omega, \mathcal{F}_0, \mathbb{P}, H)$ with some $p > 2$. By such a solution we mean an adapted process $u(\cdot)$ taking values in H and such that:

(1) $u(\cdot, \omega)$ is continuous for $\mathbb{P}$-a.a. $\omega \in \Omega$.
(2) For any $T > 0$, $u(\cdot)$ is a mild solution, on the interval $[0, T]$, of the problem

$$\begin{cases} du(t) = Au(t)\, dt + f(t)\, dt + \sum_{i=1}^d B_i u(t) dW_t^i, \\ u(0) = u_0, \end{cases} \tag{5.4}$$

i.e., , we have for all $t \in [0, T]$,

$$u(t) = S(t)u_0 + \int_0^t S(t-s)f(s)\, ds + \sum_{i=1}^d \int_0^t S(t-s)B_i u(s)\, dW_s^i,$$

where $f(\cdot)$ an adapted process such that
$$f(s,\omega) \in F(u(s,\omega)), \text{ for a.a. } (s,\omega) \in (0,T) \times \Omega,$$
with
$$\mathbb{E}\left(\int_0^T |f(s)|^2 ds\right) < \infty.$$

Observe that, for the selection $f(\cdot)$, the unique mild solution to (5.4) is given by $u(\cdot)$.

However, in order to apply Itô's formula (or to make an appropriate change of variable) we need to handle a stronger concept of solution, say, either the so-called strong solution or the variational concept of solution. We will consider the latter here.

In addition to the assumptions just used we now also assume the coercivity (5.2) and boundedness (5.3) assumptions. Then, (see Pardoux[45]), for any $u_0 \in L^p(\Omega, \mathcal{F}_0, \mathbb{P}, H)$ ($p > 2$), there exists a unique *variational solution* of problem (5.4). In other words, there exists a stochastic process $v(\cdot)$ which belongs to $L^p(\Omega \times (0,T); V) \cap L^2(\Omega; C(0,T;H))$ and that satisfies the equation in (5.4) in the sense of V', i.e.,

$$v(t) = u_0 + \int_0^t (Av(s) + f(s))\, ds + \sum_{i=1}^d \int_0^t B_i v(s)\, dW_s^i, \quad \mathbb{P}-\text{a.a. } \omega \in \Omega,$$

for all $t \in [0,T]$, where the equality is understood in the sense of V'. Now, taking into account that the variational solution (when it exists) is also a mild solution (see, e.g. Caraballo[9]) it follows that for an initial datum $u_0 \in L^p(\Omega, \mathcal{F}_0, \mathbb{P}, H)$ and for the selection $f(\cdot)$ in (5.4), there exists a unique variational solution which is also a solution of (5.1) in the sense of Da Prato and Frankowska. Henceforth, when we talk about a solution of (5.1) we will be always referring to this one.

As was seen in the single-valued case, in order to produce a stabilization effect on deterministic (and even stochastic) systems one does not need to perturb the model with a very general noise (provided it is considered in the Itô sense). In fact, a very simple multiplicative one is enough and the following stabilisation result holds (see Caraballo et al.[21]).

Theorem 5.1. *Assume that $B_2 = \cdots = B_d = 0$ and B_1 is given by $B_1 v = \sigma v$ for $v \in H$ and $\sigma \in \mathbb{R}$. Under the preceding assumptions, for σ^2 large enough so that*
$$\gamma = \sigma^2 - \lambda - 2C > 0$$
(whereλ and C are the constants appearing in (5.2) and (F2)), there exists $\Omega_0 \subset \Omega$ with $\mathbb{P}(\Omega_0) = 0$ and a random variable $T(\omega) \geq 0$ such that for any initial datum $u_0 \in L^p(\Omega, \mathcal{F}_0, \mathbb{P}, H)$ ($p > 2$), any of its corresponding solutions $u(\cdot)$ of problem (5.1) satisfies

$$|u(t,\omega)|^2 \leq e^{-\gamma t/2}|u_0(\omega)|^2 \quad \text{for all } t \geq T(\omega), \text{ a.s.} \tag{5.5}$$

Remark 5.2. It is worth mentioning that the stabilization of this evolution inclusion by Stratonovich noise has not been proved yet.

One can also find a result on the stabilization of set-valued dynamical systems generated by evolution partial differential equations without uniqueness. The main tool for the proof is the theory of random dynamical systems and random attractors. Specifically, in Caraballo et al.[21] is proved that some kind of high intensity noise, in the sense of Itô, will ensure the existence of a random attractor for the perturbed problem, while the deterministic one does not have an attractor (or it is not known whether the deterministic attractor exists). We will not include more details on this problem, since it is not our aim to introduce the theory of random attractors here.

5.3. *Stabilization of stationary solutions of a stochastic PDE*

The analysis in the preceding sections was concerned with the stabilization of the null solution of an evolution equation or inclusion by introducing some kind of noise in the deterministic model. However, it may happens that zero is not solution of the unperturbed equation, or may not be solution of the stochastic model when the noise is present. In this case, Caraballo et al.[17] showed the existence of an exponentially stable stationary stochastic solution when a suitable Itô noise was included, i.e., where the stationarity is understood in the sense of stochastic processes. Again, it is an open problem to analyze the same stabilization effect produced by Stratonovich noise (see Caraballo et al.[16,17] for more details).

5.4. *Other types of problems*

Another problem that may be even closer to reality is related to the effect produced by the noise when it acts only on (part of) the boundary of the domain and not in the forcing term in the equation. For instance, if we are considering an oceanic model, the stochastic disturbances may appear on the ocean surface and not in the equations driving the system. To the best of our knowledge, no papers have been published on this topic, although we have started to investigate this problem and some preliminary results will appear in the future.

On the other hand, the synchronization of systems modelled by differential equations has received much attention over the last years, for example, see the review articles Caraballo et al.[15] and Kloeden & Pavani.[35] For instance, Chueshov & Rekalo[27,27] analyzed the synchronization of a deterministic model concerning a reaction diffusion equation on either side of a permeable membrane. The synchronization proved by Chueshov and Rekalo is at the level of global attractors. However, the addition of a non-degenerate noise reduces the attractors to pathwise asymptotically stable stochastic stationary solutions and, in this way, almost all solutions starting in two points on both sides of the membrane, have to evolve in a synchronized way which is determined by the random attractor (which is given by a stationary solution) on the membrane. This can be regarded as a form of stabilization (see see Caraballo et al.[12] for more details).

Acknowledgement

This work has been partly supported by Ministerio de Ciencia e Innovación, Spain, under the grant MTM2008-00088, and Junta de Andalucía grant P07-FQM-02468.

References

1. J. Appleby and X. Mao, Stochastic stabilization of functional differential equations, *Systems and Control Letters* 54(11) (2005), 1069–1081.
2. J. Appleby, X. Mao and A. Rodkina, Stabilization and destabilization of nonlinear differential equations by noise, *IEEE Trans. Automatic Control* 53(3) (2008), 683–691.
3. L. Arnold, *Stochastic Differential Equations: Theory and Applications*, Wiley & Sons, New York, (1974).
4. L. Arnold, *Random Dynamical Systems*. Springer, New York, 1998.
5. L. Arnold, Stabilization by noise revisited, *Z. Angew. Math. Mech.* 70(1990), 235-246.
6. L. Arnold and I. Chueshov, Order-preserving random dynamical systems: Equilibria, attractors, applications, *Dyn. Stab. Sys.* 13 (1998), 265–280.
7. L. Arnold, H. Crauel and V. Wihstutz, Stabilization of linear systems by noise, *SIAM J. Control Optim.* 21(1983), 451-461.
8. L. Arnold and P. E. Kloeden, Lyapunov exponents and rotation number of two-dimensional systems with telegraphic noise, *SIAM J. Applied Math.* 49 (1989), 1242-1274.
9. T. Caraballo, PhD. Thesis, University of Sevilla, Spain (1988).
10. T. Caraballo, Recent results on stabilization of PDEs with noise, *Bol. Soc. Esp. Mat. Apl.* 37 (2006), 47-70.
11. T. Caraballo and J.A. Langa, Comparison of the long-time behavior of linear Ito and Stratonovich partial differential equations, *Stoch. Anal. Appl.* 19(2) (2001), 183-195.
12. T. Caraballo, I. Chueshov and P.E. Kloeden, Synchronization of a stochastic reaction-diffusion system on a thin two-layer domain, *SIAM J. Math. Anal.* **38** (2007), 1489–1507.
13. T. Caraballo, H. Crauel, J.A. Langa and J.C. Robinson, The effect of noise on the Chafee-Infante equation: a nonlinear case study, *Proc. Amer. Math. Soc.*, in press.
14. T. Caraballo, M.J. Garrido-Atienza and J. Real, Stochastic stabilization of differential systems with general decay rate, *Systems & Control Letters* 48(5) (2003), 397-406.
15. T. Caraballo, P.E. Kloeden, A. Neuenkirch and R. Pavani, Synchronization of dissipative systems with additive and linear noise, in *Festschrift in Celebration of Prof. Dr. Wilfried Grecksch's 60th Birthday*, Christiane Tammer and Frank Heyde (Editors), Shaker-Verlag, Aachen, 2008, pp. 25–48. (ISBN 978-3-8322-7500-6 ISSN 0954-0882)
16. T. Caraballo, P.E. Kloeden and B. Schmalfuss, Exponentially stable starionary solutions for stochastic evolution equations and their perturbations, *Appl. Math. Optim.* 20(2004), 183–207
17. T. Caraballo, P.E. Kloeden and B. Schmalfuß, Stabilization of stationary solutions of evolution equations by noise, *Discrete Conts. Dyn. Systems, Series B.* **6** (2006) 1199-1212.
18. T. Caraballo, J.A. Langa and J.C. Robinson, Stability and random attractors for

a reaction-diffusion equation with multiplicative noise, *Discrete Cont. Dyn. Sys.* 6 (2000), 875–892.
19. T. Caraballo, J.A. Langa and J.C. Robinson, A stochastic pitchfork bifurcation in a reaction-diffusion equation, *R. Soc. Lond. Proc. Ser. A* 457 (2001), 2041–2061.
20. T. Caraballo, J.A. Langa and T. Taniguchi, The exponential behaviour and stabilizability of stochastic 2D-Navier-Stokes equations, *J. Diff. Eqns.* 179(2002), 714-737.
21. T. Caraballo, J.A. Langa and J. Valero, Stabilisation of differential inclusions and PDEs without uniqueness by noise, *Comm. Pure and Applied Analysis* 7 (2008), 1375-1392.
22. T. Caraballo and K. Liu, On exponential stability criteria of stochastic partial differential equations, *Stoch. Proc. & Appl.* 83 (1999), 289-301.
23. T. Caraballo, K. Liu and X.R. Mao, On stabilization of partial differential equations by noise, *Nagoya Math. J.* 161(2) (2001), 155-170.
24. T. Caraballo, A.M. Márquez-Durán and J. Real, On the asymptotic behaviour of a stochastic 3D-Lans-alpha model, *Appl. Math. Optim.* 53(2006), 141-161.
25. T. Caraballo, J. Real and T. Taniguchi, On the existence and uniqueness of solutions to stochastic 3-dimensional Lagrangian averaged Navier-Stokes equations, *R. Soc. Lond. Proc. Ser. A* 462(2006), 459-479.
26. T. Caraballo and J.C. Robinson, stabilization of linear PDEs by Stratonovich noise, *Systems & Control Letters* 53(2004), 41-50.
27. I.D. Chueshov and A.M. Rekalo, Global attractor of contact parabolic problem on thin two-layer domain, *Sbornik: Mathematics,* **195** No. 1 (2004), 103–128.
28. I.D. Chueshov and A.M. Rekalo, Long-time dynamics of reaction-diffusion equations on thin two-layer domains, in *EQUADIFF-2003 Proceedings*, edited by F. Dumortier, H. Broer, J. Mawhin, A. Vanderbauwhede and S.V. Lunel, World Scientific Publishing, Singapore 2005, pp. 645–650.
29. G. Da Prato, A. Debussche, and B. Goldys, Some properties of invariant measures of non symmetric dissipative stochastic systems, *Prob. Theor. Relat. Fields* **123** (2002), 355-380.
30. G. Da Prato and H. Frankowska, *A stochastic Filippov theorem*, Stochastic Anal. Appl. **12** (4) (1994), 409-426.
31. G. Da Prato and J. Zabczyk, *Stochastic Equations in Infinite Dimensions*, Cambridge University Press, (1992).
32. R. Dautray and J.L. Lions, "Analyse Mathèmatique et Calcul Numérique por les Sciences et les Techniques", Masson, Paris (1984).
33. M. Hairer, Exponential mixing properties of stochastic PDEs through asymptotic coupling, *Prob. Theory Relat. Fields* 124 (2002), 345–380.
34. R. Has'minskii, *Stochastic Stability of Differential Equations*, Sijthoff and Noordhoff, Netherlands, (1980).
35. P.E. Kloeden and R. Pavani, Dissipative synchronization of nonautonomous and random systems, *GAMM-Mitt.* **32** (2009), No. 1, 80 92.
36. P.E. Kloeden and E. Platen, *The Numerical Solution of Stochastic Differential Equations*, Springer–Verlag, 1992 (revised reprinting 1995, 3rd revised and updated printing 1999)
37. H. Kunita, Stochastic Partial Differential Equations connected with Non-Linear Filtering, in Lecture Notes in Mathematics 972, pp. 100-169, (1981).
38. A.A. Kwiecinska, Stabilization of partial differential equations by noise, *Stoch. Proc.& Appl.* 79 (1999), no. 2, 179–184
39. A.A. Kwiecinska, Stabilization of evolution equations by noise, *Proc. Amer. Math. Soc.* 130(2002), No. 10, 3067-3074.

40. J.A. Langa and J.C. Robinson, Upper box-counting dimension of a random invariant set, *J. Math. Pures App.*, 85 (2006), no. 2, 269–294.
41. G. Leha, B. Maslowski and G. Ritter, Stability of solutions to semilinear stochastic evolution equations, *Stoch. Anal. Appl.* 17(1999), No. 6, 1009-1051.
42. J.L. Lions, *Quelque méthodes de résolution des problèmes aux limites non linéaires*, Dunod, Gauthier- Villars, Paris, 1969.
43. X.R. Mao, Stochastic stabilization and destabilization, *Systems & Control Letters* 23 (1994), 279-290.
44. A. Pazy, *Semigroups of Linear Operators and Applications to Partial Differential Equations*, Springer-Verlag, New York Inc., 1983.
45. E. Pardoux, *Équations aux Dérivées Partielles Stochastiques non Linéaires Monotones*, Thesis Univ. Paris XI, 1975.
46. M. Scheutzow, Stabilization and destabilization by noise in the plane, *Stoch. Anal. Appl.* 11(1) (1993), 97-113.
47. H.J. Sussmann, On the gap between deterministic and stochastic ordinary differential equations, *The Annals of Probability* 6(1978) , No. 1, 19-41.
48. E. Wong & M. Zakai, On the relationship between ordinary and stochastic differential equations and applications to stochastic problems in control theory, *Proc. Third IFAC Congress*, paper 3B, 1966.

Chapter 4

Stochastic Quantification of Missing Mechanisms in Dynamical Systems

Baohua Chen and Jinqiao Duan*

Department of Applied Mathematics, Illinois Institute of Technology,
Chicago, IL 60616, USA
duan@iit.edu

Complex systems often involve multiple scales and multiple physical, chemical and biological mechanisms. Due to the lack of scientific understanding, some mechanisms are not represented (i.e., "unresolved") in mathematical models for these complex systems. The impact of these unresolved processes on the resolved ones may be delicate and needs to be quantified. A stochastic dynamical approach is proposed to parameterize these missing mechanisms or unresolved scales. An example is briefly discussed to demonstrate this stochastic approach.

Key Words: Stochastic parameterization; impact of unresolved scales or missing mechanisms; fractional Brownian motion; correlated noise; stochastic partial differential equations

Contents

1. Introduction 67
2. Stochastic analysis and stochastic parameterizations 68
3. An example 70
References 75

1. Introduction

In building mathematical models for complex dynamical systems in science and engineering, not all mechanisms are taken into account, due to the lack of scientific understanding of or difficulty in representing such mechanisms. These missing mechanisms, although perhaps small in spatial scales or fast in temporal scales, may have delicate impact on the overall, either in transient or in long time, evolution of the systems. This is especially the case when a deterministic model could predict some dynamical behaviors of a complex phenomenon, but fails to capture other system features. The missing mechanisms may appear as random fluctuations, which are thus more amenable to stochastic representations.

---

*Partially supported by NSF grants 0620539 and 0731201, the Cheung Kong Scholars Program and the K. C. Wong Education Foundation

We propose a stochastic approach for representing missing mechanisms in dynamical systems, as mathematical models for complex phenomena. This is a stochastic analysis-based, data-driven method for stochastic parameterizations.

In §2, we present a stochastic parameterization approach, and then apply it to a geophysical model in §3.

2. Stochastic analysis and stochastic parameterizations

We review a few stochastic concepts.

- *Brownian motion*: Brownian motion (also known as Wiener process) is among the simplest of the continuous-time stochastic processes.
  Brownian motion B_t on a probability space $(\Omega, \mathcal{F}, \mathbb{P})$ is characterized by the following facts:
  - $B_0 = 0$ almost surely.
  - B_t is almost surely continuous.
  - B_t has independent and stationary increments with distribution $B_t - B_s \sim N(0, t-s)$ for $0 \leq s < t$.

  The most important property of Brownian motion is that its successive increments are uncorrelated: each displacement is independent of the former, in direction as well as in amplitude.

- *Fractional Brownian motion*: Mandelbrot and van Ness (1968) defined a family of stochastic processes they called fractional Brownian motion. The main difference with ordinary Brownian motion is that in a fractional Brownian motion successive increments are correlated. Fractional Brownian motion has for long served as the archetype of a process with long range dependence (LRD).[18]

A fractional Brownian motion $B^H(t)$, $t \in \mathbb{R}$, is a continuous-time Gaussian process starting at zero, with mean zero, and having the following correlation function:

$$E[B^H(t) B^H(s)] = \frac{1}{2}(|t|^{2H} + |s|^{2H} - |t-s|^{2H}),$$

where $0 < H < 1$, called the self-similar parameter or Hurst parameter. The Hurst parameter may be interpreted as a measure of the roughness.

Figure 4.1 show the sample paths of fractional Brownian motion with various parameter values H. The larger the parameter H is, the smoother the path is. Some properties of the fractional Brownian motion are as follows.

- Fractional Brownian motion is a Gaussian process.
- Fractional Brownian motion is statistically self-similar

$$B^H(\lambda t) = |\lambda|^H B^H(t), \quad 0 < H < 1, \quad \lambda \in \mathbb{R}.$$

- Fractional Brownian motion has dependent and stationary increments.
- For $H = \frac{1}{2}$, the process is the ordinary Brownian motion, which has independent increments.
- For $H > \frac{1}{2}$, the increments of the process are positively correlated.
- For $H < \frac{1}{2}$, the increments of the process are negatively correlated.

The Hurst parameter H characterizes the important properties of fractional Brownian motion. Several statistics have been introduced to estimate H, such as wavelets, k-variations, maximum likelihood estimator, and spectral methods. A recent reference for these various approaches is represented by Ref. 2.

- *Colored noise:* Fractional Brownian motion has dependent and stationary increments. Its generalized time derivative is a mathematic model of colored noise. By the way, white noise is modeled as the generalized time derivative of Brownian motion.

Consider a dynamical system modeled by the following deterministic partial differential equation for a state variable $q(\vec{x}, t)$

$$q_t = Aq + N(q), \tag{2.1}$$

where A is a (unbounded) linear differential operator and $N(\cdot)$ is a nonlinear operator. If the model output matches well (in certain metric) with observation for q, there is not much need to go to stochastic dynamical modeling. However, when there is no such good match, this deterministic model (sometimes, a simplified or idealized conceptual model) needs to be improved. In fact, mathematical modeling is a process of obtaining more and more accurate descriptions for complex systems.

To improve this deterministic model (2.1), we first figure out the model error or model uncertainty. Namely, we use observational data for q to feed into an appropriately space-time discretized version of (2.1). Because of the mis-match mentioned above, there will be a residual term F in the discretized version of (2.1). This residual term, defined on a space-time grid, is the model uncertainty. It contains information for the unaccounted, missing dynamical mechanisms in the original deterministic model. Note that the observational data is usually fluctuating or random, the missing mechanisms $F(\vec{x}, t, \omega)$ as represented as data on the space-time grid are also random, i.e., with different realizations or samples. This calls for a stochastic parameterization for the missing mechanisms $F(\vec{x}, t, \omega)$.

For example, we may try this approximation

$$F(\vec{x}, t, \omega) = \sigma(\vec{x}, t, q) \, \dot{B}_t^H, \tag{2.2}$$

where σ is an empirical formula describing fluctuating amplitude or noise intensity for F, and B_t^H is a fractional Brownian motion with Husrt parameter H. Here σ

may depend on parameters. These parameters and H may be estimated using data for F, with help of stochastic analysis. Thus we obtain a new, hopefully improved, dynamical model, i.e., a stochastic partial differential equation

$$q_t = Aq + N(q) + \sigma(\vec{x}, t, q)\, \dot{B}_t^H. \qquad (2.3)$$

For more information on stochastic parameterization, see Refs. 8–11,16,23,30 and references therein.

3. An example

We now consider an example to demonstrate the above stochastic parameterization method. We consider a model for water vapor dynamics in the climate system. An advection-diffusion-condensation equation is an idealized model for specific humidity.[20,21,25] In the absence of diabetic effects, parcel motion is restricted to two dimensional surfaces of the constant potential temperature, i.e., to isentropic surface (We consider $\theta = 315K$ here).[14] The specific humidity q, subject to an advection, diffusion and condensation, is governed by the following equation:

$$q_t + \mathbf{v} \cdot \nabla q = \eta \Delta q + S(q, q_s) + F, \qquad (3.1)$$

where $\mathbf{v} = (u, v)$ is advection velocity field, η is diffusivity of condensable substance, and $S(q, q_s)$ represents the sources and sinks of water vapor for the parcel of air considered. A frequent simplification is made in modeling that the water vapor falls out in the form of precipitation as soon as it condenses. The sink is thus represented as

$$S = -\frac{1}{\tau}(q - q_s) \quad \text{if } q > q_s,$$
$$= 0 \quad \text{if } q \le q_s.$$

Saturation specific humidity is $q_s = \epsilon e_s / p$, with water vapor saturation vapor pressure e_s, which is a function of temperature. Saturation vapor pressure is calculated according to the Clausius-Clapeyron relation. The empirical calculating formula is introduced in Ref. 3.

The missing mechanism F includes the small scale convective moistening, which is also called "convective forcing" thereinafter. It consists of contributions owing to subgrid-scale moist convection and to other subgrid scale turbulence. They are vital to the water vapor distribution but are not resolved explicitly in the model.

We analyzed the mean fields of water vapor in ECMWF Re-Analysis (ERA-40) 4-time daily data for the years 1975-2000, level is fixed at one isentropic surface $\theta = 315K$. Water vapor content is measured by specific humidity q. Dynamical fields in the reanalysis data are represented spectrally, with triangular truncation at total wavenumber 159. Specific humidity is represented on a corresponding reduced Gaussian grid at N80 resolution. For the computation of water vapor and other flow fields, we transformed spectral fields to the corresponding regular Gaussian grid and interpolated specific humidity to the same grid. The resolution of the data is $2.5° \times 2.5°$ in longitude and latitude. The ERA-40 data currently represents the higher resolution data set that provides water vapor and dynamical fields with continuous long-term coverage.

Convective moistening is deduced by equation (3.1):

$$F = q_t + \mathbf{v} \cdot \nabla q - \eta \Delta q - S(q, q_s). \tag{3.2}$$

Discritizing this equation by utilizing up-wind scheme, we obtain

$$F_{i,j,k} = \frac{q_{i,j,k+1} - q_{i,j,k}}{\Delta t} + (u_{i,j,k} \frac{q_{i+1,j,k} - q_{i,j,k}}{\Delta x} + v_{i,j,k} \frac{q_{i,j+1,k} - q_{i,j,k}}{\Delta y})$$
$$- \eta(\frac{q_{i+1,j,k} + q_{i-1,j,k} - 2q_{i,j,k}}{(\Delta x)^2} + \frac{q_{i,j+1,k} + q_{i,j-1,k} - 2q_{i,j,k}}{(\Delta y)^2})$$
$$- S_{i,j,k}. \tag{3.3}$$

Plugging data of $\mathbf{v} = (u, v)$, q and $S(q, q_s)$ into this discretized equation, convective forcing $F(x, y, t)$ can be estimated at each time and each grid, for various realizations or samples.

Autocorrelation of F shows that it is temporally correlated, so colored noise is needed to parameterize it:[4]

$$F(\vec{x}, t) = \sigma(\vec{x})(q - \alpha \cdot q_s) \frac{dB_t^H}{dt}, \tag{3.4}$$

where B_t^H is the fractional Brownian motion with Hurst parameter H, and $\sigma(\vec{x}) \geq 0$ is the deterministic noise intensity which usually depends on space. The noise $\sigma(\vec{x})(q - \alpha \cdot q_s)\dot{B}_t^H$ is a multiplicative colored noise with tuning parameter α, which is determined by empirical testing.

Now we estimate the Hurst parameter H and noise intensity σ. Integrating the equation (3.4) on the observational time interval $[0, T]$, we get

$$\int_0^T F(\vec{x}, t) dt = \sigma(\vec{x}) \int_0^T (q(\vec{x}, t) - \alpha \cdot q_s(\vec{x}, t)) dB_t^H. \tag{3.5}$$

In this way, the mean of this integral is not zero (different from Ito integral [a]), but in consistent with the mean of the convective forcing F. Therefore, there is no need to find an expression for the mean of F, as its impact is included in F.

Let us denote $Z_t = \int_0^t F ds$, then equation (3.5) can be written:

$$Z_T = \sigma(\vec{x}) \int_0^T (q(\vec{x}, t) - \alpha \cdot q_s(\vec{x}, t)) dB_t^H. \tag{3.6}$$

We have the following property for fractional stochastic integrals:[6]

---

[a] Ito Integral's properties: if $f \in \nu(0, T)$, $\mathbb{E}[\int_0^T f dB_t] = 0$[19]

Theorem 3.1 (Convergence of Variation[6]). For a given $p < \frac{1}{1-H}$, as $n \to \infty$, we have the (uniform) convergence in probability:

$$\triangle_n^{1-pH} V_p^n(Z)_T \to c_p \, \sigma^p(\vec{x}) \int_0^T |q(\vec{x},\tau) - \alpha \cdot q_s(\vec{x},\tau)|^p d\tau, \qquad (3.7)$$

where $c_p = \mathbb{E}(|B_1^H|^p) = \frac{2^{p/2}\Gamma(\frac{p+1}{2})}{\Gamma(\frac{1}{2})}$ with Γ the Gamma function, the p-power variation $V_p^n(Z)_T$ is defined as

$$V_p^n(Z)_T = \sum_{i=1}^n |Z_{i/n} - Z_{(i-1)/n}|^p,$$

and finally $\triangle_n = \frac{T}{n}$ and $t_i^n = i\triangle_n$.

Thus, we have an estimator for the noise intensity when n is large.

$$\tilde{\sigma}(\vec{x}) \approx \left\{ \frac{\triangle_n^{1-pH} V_p^n(Z)_T}{c_p \int_0^T |q(\vec{x},\tau) - \alpha \cdot q_s(\vec{x},\tau)|^p d\tau} \right\}^{\frac{1}{p}}. \qquad (3.8)$$

Since $\tilde{\sigma}(\vec{x})$ is sample-wise, we take the mean

$$\sigma(\vec{x}) \approx \mathbb{E}\left[\left\{ \frac{\triangle_n^{1-pH} V_p^n(Z)_T}{c_p \int_0^T |q(\vec{x},\tau) - \alpha \cdot q_s(\vec{x},\tau)|^p d\tau} \right\}^{\frac{1}{p}} \right]. \qquad (3.9)$$

How to find the estimator $\hat{H}_n$ for H? Estimator $\hat{H}_n$ is taken as the minimizer for the optimization problem:

$$\min_{0 < H < 1} \left[\triangle_n^{1-pH} V_p^n(Z)_T - c_p \, \sigma^p(\vec{x}) \int_0^T |q(\vec{x},\tau) - \alpha \cdot q_s(\vec{x},\tau)|^p d\tau \right], \qquad (3.10)$$

where σ is estimated as (3.9). See Ref. 4 for more details.

Thus we obtain a stochastic model for water vapor evolution:

$$q_t + \mathbf{v} \cdot \nabla q = \eta \Delta q + S(q, q_s) + \sigma \cdot (q - \alpha \cdot q_s) \dot{B}_t^H. \qquad (3.11)$$

Figure 4.2 shows a reasonable agreement between observation, and simulation of the stochastic model (3.11), for the El Niño-associated eastward shift in the moister regions from the western Pacific warm pool to the central eastern Pacific during 1997/1998 winter.

Acknowledgements

We are very grateful to Ray Pierrehumbert, Joe Tribbia, and Grant Branstator for their constructive suggestions and insights. We also thank Xiaofan Li for numerous valuable comments with numerical simulations, and thank Chi-fan Shih for providing access to data sets.

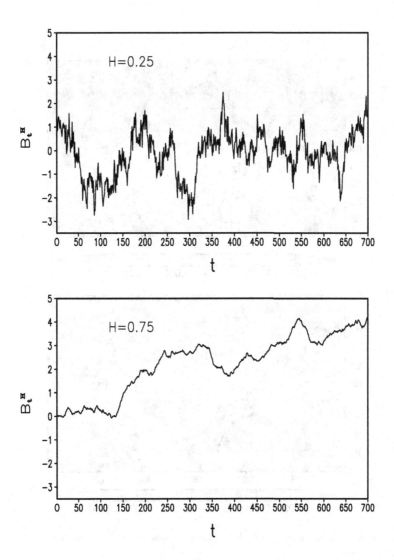

Figure 4.1. Fractional Brownian motion with Hurst parameters $H = 0.25$ (top) and $H = 0.75$ (bottom)

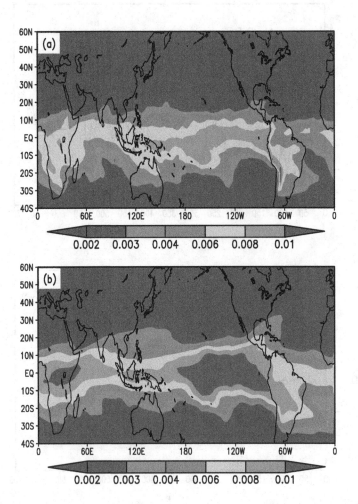

Figure 4.2. Seasonally-averaged specific humidity in El Niño winter: (a) Observation; (b) Stochastic model simulation

References

1. Bender, C. "An Ito Formula for Generalized Functionals of a Fractional Brownian Motion with Arbitrary Hurst Parameter". *Stoch. Proc. Appl.* **104**(2003): 81-106.
2. Beran, J. "Statistics for Long-Memory Processes". Chapman and Hall, 1994.
3. Buck, A. L. "New Equations for Computing Vapor Pressure and Enhancement Factor". *Journal of Applied Meteorology* **20**(1981): 1527-1532.
4. Chen, B. *Stochastic Dynamics of Water Vapor in the Climate System*. Ph.D. Thesis, 2009, Illinois Institute of Technology, Chicago, USA.
5. Coeurjolly, J.-P. "Estimating the Parameters of a Fractional Brownian Motion by Discrete Variations of Its Sample Paths". *Stat. Inference for Stoch. Proc.* **4**(2001): 199-227.
6. Corcuera, J. M., D. Nualart, and J. H. C. Woerner. "Power Variation of Some Integral Fractional Processes". *Bernoulli* **12**(2006): 713-735.
7. Dai, A. "Recent Climatology, Variability, and Trends in Global Surface Humidity". *J. Clim.* **19**(2006): 3589-3606.
8. Du, A. and J. Duan. "A Stochastic Approach for Parametering Unresolved Scales in a System with Memory". *Journal of Algorithms and Computational Technology* **3**(2009): 393-405.
9. Duan, J. and B. Nadiga. "Stochastic Parameterization of Large Eddy Simulation of Geophysical Flows". *Proc. American Math. Soc.* **135**(2007): 1187-1196.
10. Duan, J. and B. Chen. "Quantifying Model Uncertainty by Fractional Brownian Motions". *Oberwolfach Reports* **5**(2008).
11. Duan, J. "Stochastic Modeling of Unresolved Scales in Complex Systems". *Frontiers of Math. in China* **4** (2009), 425-436.
12. Duan, J. "Predictability in Nonlinear Dynamical Systems with Model Uncertainty". In *Stochastic Physics and Climate Modeling*, T. N. Palmer and P. Williams (eds.), Cambridge Univ. Press, 2009.
13. Held I. M. and B. J. Soden. "Water Vapor Feedback and Global Warming". *Annual Review Energy Environment* **25**(2000): 441-475.
14. Hoskins, B. "Towards a PV-θ View of the General Circulation". *Tellus* **43AB**(1991): 27-35.
15. Kolmogorov, A. N. "The Local Structure of Turbulence in Incompressible Fluid at Very High Reymond Numbers". *Dokl. Acad. Sci., USSR*, **30**(1941): 299-303.
16. Lin, J. W.-B. and J. D. Neelin. Considerations for Stochastic Convective Parameterization. *J. Atmos. Sci.* **59**(2002): 959-975.
17. Marshak, A., A. Davis, R. Sahalan, and W. Wiscombe. "Bounded Cascade Models as Nonstationaty Multifractals". *Phys. Rev.* **E49**(1994).
18. Nualart, D. "Stochastic Calculus with respect to the Fractional Brownian Motion and Applications". *Contemporary Mathematics* **336** (2003), 3-39.
19. Oksendal, B. " Stochastic Differenntial Equations". Sixth Ed., Springer-Verlag, New York, 2003.
20. Pierrehumbert, R. T., Brogniez H, and Roca R, 2005: "'On the Relative Humidity of the Earth's Atmosphere". In *The General Circulation of the Atmosphere*, 143-185, T. Schneider and A. Sobel, eds. Princeton University Press, 2007.
21. Pierrehumbert, R. T. "Subtropical Water Vapor As a Mediator of Rapid Global Climate Change". In Clark PU, Webb RS and Keigwin LD eds. *Mechanisms of Global Change at Millennial Time Scales*. American Geophysical Union: Washington, D.C. Geophysical Monograph Series **112**: 177-201.
22. Pierrehumbert, R. T. "Anomalous Scaling of High Cloud Variability in the Tropical

Pacific". *Geophysical Research Letters*, **23**(1996): 1095-1098.
23. Palmer, T. N., G. J. Shutts, R. Hagedorn, F. J. Doblas-Reyes, T. Jung and M. Leutbecher. "Representing Model Uncertainty in Weather and Climate Prediction". *Annu. Rev. Earth Planet. Sci.* **33**(2005): 163-193.
24. Sui, H., W. G. Read, J. H. Jiang, and et.al, "Enhanced Positive Water Vapor Feedback Associated With Tropical Deep Convection: New Evidence From Aura MLS". *Gephys. Res. Lett.* **33**(2006).
25. Sukhatme, J. and R. T. Pierrehumbert. "Statistical Equilibria of Uniformly Forced Advection Condensation". (posted on ArXiV), 2005.
26. Sun, D.-Z. and R. S. Lindzen, "Distribution of Tropical Tropospheric Water Vapor". *J. Atmos. Sci.* **50**(1993a): 1643-1660.
27. Teixeira, J. and Carolyn A. Reynolds. "Stochastic Nature of Physical Parameterizations in Ensemble Predictions: A stochsatic Convection Approach". *Monthly Weather Review* **136**(2008): 483-496.
28. Tudor, C. A. and F. G. Viens. "Variations and Estimators for the Self-similarity Order through Malliavan Calculus". *Submitted*, 2007.
29. Vainshtein, S. I., K. R. Sreenivasan, R. T. Pierrehumbert and et al, "Scaling Exponents for Turbulence and Other Random Process and Their Relationships with Multifractal Structure". *Phys. Rev. E* **50**(1994).
30. Williams, P. D. "Modelling Climate Change: the Role of Unresolved Processes". *Phil. Trans. R. Soc. A* **363**(2005): 2931-2946.

Chapter 5

Banach Space-Valued Functionals of White Noise

Yin Chen and Caishi Wang*

*School of Mathematics and Information Science, Northwest Normal University,
Lanzhou 730070, China*

Let X be a complex Banach space (not necessary to be reflexive). In this chapter, we prove a moment characterization theorem and a convergent theorem for X-valued generalized functionals of white noise, which refine the corresponding results recently obtained by Wang and Huang (Acta Math. Sin. Engl. Ser. 22 (2006), 157–168).

Contents

1 Introduction 77
2 Kernel theorems 78
3 Main theorems 81
References 89

1. Introduction

Let X be a complex Banch space. By X-valued generalized functionals of white noise we mean continuous linear mappings defined on the testing functionals of white noise and valued in X. Such mappings form an important subset of vector-valued functionals of white noise and may find their applications in the study of stochastic evolution equations in Banach spaces.

Kondratiev and Streit[5] introduced the moment approach to characterize scalar-valued generalized functionals of white noise. It turns out that the moment approach is independent of the choice of the framework of white noise analysis. In Ref. 9, the authors applied the moment approach to X-valued generalized functionals of white noise and obtained some interesting results under the condition that X is reflexive. The reflexive condition on X, however, appears somewhat unnatural.

The main purpose of the present paper is to refine the work of Ref. 9. More precisely, we will show that the moment characterization theorem and the convergent theorem given in Ref. 9 remain true without the reflexive condition on X.

---

*Corresponding author (wangcs@nwnu.edu.cn)

2. Kernel theorems

Let X be a complex Banach space (not necessary to be reflexive). In this section, we prove some kernel theorems for X-valued multilinear mappings. These theorems will play a crucial role in proving our main theorems.

Let $\mathcal{H}$ be a separable complex Hilbert space with inner product $\langle \cdot, \cdot \rangle$ and norm $|\cdot|$. We denote by $\langle \cdot, \cdot \rangle_{X^* \times X}$ the canonical bilinear form on $X^* \times X$, where X^* is the dual of X.

Let $n \geq 1$ and $M: \mathcal{H}^n \longmapsto X$ an n-linear mapping. M is called bounded if $\|M\| < \infty$, where $\|M\|$ is defined by

$$\|M\| = \sup \Big\{ \|M(h_1, h_2, \cdots, h_n)\|_X \mid |h_1| \leq 1, |h_2| \leq 1, \cdots, |h_n| \leq 1,$$
$$(h_1, h_2, \cdots, h_n) \in \mathcal{H}^n \Big\}.$$

In that case, $\|M\|$ is called the norm of M.

Definition 2.1. Let $n \geq 1$. A bounded n-linear mapping $M: \mathcal{H}^n \longmapsto X$ is said to be strongly bounded if there exists an orthonormal basis $\{e_k\}_{k \geq 1}$ of $\mathcal{H}$ such that

$$\|M\|_s^2 \equiv \sup_{\|g\|=1, g \in X^*} \sum_{j_1, j_2, \cdots, j_n} |\langle g, M(e_{j_1}, e_{j_2}, \cdots, e_{j_n}) \rangle_{X^* \times X}|^2 < \infty \qquad (2.1)$$

where $\sum_{j_1, j_2, \cdots, j_n} \equiv \sum_{j_1, j_2, \cdots, j_n = 1}^{\infty}$. In that case, $\|M\|_s$ is called the strong norm of M.

As is shown below, $\|M\|_s$ is actually independent of the choice of the orthonormal basis $\{e_k\}_{k \geq 1}$.

Let $\mathcal{H}^{\otimes n}$ be the n-fold Hilbert tensor product of $\mathcal{H}$. By convention, the inner product and norm of $\mathcal{H}^{\otimes n}$ are still denoted by $\langle \cdot, \cdot \rangle$ and $|\cdot|$, respectively.

Theorem 2.1. Let $n \geq 1$. If $M: \mathcal{H}^n \longmapsto X$ is a strongly bounded n-linear mapping, then there exists a unique bounded linear operator $T_M: \mathcal{H}^{\otimes n} \longmapsto X$ such that

$$M(h_1, h_2, \cdots, h_n) = T_M(h_1 \otimes h_2 \otimes \cdots \otimes h_n), \quad (h_1, h_2, \cdots, h_n) \in \mathcal{H}^n, \qquad (2.2)$$

and moreover, $\|T_M\| = \|M\|_s$, where $\|T_M\|$ stands for the usual operator norm.

Proof. Obviously, T_M is unique if it exists. To prove the existence, we define a mapping $M_+: X^* \longmapsto (\mathcal{H}^{\otimes n})^*$ as follows

$$M_+ g = \sum_{j_1, j_2, \cdots, j_n} \langle g, M(e_{j_1}, e_{j_2}, \cdots, e_{j_n}) \rangle_{X^* \times X} R(e_{j_1} \otimes e_{j_2} \otimes \cdots \otimes e_{j_n}), \quad g \in X^*,$$
$$(2.3)$$

where $\{e_k\}_{k \geq 1}$ is an orthonormal basis of $\mathcal{H}$ and $R: \mathcal{H}^{\otimes n} \longmapsto (\mathcal{H}^{\otimes n})^*$ is the Riesz mapping. It can be easily verified that $M_+: X^* \longmapsto (\mathcal{H}^{\otimes n})^*$ is a bounded linear operator and

$$\|M_+ g\|_{(\mathcal{H}^{\otimes n})^*}^2 = \sum_{j_1, j_2, \cdots, j_n} |\langle g, M(e_{j_1}, e_{j_2}, \cdots, e_{j_n}) \rangle_{X^* \times X}|^2, \quad g \in X^*, \qquad (2.4)$$

which means $\|M_+\| = \|M\|_s$.

For $(h_1, h_2, \cdots, h_n) \in \mathcal{H}^n$ and $g \in X^*$, it follows that

$$\langle M_+ g, h_1 \otimes h_2 \otimes \cdots \otimes h_n \rangle_{(\mathcal{H}^{\otimes n})^* \times \mathcal{H}^{\otimes n}}$$
$$= \sum_{j_1, j_2, \cdots, j_n} \langle g, M(e_{j_1}, e_{j_2}, \cdots, e_{j_n}) \rangle_{X^* \times X} \langle h_1, e_{j_1} \rangle \langle h_2, e_{j_2} \rangle \cdots \langle h_n, e_{j_n} \rangle$$
$$= \sum_{j_1, j_2, \cdots, j_n} \langle g, M(\langle h_1, e_{j_1} \rangle e_{j_1}, \langle h_2, e_{j_2} \rangle e_{j_2}, \cdots, \langle h_n, e_{j_n} \rangle e_{j_n}) \rangle_{X^* \times X}$$
$$= \langle g, M(h_1, h_2, \cdots, h_n) \rangle_{X^* \times X}.$$

Now let $J_1: \mathcal{H}^{\otimes n} \longmapsto (\mathcal{H}^{\otimes n})^{**}$ and $J_2: X \longmapsto X^{**}$ be the natural embedding mappings and denote by M_+^* the adjoint of M_+. Then, for $(h_1, h_2, \cdots, h_n) \in \mathcal{H}^n$ and $g \in X^*$, we have

$$\langle M_+^* J_1(h_1 \otimes h_2 \otimes \cdots \otimes h_n), g \rangle_{X^{**} \times X^*}$$
$$= \langle J_1(h_1 \otimes h_2 \otimes \cdots \otimes h_n), M_+ g \rangle_{(\mathcal{H}^{\otimes n})^{**} \times (\mathcal{H}^{\otimes n})^*}$$
$$= \langle M_+ g, h_1 \otimes h_2 \otimes \cdots \otimes h_n \rangle_{(\mathcal{H}^{\otimes n})^* \times \mathcal{H}^{\otimes n}}$$
$$= \langle g, M(h_1, h_2, \cdots, h_n) \rangle_{X^* \times X}$$
$$= \langle J_2 M(h_1, h_2, \cdots, h_n), g \rangle_{X^{**} \times X^*},$$

which implies that for each $(h_1, h_2, \cdots, h_n) \in \mathcal{H}^n$, it holds that

$$M_+^* J_1(h_1 \otimes h_2 \otimes \cdots \otimes h_n) = J_2 M(h_1, h_2, \cdots, h_n) \in J_2(X). \quad (2.5)$$

Since $\{ h_1 \otimes h_2 \otimes \cdots \otimes h_n \mid (h_1, h_2, \cdots, h_n) \in \mathcal{H}^n \}$ is total in $\mathcal{H}^{\otimes n}$ and $J_2(X)$ is a closed subspace of X^{**}, it follows that

$$M_+^* J_1(\mathcal{H}^{\otimes n}) \subset J_2(X).$$

Hence $T_M \equiv J_2^{-1} M_+^* J_1$ is a bounded linear operator from $\mathcal{H}^{\otimes n}$ to X. It follows from (2.5) that

$$M(h_1, h_2, \cdots, h_n) = T_M(h_1 \otimes h_2 \otimes \cdots \otimes h_n), \quad (h_1, h_2, \cdots, h_n) \in \mathcal{H}^n.$$

Finally,

$$\|T_M\| = \sup_{|u|=1, u \in \mathcal{H}^{\otimes n}} \|T_M u\|_X = \sup_{|u|=1, u \in \mathcal{H}^{\otimes n}} \|J_2^{-1} M_+^* J_1 u\|_X$$
$$= \sup_{|u|=1, u \in \mathcal{H}^{\otimes n}} \|M_+^* J_1 u\|_{X^{**}} = \sup_{\|v\|=1, v \in (\mathcal{H}^{\otimes n})^{**}} \|M_+^* v\|_{X^{**}}$$
$$= \|M_+^*\| = \|M_+\| = \|M\|_s.$$

This completes the proof. $\square$

Remark 2.1. According to Theorem 2.1, if $M: \mathcal{H}^n \longmapsto X$ is a strongly bounded n-linear mapping, then $\|M\| \leq \|M\|_s$.

Let $\mathcal{H}^{\widehat{\otimes}n}$ be the n-fold symmetric Hilbert tensor product of $\mathcal{H}$, which is a closed subspace of $\mathcal{H}^{\otimes n}$. Note that $\mathcal{H}^{\widehat{\otimes}0} = \mathbb{C}$. By convention, $\mathcal{H}^{\widehat{\otimes}n}$ is endowed with the inner product $n!\langle \cdot, \cdot \rangle$ instead, which is equivalent to the inner product $\langle \cdot, \cdot \rangle$ of $\mathcal{H}^{\otimes n}$. Hence $\| \cdot \|_{\mathcal{H}^{\widehat{\otimes}n}} = \sqrt{n!} |\cdot|$.

Theorem 2.2. Let $n \geq 1$. If $M: \mathcal{H}^n \longmapsto X$ be a strongly bounded symmetric n-linear mapping, then there exists a unique bounded linear operator $L_M: \mathcal{H}^{\widehat{\otimes}n} \longmapsto X$ such that

$$M(h_1, h_2, \cdots, h_n) = L_M(h_1 \widehat{\otimes} h_2 \widehat{\otimes} \cdots \widehat{\otimes} h_n), \quad (h_1, h_2, \cdots, h_n) \in \mathcal{H}^n, \qquad (2.6)$$

and moreover,

$$\|L_M\| = \frac{1}{\sqrt{n!}} \|M\|_s. \qquad (2.7)$$

Proof. By Theorem 2.1, there exists a unique bounded linear operator $T_M: \mathcal{H}^{\otimes n} \longmapsto X$ such that

$$M(h_1, h_2, \cdots, h_n) = T_M(h_1 \otimes h_2 \otimes \cdots \otimes h_n), \quad (h_1, h_2, \cdots, h_n) \in \mathcal{H}^n.$$

Put $L_M = T_M|_{\mathcal{H}^{\widehat{\otimes}n}}$. Then, it is easy to verify that $L_M: \mathcal{H}^{\widehat{\otimes}n} \longmapsto X$ is a bounded linear operator and moreover, L_M satisfies equality (2.6).

Taking an orthonormal basis $\{e_k\}_{k \geq 1}$ of $\mathcal{H}$, we have

$$\|M\|_s^2 = \sup_{\|g\|=1, g \in X^*} \sum_{j_1, j_2, \cdots, j_n} |\langle g, M(e_{j_1}, e_{j_2}, \cdots, e_{j_n})\rangle_{X^* \times X}|^2.$$

It is known that with $1 \leq i_1 < i_2 < \cdots < i_k$, $1 \leq r_1, r_2, \cdots, r_k \leq n$, $r_1 + r_2 + \cdots + r_k = n$ and $1 \leq k \leq n$, the following vector set constitutes an orthonormal basis of the symmetric Hilbert tensor $\mathcal{H}^{\widehat{\otimes}n}$

$$\left\{ \frac{e_{i_1}^{\otimes r_1} \widehat{\otimes} e_{i_2}^{\otimes r_2} \widehat{\otimes} \cdots \widehat{\otimes} e_{i_k}^{\otimes r_k}}{\sqrt{r_1! r_2! \cdots r_k!}} \right\}.$$

Hence

$$\|L_M\|^2 = \|L_M^*\|^2 = \sup_{\|g\|=1, g \in X^*} \|L_M^* g\|_{(\mathcal{H}^{\widehat{\otimes}n})^*}^2$$

$$= \sup_{\|g\|=1, g \in X^*} \sum_{\Delta_n} \left| \left\langle L_M^* g, \frac{e_{i_1}^{\otimes r_1} \widehat{\otimes} e_{i_2}^{\otimes r_2} \widehat{\otimes} \cdots \widehat{\otimes} e_{i_k}^{\otimes r_k}}{\sqrt{r_1! r_2! \cdots r_k!}} \right\rangle_{(\mathcal{H}^{\widehat{\otimes}n})^* \times \mathcal{H}^{\widehat{\otimes}n}} \right|^2$$

$$= \sup_{\|g\|=1, g \in X^*} \frac{1}{n!} \sum_{j_1, j_2, \cdots, j_n} |\langle L_M^* g, e_{j_1} \widehat{\otimes} e_{j_2} \widehat{\otimes} \cdots \widehat{\otimes} e_{j_n}\rangle_{(\mathcal{H}^{\widehat{\otimes}n})^* \times \mathcal{H}^{\widehat{\otimes}n}}|^2$$

$$= \sup_{\|g\|=1, g \in X^*} \frac{1}{n!} \sum_{j_1, j_2, \cdots, j_n} |\langle g, L_M(e_{j_1} \widehat{\otimes} e_{j_2} \widehat{\otimes} \cdots \widehat{\otimes} e_{j_n})\rangle_{X^* \times X}|^2$$

$$= \sup_{\|g\|=1, g \in X^*} \frac{1}{n!} \sum_{j_1, j_2, \cdots, j_n} |\langle g, M(e_{j_1}, e_{j_2}, \cdots, e_{j_n})\rangle_{X^* \times X}|^2$$

$$= \frac{1}{n!} \|M\|_s^2,$$

where $\triangle_n$ denotes the following relation:

$$1 \leq i_1 < i_2 < \cdots < i_k, \ 1 \leq r_1, r_2, \cdots, r_k \leq n, \ r_1 + r_2 + \cdots + r_k = n$$

and $1 \leq k \leq n$. Hence $\|L_M\| = \|M\|_s/\sqrt{n!}$. □

3. Main theorems

Let X be a complex Banach space (not necessary to be reflexive). In the present section, we present a moment characterization theorem and a convergent theorem for X-valued generalized functionals of white noise, which refine that of Ref. 9. As is seen, their proofs depend on the kernel theorems given in Section 2

We first outline the framework of white noise analysis where we work. Let H be a real separable Hilbert space with norm $|\cdot|_0$ and inner product $\langle \cdot, \cdot \rangle$. Let A be a positive self-adjoint operator in H such that there exists an orthonormal basis $\{e_i\}_{i \geq 1}$ for H satisfying the following conditions

(1) $A e_i = \lambda_i e_i$, $i = 1, 2, \cdots$,
(2) $1 < \lambda_1 \leq \lambda_2 \leq \cdots \leq \lambda_n \leq \cdots$,
(3) $\sum_{i=1}^{\infty} \lambda_i^{-\alpha} < \infty$ for some positive constant α.

For each $p \in \mathbb{R}$, define $|\cdot|_p \equiv |A^p \cdot |_0$ and let E_p be the completion of $\text{Dom} A^p$ with respect to $|\cdot|_p$. Then E_p is a real Hilbert space for each $p \in \mathbb{R}$ and moreover E_p and E_{-p} can be viewed as each other's topological dual.

Let E be the projective limit of $\{E_p \mid p \geq 0\}$ and E^* the inductive limit of $\{E_{-p} \mid p \geq 0\}$. Then we get the following inclusion relation

$$E \subset E_q \subset E_p \subset H \subset E_{-p} \subset E_{-q} \subset E^*$$

where $0 \leq p \leq q$. Moreover E and E^* can be regarded as each other's topological dual and $E \subset H \subset E^*$ constitutes a Gel'fand triple. We denote by $\langle \cdot, \cdot \rangle$ the canonical bilinear form on $E^* \times E$ which is consistent with the inner product of H. By the Minlos theorem,[3] there exists a Gaussian measure μ on E^* such that

$$\int_{E^*} e^{i\langle x, \xi \rangle} \mu(dx) = e^{-|\xi|_0^2/2}, \quad \xi \in E. \tag{3.1}$$

The measure space (E^*, μ) is known as the *white noise space*. Let $(L^2) \equiv L^2(E^*, \mu)$ be the complex Hilbert space of μ-square integrable functions on E^* with inner product $((\cdot, \cdot))$ and norm $\|\cdot\|_0$.

Let H_c be the complexification of H and $(\cdot, \cdot)$ the inner product of H_c. For $n \geq 1$, let $H_c^{\widehat{\otimes} n}$ denote the symmetric n-fold Hilbert tensor product of H_c with norm $\sqrt{n!} |\cdot|_0$, where $|\cdot|_0$ stands for the norm of $H_c^{\otimes n}$. It is known that, for each integer $n \geq 0$, there exists a linear isometry $I_n : H_c^{\widehat{\otimes} n} \longmapsto (L^2)$ such that

$$I_n(\xi^{\otimes n}) = \langle :\cdot^{\otimes n}:, \xi^{\otimes n} \rangle, \quad \xi \in E_c \tag{3.2}$$

where E_c means the complexification of E and $\langle \cdot, \cdot \rangle$ the canonical bilinear form on $(E_c^{\otimes n})^* \times E_c^{\otimes n}$. In addition, $I_m(H_c^{\widehat{\otimes}m}) \perp I_n(H_c^{\widehat{\otimes}n})$ whenever $m, n \geq 0$ and $m \neq n$. The next lemma is known as the *Wiener-Itô-Segal isomorphism theorem*.

Lemma 3.1. *Let $\Gamma(H_c)$ be the symmetric Fock space over H_c. Then there exists an isometric isomorphism $I \colon \Gamma(H_c) \longmapsto (L^2)$ such that*

$$I\left(\bigoplus_{n=0}^{\infty} f_n\right) = \sum_{n=0}^{\infty} I_n(f_n), \quad \bigoplus_{n=0}^{\infty} f_n \in \Gamma(H_c) \tag{3.3}$$

where the series on the righthand side converges in the norm of (L^2).

Now let $\Gamma(A)$ be the second quantization operator of A defined by

$$\Gamma(A)\varphi = \sum_{n=0}^{\infty} I_n(A^{\otimes n} f_n), \quad \varphi \in \mathrm{Dom}\Gamma(A) \tag{3.4}$$

where $\varphi = I(\bigoplus_{n=0}^{\infty} f_n)$. Then $\Gamma(A)$ is a self-adjoint operator in (L^2) with inverse $\Gamma(A^{-1})$. Similarly, for each $p \in \mathbb{R}$, define $\|\cdot\|_p \equiv \|\Gamma(A)^p \cdot \|_0$ and let (E_p) be the completion of $\mathrm{Dom}\Gamma(A)^p$ with respect to norm $\|\cdot\|_p$. Then (E_p) is complex Hilbert space for each $p \in \mathbb{R}$ and (E_p) and (E_{-p}) can be viewed as each other's dual. Let (E) be the projective limit of $\{(E_p) \mid p \geq 0\}$ and $(E)^*$ the inductive limit of $\{(E_{-p}) \mid p \geq 0\}$. Then we have the following inclusion relation

$$(E) \subset (E_q) \subset (E_p) \subset (L^2) \subset (E_{-p}) \subset (E_{-q}) \subset (E)^* \tag{3.5}$$

where $0 \leq p \leq q$. Moreover (E) is a countably Hilbertian nuclear space and $(E)^*$ can be regarded as the topological dual of (E). Hence we come to a second Gel'fand triple

$$(E) \subset (L^2) \subset (E)^* \tag{3.6}$$

which is known as the *framework of white noise analysis* over $E \subset H \subset E^*$. Usually, elements of (E) are called *testing functionals* while elements of $(E)^*$ are referred to as *generalized functionals*. We denote by $\langle\!\langle \cdot, \cdot \rangle\!\rangle$ the canonical bilinear form on $(E)^* \times (E)$.

Let $\varphi \in (L^2)$ with $\varphi = I(\bigoplus_{n=0}^{\infty} f_n)$, $\bigoplus_{n=0}^{\infty} f_n \in \Gamma(H_c)$. Then $\varphi \in (E)$ if and only if $f_n \in E_c^{\otimes n}$ for each $n \geq 0$ and

$$\sum_{n=0}^{\infty} n! |f_n|_p^2 < \infty, \quad \forall p \geq 0. \tag{3.7}$$

For $\varphi, \psi \in (E)$, their Wick product $\varphi \diamond \psi$ is defined as

$$\varphi \diamond \psi = I\left(\bigoplus_{n=0}^{\infty} \sum_{l+m=n} f_l \widehat{\otimes} g_m\right) \tag{3.8}$$

where $\varphi = I(\bigoplus_{n=0}^{\infty} f_n)$ and $\psi = I(\bigoplus_{n=0}^{\infty} g_n)$. It is known that $\varphi \diamond \psi \in (E)$ whenever $\varphi, \psi \in (E)$. Moreover the mapping $(\varphi, \psi) \in (E) \times (E) \longmapsto \varphi \diamond \psi \in (E)$ is continuous.

For $\varphi, \psi \in (E)$, let $\varphi\psi$ be the usual product of φ and ψ. Then $\varphi\psi \in (E)$ and moreover the mapping $(\varphi, \psi) \in (E) \times (E) \longmapsto \varphi\psi \in (E)$ is continuous.

The following lemma shows the relationship between the Wick and the usual products. See Ref. 4 or Ref. 6 for its proof.

Lemma 3.2. *There exists a linear homeomorphism* $\Theta \colon (E) \longmapsto (E)$ *such that for any* $\xi \in E_c$, $\Theta[I_1(\xi)] = I_1(\xi)$ *and moreover*

$$\Theta(\varphi\psi) = \Theta(\varphi) \diamond \Theta(\psi), \quad \varphi, \psi \in (E). \tag{3.9}$$

The linear homeomorphism $\Theta \colon (E) \longmapsto (E)$ *is known as the* renormalization operator

Recall that X is a complex Banach space (not necessary to be reflexive). By an X-valued generalized functional we mean a continuous linear mapping from (E) into X. As usual we denote by $\mathcal{L}[(E), X]$ the space of X-valued generalized functionals.

Definition 3.2. Let $T \in \mathcal{L}[(E), X]$ be an X-valued generalized functional. Define $M_0^T = T1$ and

$$M_n^T(\xi_1, \xi_2, \cdots, \xi_n) = T[I_1(\xi_1)I_1(\xi_2)\cdots I_1(\xi_n)], \quad \xi_1, \xi_2, \cdots, \xi_n \in E_c \tag{3.10}$$

for $n \geq 1$. We call M_n^T the moment of order n of the X-valued generalized functional T.

It is easy to see that the moment of order n of an X-valued generalized functional is a symmetric n-linear mapping from E_c^n to X. In the following, for an n-linear mapping $M \colon E_c^n \longmapsto X$, we use the following notation

$$\widehat{M}(\xi) = M(\xi, \xi, \cdots, \xi), \quad \xi \in E_c \tag{3.11}$$

where ξ appears n times on the righthand side.

The next proposition shows that an X-valued generalized functional is uniquely determined by its moment sequence.

Proposition 3.1. *Let* $T_1, T_2 \in \mathcal{L}[(E), X]$. *Then* $T_1 = T_2$ *if and only if*

$$\widehat{M_n^{T_1}}(\xi) = \widehat{M_n^{T_2}}(\xi), \quad \xi \in E_c, \; n \geq 0. \tag{3.12}$$

Proof. We need only to prove the *if* part. Let $\Lambda_n = \{I_n(\xi^{\otimes n}) \mid \xi \in E_c\}$. Then the set $\mathrm{span}\{\cup_{n \geq 0}\Lambda_n\}$ is dense in (E). For each $n \geq 0$ and any $\xi \in E_c$, we have

$$T_1\Theta^{-1}[I_n(\xi^{\otimes n})] = \widehat{M_n^{T_1}}(\xi) = \widehat{M_n^{T_2}}(\xi) = T_2\Theta^{-1}[I_n(\xi^{\otimes n})]$$

which implies that

$$T_1\Theta^{-1}(\varphi) = T_2\Theta^{-1}(\varphi), \quad \varphi \in \mathrm{span}\{\cup_{n \geq 0}\Lambda_n\}.$$

Hence, by the continuity of $T_1\Theta^{-1}$ and $T_2\Theta^{-1}$, we come to that $T_1\Theta^{-1} = T_2\Theta^{-1}$ which implies that $T_1 = T_2$. $\square$

Theorem 3.1. *Let $T \in \mathcal{L}[(E), X]$ be an X-valued generalized functional and $\{M_n^T\}_{n\geq 0}$ the moment sequence of T. Then there exist constants $K > 0$ and $p \geq 0$ such that*

$$\|\widehat{M_n^T}(\xi)\|_X \leq K\sqrt{n!}\,|\xi|_p^n, \quad \xi \in E_c,\ n \geq 0. \tag{3.13}$$

Proof. By Lemma 3.2, we have

$$\widehat{M_n^T}(\xi) = T[(I_1(\xi))^n] = T\Theta^{-1}(I_n(\xi^{\otimes n})), \quad \xi \in E_c,\ n \geq 0.$$

On the other hand, we see that $T\Theta^{-1} \in \mathcal{L}[(E), X]$, which implies that there exist constants $K > 0$ and $p \geq 0$ such that

$$\|T\Theta^{-1}(\varphi)\|_X \leq K\|\varphi\|_p, \quad \varphi \in (E).$$

Hence, for any $\xi \in E_c$ and $n \geq 0$, it holds that

$$\|\widehat{M_n^T}(\xi)\|_X \leq K\|I_n(\xi^{\otimes n})\|_p = K\sqrt{n!}\,|\xi|_p^n.$$

This completes the proof. $\square$

Theorem 3.2. *Let $M_0 \in X$. For each $n \geq 1$, let $M_n\colon E_c^n \longmapsto X$ be a symmetric n-linear mapping. Assume that there exist constants $K > 0$ and $p \geq 0$ such that*

$$\|\widehat{M_n}(\xi)\|_X \leq K\sqrt{n!}\,|\xi|_p^n, \quad \xi \in E_c,\ n \geq 0. \tag{3.14}$$

Then there exists a unique X-valued generalized functional $T \in \mathcal{L}[(E), X]$ such that

$$M_n^T(\xi_1, \xi_2, \cdots, \xi_n) = M_n(\xi_1, \xi_2, \cdots, \xi_n), \quad \xi_1, \xi_2, \cdots, \xi_n \in E_c. \tag{3.15}$$

Moreover for $q \geq p + \frac{\alpha}{2}$ with $e^2\|A^{-(q-p)}\|_{HS}^2 < 1$ we have

$$\|T\Theta^{-1}(\varphi)\|_X \leq \frac{K}{\sqrt{1 - e^2\|A^{-(q-p)}\|_{HS}^2}}\|\varphi\|_q, \quad \varphi \in (E). \tag{3.16}$$

Proof. T is obviously unique if it exists. Below we verify the existence of T. Take $q \geq p + \frac{\alpha}{2}$ such that $e^2\|A^{-(q-p)}\|_{HS}^2 < 1$. Firstly, by (3.14) and the polarization formula, we get

$$\|M_n(\xi_1, \xi_2, \cdots, \xi_n)\|_X \leq K\frac{n^n}{\sqrt{n!}}|\xi_1|_p|\xi_2|_p\cdots|\xi_n|_p, \quad \xi_1, \xi_2, \cdots, \xi_n \in E_c,\ n \geq 0. \tag{3.17}$$

Since $|\cdot|_p \leq |\cdot|_q$, we then come to

$$\|M_n(\xi_1, \xi_2, \cdots, \xi_n)\|_X \leq K\frac{n^n}{\sqrt{n!}}|\xi_1|_q|\xi_2|_q\cdots|\xi_n|_q, \quad \xi_1, \xi_2, \cdots, \xi_n \in E_c,\ n \geq 0 \tag{3.18}$$

which imply that each M_n has a unique bounded extension to $E_{q,c}^n$, where $E_{q,c}$ is the complexification of E_q.

Let $M_n^{(q)}$ be the bounded extension of M_n to $E_{q,c}^n$ for $n \geq 0$. Then $M_n^{(q)}$ remains symmetric. It is known that $\{A^{-q}e_k\}_{k\geq 1}$ is an orthonormal basis of $E_{q,c}$. And moreover, for $n \geq 1$ we have

$$\sum_{j_1,j_2,\cdots,j_n} \|M_n^{(q)}(A^{-q}e_{j_1}, A^{-q}e_{j_2}, \cdots, A^{-q}e_{j_n})\|_X^2$$

$$= \sum_{j_1,j_2,\cdots,j_n} \|M_n(A^{-q}e_{j_1}, A^{-q}e_{j_2}, \cdots, A^{-q}e_{j_n})\|_X^2$$

$$\leq \sum_{j_1,j_2,\cdots,j_n} K^2 \frac{n^{2n}}{n!} |A^{-q}e_{j_1}|_p^2 |A^{-q}e_{j_2}|_p^2 \cdots |A^{-q}e_{j_n}|_p^2$$

$$= K^2 \frac{n^{2n}}{n!} \|A^{-(q-p)}\|_{HS}^{2n}$$

$$< \infty.$$

Hence, for $n \geq 1$, the symmetric n-linear mapping $M_n^{(q)}: E_{q,c}^n \longmapsto X$ is strongly bounded and

$$\|M_n^{(q)}\|_s^2 \leq K^2 \frac{n^{2n}}{n!} \|A^{-(q-p)}\|_{HS}^{2n}. \tag{3.19}$$

By Theorem 2.2, for each $n \geq 1$ there exists an $L_n^{(q)} \in \mathcal{L}[E_{q,c}^{\hat{\otimes} n}, X]$ such that

$$M_n^{(q)}(\xi_1, \xi_2, \cdots, \xi_n) = L_n^{(q)}(\xi_1 \hat{\otimes} \xi_2 \hat{\otimes} \cdots \hat{\otimes} \xi_n), \quad \xi_1, \xi_2, \cdots, \xi_n \in E_{q,c} \tag{3.20}$$

and $\|L_n^{(q)}\|^2 = \frac{1}{n!}\|M_n^{(q)}\|_s^2$, where $\|L_n^{(q)}\|$ is the usual operator norm of $L_n^{(q)}$ as a bounded operator from $E_{q,c}^{\hat{\otimes} n}$ to X.

Define a mapping $L^{(q)}: \Gamma(E_{q,c}) \longmapsto X$ as follows

$$L^{(q)}(F) = \sum_{n=0}^{\infty} L_n^{(q)}(f_n), \quad F = \bigoplus_{n=0}^{\infty} f_n \in \Gamma(E_{q,c}) \tag{3.21}$$

where $L_0^{(q)}: \mathbb{C} \longmapsto X$ is defined as $L_0^{(q)}(z) = zM_0$. We assert that $L^{(q)}$ is well defined and moreover $L^{(q)} \in \mathcal{L}[\Gamma(E_{q,c}), X]$.

In fact, for $F = \bigoplus_{n=0}^{\infty} f_n \in \Gamma(E_{q,c})$, we have

$$\sum_{n=0}^{\infty} \|L_n^{(q)}(f_n)\|_X \leq \sum_{n=0}^{\infty} \|L_n^{(q)}\| \sqrt{n!} |f_n|_q$$

$$\leq \left\{ \sum_{n=0}^{\infty} \|L_n^{(q)}\|^2 \right\}^{1/2} \left\{ \sum_{n=0}^{\infty} n! |f_n|_q^2 \right\}^{1/2}$$

$$= \left\{ \sum_{n=0}^{\infty} \frac{1}{n!} \|M_n^{(q)}\|_s^2 \right\}^{1/2} \left\| \bigoplus_{n=0}^{\infty} f_n \right\|_{\Gamma(E_{q,c})}$$

$$\leq K \left\{ \sum_{n=0}^{\infty} \frac{n^{2n}}{n!n!} \|A^{-(q-p)}\|_{HS}^{2n} \right\}^{1/2} \|F\|_{\Gamma(E_{q,c})}$$

$$\leq K \left\{ \sum_{n=0}^{\infty} e^{2n} \|A^{-(q-p)}\|_{HS}^{2n} \right\}^{1/2} \|F\|_{\Gamma(E_{q,c})}$$

$$\leq \frac{K}{\sqrt{1 - e^2 \|A^{-(q-p)}\|_{HS}^2}} \|F\|_{\Gamma(E_{q,c})}$$

$$< \infty$$

which implies that the series $\sum_{n=0}^{\infty} L_n^{(q)}(f_n)$ is absolutely convergent in X. Hence the mapping $L^{(q)}: \Gamma(E_{q,c}) \longmapsto X$ is well defined. Clearly $L^{(q)}$ is a linear mapping. Moreover, from the above estimate, we see that

$$\|L^{(q)}(F)\|_X \leq \frac{K}{\sqrt{1 - e^2 \|A^{-(q-p)}\|_{HS}^2}} \|F\|_{\Gamma(E_{q,c})}, \quad F \in \Gamma(E_{q,c}) \tag{3.22}$$

which implies that $L^{(q)} \in \mathcal{L}[\Gamma(E_{q,c}), X]$.

Now let $T = L^{(q)} I^{(q)^{-1}} \Theta$, where $I^{(q)} = I|_{\Gamma(E_{q,c})}$ which is an isometric isomorphism from $\Gamma(E_{q,c})$ to (E_q). Then $T \in \mathcal{L}[(E), X]$ and moreover for any $\xi_1, \xi_2, \cdots, \xi_n \in E_c$ and $n \geq 0$

$$M_n^T(\xi_1, \xi_2, \cdots, \xi_n) = T[I_1(\xi_1) I_1(\xi_2) \cdots I_1(\xi_n)]$$
$$= L^{(q)} I^{(q)^{-1}} [I_n(\xi_1 \widehat{\otimes} \xi_2 \widehat{\otimes} \cdots \widehat{\otimes} \xi_n)]$$
$$= L_n^{(q)}(\xi_1 \widehat{\otimes} \xi_2 \widehat{\otimes} \cdots \widehat{\otimes} \xi_n)$$
$$= M_n^{(q)}(\xi_1, \xi_2, \cdots, \xi_n)$$
$$= M_n(\xi_1, \xi_2, \cdots, \xi_n).$$

Clearly (3.22) implies (3.16). $\square$

Remark 3.2. The combination of Theorems 3.1 and 3.2 forms a moment characterization theorem for X-valued generalized functionals of white noise. It was originally proved in Ref. 9 under the condition of X being reflexive.

As an immediate consequence of Proposition 3.1 and Theorem 3.2, the next theorem gives a useful norm estimate for X-valued generalized functionals.

Theorem 3.3. Let $T \in \mathcal{L}[(E), X]$ and $K > 0$, $p \geq 0$. Assume that

$$\|\widehat{M_n^T}(\xi)\|_X \leq K\sqrt{n!}\,|\xi|_p^n, \quad \xi \in E_c,\ n \geq 0. \tag{3.23}$$

Then for $q \geq p + \frac{\alpha}{2}$ with $e^2\|A^{-(q-p)}\|_{HS}^2 < 1$ it holds that

$$\|T\Theta^{-1}(\varphi)\|_X \leq \frac{K}{\sqrt{1 - e^2\|A^{-(q-p)}\|_{HS}^2}}\|\varphi\|_q, \quad \varphi \in (E). \tag{3.24}$$

Let $T \in \mathcal{L}[(E), X]$ and $\{T_k \mid k \geq 1\} \subset \mathcal{L}[(E), X]$. The sequence $\{T_k \mid k \geq 1\}$ is said to converge strongly to T if for each $\varphi \in (E)$ we have $T_k\varphi \longrightarrow T\varphi$ in the norm of X.

Using the above results, we can prove the next theorem, which offers a necessary and sufficient condition for a sequence of X-valued generalized functionals to converge strongly.

Theorem 3.4. $T \in \mathcal{L}[(E), X]$ and $\{T_k \mid k \geq 1\} \subset \mathcal{L}[(E), X]$. Then $\{T_k \mid k \geq 1\}$ converges strongly to T if and only if the following two conditions are satisfied

(1) For each $\xi \in E_c$ and each $n \geq 0$, $T_k(W_\xi^n) \longrightarrow T(W_\xi^n)$ $(k \longrightarrow \infty)$ in the norm of X.

(2) There exist $K > 0$, $p \geq 0$ such that

$$\sup_{k \geq 1}\|T_k(W_\xi^n)\|_X \leq K\sqrt{n!}\,|\xi|_p^n, \quad \xi \in E_c,\ n \geq 0. \tag{3.25}$$

Here, by convention, $W_\xi = I_1(\xi)$ for $\xi \in E_c$.

Proof. We first prove the *if* part. Let $S_k = T_k\Theta^{-1}$ for $k \geq 1$ and $S = T\Theta^{-1}$. Then we see that $S, S_k \in \mathcal{L}[(E), X]$ and moreover $\{T_k\}$ converges strongly to T if and only if $\{S_k\}$ converges strongly to S. Hence it suffices to show that $\{S_k\}$ converges strongly to S.

Take $q \geq p + \frac{\alpha}{2}$ with $e^2\|A^{-(q-p)}\|_{HS}^2 < 1$. Write $T_0 = T$ and $S_0 = S$. Then by (3.25) we have

$$\sup_{k \geq 0}\|\widehat{M_n^{T_k}}(\xi)\|_X \leq K\sqrt{n!}\,|\xi|_p^n, \quad \xi \in E_c,\ n \geq 0.$$

Hence by Theorem 3.3 we get

$$\sup_{k \geq 0}\|S_k(\varphi)\|_X = \sup_{k \geq 1}\|T_k\Theta^{-1}(\varphi)\|_X \leq \frac{K}{\sqrt{1 - e^2\|A^{-(q-p)}\|_{HS}^2}}\|\varphi\|_q, \quad \varphi \in (E).$$

Let $S_k^{(q)}$ be the bounded extension of S_k to (E_q) for $k \geq 0$. Then from the above inequality we get

$$\sup_{k \geq 0}\|S_k^{(q)}\| \leq \frac{K}{\sqrt{1 - e^2\|A^{-(q-p)}\|_{HS}^2}}$$

where $\|S_k^{(q)}\|$ denotes the usual operator norm of $S_k^{(q)}$ as an element of $\mathcal{L}[(E_q), X]$.

On the other hand, for any $n \geq 0$ and $\xi \in E_c$, we have

$$S_k[I_n(\xi^{\otimes n})] = T_k(W_\xi^n) \longrightarrow T_0(W_\xi^n) = S_0[I_n(\xi^{\otimes n})]$$

which implies that

$$S_k^{(q)}(\varphi) = S_k(\varphi) \longrightarrow S_0(\varphi) = S_0^{(q)}(\varphi), \quad \varphi \in \text{span}\{\cup_{n \geq 0}\Lambda_n\}$$

where $\Lambda_n = \{I_n(\xi^{\otimes n}) \mid \xi \in E_c\}$.

It is known that the set $\text{span}\{\cup_{n \geq 0}\Lambda_n\}$ is dense in (E_q). Therefore, by the Banach-Steinhaus theorem,[2] we know that $\{S_k^{(q)}\}$ converges strongly to $S_0^{(q)} = S^{(q)}$. In particular $\{S_k\}$ converges strongly to $S_0 = S$.

We now prove the *only if* part. Obviously, the property that T_k converges strongly to T implies that $T_k(W_\xi^n) \longrightarrow T(W_\xi^n)$ ($k \longrightarrow \infty$) in the norm of X for each $\xi \in E_c$ and each $n \geq 0$. To verify what remains, we put

$$U_N = \{\varphi \mid \varphi \in (E), \sup_{k \geq 1} \|T_k \Theta^{-1}(\varphi)\|_X \leq N\}, \quad N \geq 1$$

We can see that each U_N is a closed subset of (E) since

$$U_N = \bigcap_{k=1}^{\infty} \{\varphi \mid \varphi \in (E), \|T_k \Theta^{-1}(\varphi)\|_X \leq N\}.$$

Moreover $(E) = \bigcup_{N \geq 1} U_N$ since $\{T_k \Theta^{-1}\}$ converges strongly to $T\Theta^{-1}$.

It is known that (E) is a Fréchet space whose distance ρ can be taken as

$$\rho(\varphi, \psi) = \sum_{p=0}^{\infty} \frac{1}{2^p} \frac{\|\varphi - \psi\|_p}{1 + \|\varphi - \psi\|_p}.$$

Hence, by the well known Baire's category theorem,[2] there exists $N_0 \geq 1$ such that U_{N_0} has an interior point $\varphi_0 \in U_{N_0}$. Noting that on (E) it holds that $\|\cdot\|_0 \leq \|\cdot\|_1 \leq \|\cdot\|_2 \leq \cdots$, we find that there exist $p \geq 0$ and $\delta > 0$ such that $B_p(\varphi_0, \delta) \subset U_{N_0}$, where

$$B_p(\varphi_0, \delta) = \{\varphi \mid \varphi \in (E), \|\varphi - \varphi_0\|_p < \delta\}.$$

It is easy to see that for any $\varphi \in (E)$ with $\|\varphi\|_p \neq 0$

$$\frac{\delta \varphi}{2\|\varphi\|_p} + \varphi_0 \in B_p(\varphi_0, \delta).$$

Hence

$$\sup_{k \geq 1} \|T_k \Theta^{-1}[\frac{\delta \varphi}{2\|\varphi\|_p} + \varphi_0]\|_X \leq N_0, \quad \varphi \in (E), \|\varphi\|_p \neq 0$$

which implies that

$$\sup_{k \geq 1} \|T_k \Theta^{-1}(\varphi)\|_X \leq \frac{4N_0}{\delta} \|\varphi\|_p, \quad \varphi \in (E).$$

In particular, we have

$$\sup_{k\geq 1}\|T_k(W_\xi^n)\|_X = \sup_{k\geq 1}\|T_k\Theta^{-1}[I_n(\xi^{\otimes n})]\|_X \leq \frac{4N_0}{\delta}\sqrt{n!}\,|\xi|_p^n, \quad \xi \in E_c,\ n \geq 0.$$

This completes the proof. $\square$

Acknowledgement

This work is supported by National Natural Science Foundation of China (10571065), Natural Science Foundation of Gansu Province (0710RJZA106) and NWNU-KJCXGC, China.

References

1. D.M. Chung, T.S. Chung and U.C. Ji, A characterization theorem for operators on white noise functionals, *J. Math. Soc. Japan* 51 (1999), 437–447.
2. J.B. Conway, *A Course in Functional Analysis*, 2nd edition, Spinger-Verlag, New York, 1990.
3. T. Hida, H.H. Kuo, J. Potthoff and L. Streit, *White Noise–An Infinite Dimensional Calculus*, Kluwer Academic, Dordrecht, 1993.
4. Z.Y. Huang and J.A. Yan, *Introduction to Infinite Dimensional Stochastic Analysis*, Kluwer Academic, Dordrecht, 1999.
5. Yu.G. Kondratiev and L. Streit, Spaces of white noise distributions: constructions, applications, I. *Rep. Math. Phys.* 33 (1993), 341–366.
6. H.H. Kuo, *White Noise Distribution Theory*, CRC Press, 1996.
7. N. Obata, Operator calculus on vector-valued white noise functionals, *J. Funct. Anal.* 121 (1994), 185–208.
8. J. Potthoff and L. Streit, A characterization of Hida distributions, *J. Funct. Anal.* 101 (1991), 212–229.
9. C.S. Wang and Z.Y. Huang, A moment characterization of B-valued generalized functionals of white noise, *Acta Math. Sin. Engl. Ser.* 22 (2006), 157–168

Chapter 6

Hurst Index Estimation for Self-Similar Processes with Long-Memory

Alexandra Chronopoulou* and Frederi G. Viens*

Department of Statistics, Purdue University
150 N. University St. West Lafayette
IN 47907-2067, USA.
achronop@purdue.edu,   viens@purdue.edu

The statistical estimation of the Hurst index is one of the fundamental problems in the literature of long-range dependent and self-similar processes. In this article, the Hurst index estimation problem is addressed for a special class of self-similar processes that exhibit long-memory, the Hermite processes. These processes generalize the fractional Brownian motion, in the sense that they share its covariance function, but are non-Gaussian. Existing estimators such as the R/S statistic, the variogram, the maximum likelihood and the wavelet-based estimators are reviewed and compared with a class of consistent estimators which are constructed based on the discrete variations of the process. Convergence theorems (asymptotic distributions) of the latter are derived using multiple Wiener-Itô integrals and Malliavin calculus techniques. Based on these results, it is shown that the latter are asymptotically more efficient than the former.

Keywords: self-similar process, parameter estimation, long memory, Hurst parameter, multiple stochastic integral, Malliavin calculus, Hermite process, fractional Brownian motion, non-central limit theorem, quadratic variation.

2000 AMS Classification Numbers: Primary 62F12; Secondary 60G18, 60H07, 62M09.

Contents

1	Introduction	92
	1.1 Motivation	92
	1.2 Mathematical Background	93
2	Most Popular Hurst Parameter Estimators	96
	2.1 Heuristic Estimators	96
	2.2 Maximum Likelihood Estimation	98
	2.3 Wavelet Estimator	99
3	Multiplication in the Wiener Chaos & Hermite Processes	101
	3.1 Basic Tools on Multiple Wiener-Itô Integrals	101
	3.2 Main Definitions	103

*Both authors' research partially supported by NSF grant 0606615.

4   Hurst Parameter Estimator Based on Discrete Variations 105
    4.1   Estimator Construction 105
    4.2   Asymptotic Properties of $\hat{H}_N$ 106
5   Comparison & Conclusions 111
    5.1   Variations Estimator vs. mle 112
    5.2   Variations' vs. Wavelet Estimator 114
References 116

1. Introduction

1.1. *Motivation*

A fundamental assumption in many statistical and stochastic models is that of independent observations. Moreover, many models that do not make this assumption have the convenient Markov property, according to which the future of the system is not affected by its previous states but only by the current one.

The phenomenon of *long memory* has been noted in nature long before the construction of suitable stochastic models: in fields as diverse as hydrology, economics, chemistry, mathematics, physics, geosciences, and environmental sciences, it is not uncommon for observations made far apart in time or space to be non-trivially correlated.

Since ancient times the Nile River has been known for its long periods of dryness followed by long periods of floods. The hydrologist Hurst ([13]) was the first one to describe these characteristics when he was trying to solve the problem of flow regularization of the Nile River. The mathematical study of long-memory processes was initiated by the work of Mandelbrot [16] on self-similar and other stationary stochastic processes that exhibit long-range dependence. He built the foundations for the study of these processes and he was the first one to mathematically define the fractional Brownian motion, the prototype of self-similar and long-range dependent processes. Later, several mathematical and statistical issues were addressed in the literature, such as derivation of central (and non-central) limit theorems ([5], [6], [10], [17], [27]), parameter estimation techniques ([1], [7], [8], [27]) and simulation methods ([11]).

The problem of the statistical estimation of the self-similarity and/or long-memory parameter H is of great importance. This parameter determines the mathematical properties of the model and consequently describes the behavior of the underlying physical system. Hurst ([13]) introduced the celebrated *rescaled adjusted range* or *R/S* statistic and suggested a graphical methodology in order to estimate H. What he discovered was that for data coming from the Nile River the R/S statistic behaves like a constant times k^H, where k is a time interval. This was called later by Mandelbrot the *Hurst effect* and was modeled by a *fractional Gaussian noise (fGn)*.

One can find several techniques related to the Hurst index estimation problem in the literature. There are a lot of graphical methods including the R/S statistic, the correlogram and partial correlations plot, the variance plot and the variogram, which are widely used in geosciences and hydrology. Due to their graphical nature they are not so accurate and thus there is a need for more rigorous and sophisticated methodologies, such as the maximum likelihood. Fox and Taqqu ([12]) introduced the Whittle approximate maximum likelihood method in the Gaussian case which was later generalized for certain non-Gaussian processes. However, these approaches were lacking computational efficiency which lead to the rise of wavelet-based estimators and discrete variation techniques.

1.2. Mathematical Background

Let us first recall some basic definitions that will be useful in our analysis.

Definition 1.1. A stochastic process $\{X_n; n \in \mathbb{N}\}$ is said to be *stationary* if the vectors $(X_{n_1}, \ldots, X_{n_d})$ and $(X_{n_1+m}, \ldots, X_{n_d+m})$ have the same distribution for all integers d, $m \geq 1$ and $n_1, \ldots, n_d \geq 0$. For Gaussian processes this is equivalent to requiring that $Cov(X_m, X_{m+n}) := \gamma(n)$ does not depend on m. These two notions are often called *strict stationarity* and *second-order stationarity*, respectively. The function $\gamma(n)$ is called the *autocovariance function*. The function $\rho(n) = \gamma(n)/\gamma(0)$ is the called *autocorrelation* function.

In this context, long memory can be defined in the following way:

Definition 1.2. Let $\{X_n; n \in \mathbb{N}\}$ be a stationary process. If $\sum_n \rho(n) = +\infty$ then X_n is said to exhibit *long memory* or *long-range dependence*. A sufficient condition for this is the existence of $H \in (1/2, 1)$ such that

$$\liminf_{n \to \infty} \frac{\rho(n)}{n^{2H-2}} > 0.$$

Typical long memory models satisfy the stronger condition $\lim_{n \to \infty} \rho(n)/n^{2H-2} > 0$, in which case H can be called the *long memory parameter* of X.

A process that exhibits long-memory has an autocorrelation function that decays very slowly. This is exactly the behavior that was observed by Hurst for the first time. In particular, he discovered that the yearly minimum water level of the Nile river had the long-memory property, as can been seen in Figure 6.1.

Another property that was observed in the data collected from the Nile river is the so-called *self-similarity* property. In geometry, a self-similar shape is one composed of a basic pattern which is repeated at multiple (or infinite) scale. The statistical interpretation of self-similarity is that the paths of the process will look the same, in distribution, irrespective of the distance from which we look at. The rigorous definition of the self-similarity property is as follows:

Definition 1.3. A process $\{X_t; t \geq 0\}$ is called *self-similar* with self-similarity parameter H, if for all $c > 0$, we have the identity in distribution

$$\{c^{-H} X_{ct} : t \geq 0\} \stackrel{\mathcal{D}}{=} \{X_t : t \geq 0\}.$$

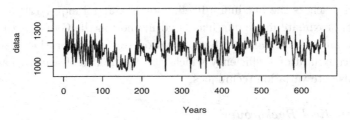

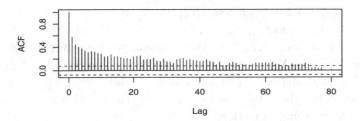

Figure 6.1. Yearly minimum water levels of the Nile River at the Roda Gauge (622-1281 A.D.). The dotted horizontal lines represent the levels $\pm 2/\sqrt{600}$. Since our observations are above these levels, it means that they are significantly correlated with significance level 0.05.

In Figure 6.2, we can observe the self-similar property of a simulated path of the fractional Brownian motion with parameter $H = 0.75$.

In this chapter, we concentrate on a special class of long-memory processes which are also self-similar and for which the self-similarity and long-memory parameters coincide, the so-called *Hermite processes*. This is a family of processes parametrized by the order q and the self-similarity parameter H. They all share the same covariance function

$$Cov(X_t, X_s) = \frac{1}{2}\left(t^{2H} + s^{2H} - |t-s|^{2H}\right). \quad (1.1)$$

From the structure of the covariance function we observe that the Hermite processes have stationary increments, they are H-self-similar and they exhibit long-range dependence as defined in Definition 1.2 (in fact, $\lim_{n\to\infty} \rho(n)\;/\!/n^{2H-2} = H(2H-1)$). The Hermite process for $q = 1$ is a standard *fractional Brownian motion* with Hurst parameter H, usually denoted by B^H, the only Gaussian process in the Hermite class. A Hermite process with $q = 2$ known as *the Rosenblatt process*. In the sequel, we will call H either long-memory parameter or self-similarity parameter or *Hurst parameter*. The mathematical definition of these processes is given in Definition 3.5.

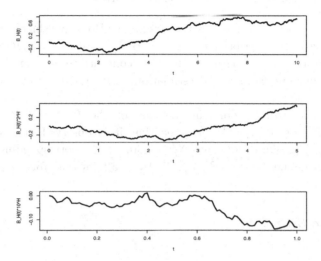

Figure 6.2. Self-similarity property for the fractional Brownian motion with $H = 0.75$. The first graph shows the path from time 0 to 10. The second and third graph illustrate the normalized sample path for $0 < t < 5$ and $0 < t < 1$ respectively.

Another class of processes used to model long-memory phenomena are the fractional ARIMA (Auto Regressive, Integrated, Moving Average) or FARIMA processes. The main technical difference between a FARIMA and a Hermite process is that the first one is a discrete-time process and the second one a continuous-time process. Of course, in practice, we can only have discrete observations. However, most phenomena in nature evolve continuously in time and the corresponding observations arise as samplings of continuous time processes. A discrete-time model depends heavily on the sampling frequency: daily observations will be described by a different FARIMA model than weekly observations. In a continuous time model, the observation sampling frequency does not modify the model. These are compelling reasons why one may choose to work with the latter.

In this article we study the Hurst parameter estimation problem for the Hermite processes. The structure of the paper is as follows: in Section 2, we provide a survey of the most widely used estimators in the literature. In Section 3 we describe the main ingredients and the main definitions that we need for our analysis. In Section 4, we construct a class of estimators based on the discrete variations of the process and describe their asymptotic properties, including a sketch of the proof of the main theoretical result, Theorem 4.2, which summarizes the series of papers [6], [7], [27] and [28]. In the last section, we compare the variations-based estimators with the existing ones in the literature, and provide an original set of practical recommendations based on theoretical results and on simulations.

2. Most Popular Hurst Parameter Estimators

In this section we discuss the main estimators for the Hurst parameter in the literature. We start with the description of three heuristic estimators: the R/S estimator, the correlogram and the variogram. Then, we concentrate on a more traditional approach: the maximum likelihood estimation. Finally, we briefly describe the wavelet-based estimator.

The description will be done in the case of the *fractional Brownian motion* (fBm) $\{B_t^H;\ t \in [0,1]\}$. We assume that it is observed in discrete times $\{0, 1, \ldots, N-1, N\}$. We denote by $\{X_t^H;\ t \in [0,1]\}$ the corresponding increment process of the fBm (i.e. $X_{\frac{i}{N}}^H = B_{\frac{i}{N}}^H - B_{\frac{i-1}{N}}^H$), also known as *fractional Gaussian noise*.

2.1. *Heuristic Estimators*

R/S Estimator:

The most famous among these estimators is the so-called R/S estimator that was first proposed by Hurst in 1951, [13], in the hydrological problem regarding the storage of water coming from the Nile river. We start by dividing our data in K non-intersecting blocks, each one of which contains $M = [\frac{N}{K}]$ elements. The *rescaled adjusted range* is computed for various values of N by

$$Q := Q(t_i, N) = \frac{R(t_i, N)}{S(t_i, n)}$$

at times $t_i = M(i-1)$, $i = 1, \ldots, K$. For

$$Y(t_i, k) := \sum_{j=0}^{k-1} X_{t_i+j}^H - k \left(\frac{1}{n} \sum_{j=0}^{n-1} X_{t_i+j}^H \right), \quad k = 1, \ldots, n$$

we define $R(t_i, n)$ and $S(t_i, n)$ to be

$$R(t_i, n) := \max\{Y(t_i, 1), \ldots, Y(t_i, n)\} - \min\{Y(t_i, 1), \ldots, Y(t_i, n)\} \text{ and}$$

$$S(t_i, n) := \sqrt{\frac{1}{n} \sum_{j=0}^{n-1} X_{t_i+j}^{H\,2} - \left(\frac{1}{n} \sum_{j=0}^{n-1} X_{t_i+j}^H \right)^2}.$$

Remark 2.1. It is interesting to note that the numerator $R(t_i, n)$ can be computed only when $t_i + n \leq N$.

In order to compute a value for H we plot the logarithm of R/S (i.e $\log Q$) with respect to $\log n$ for several values of n. Then, we fit a least-squares line $y = a + b \log n$ to a central part of the data, that seem to be nicely scattered along a straight line. The slope of this line is the estimator of H.

This is a graphical approach and it is really in the hands of the statistician to determine the part of the data that is "nicely scattered along the straight line". The problem is more severe in small samples, where the distribution of the R/S statistic is far from normal. Furthermore, the estimator is biased and has a large standard error. More details on the limitations of this approach in the case of fBm can be found in [2].

Correlogram:
Recall $\rho(N)$ the autocorrelation function of the process as in Definition 1.1. In the Correlogram approach, it is sufficient to plot the *sample autocorrelation function*

$$\hat{\rho}(N) = \frac{\hat{\gamma}(N)}{\hat{\gamma}(0)}$$

against N. As a rule of thumb we draw two horizontal lines at $\pm 2/\sqrt{N}$. All observations outside the lines are considered to be significantly correlated with significance level 0.05. If the process exhibits long-memory, then the plot should have a very slow decay.

The main disadvantage of this technique is its graphical nature which cannot guarantee accurate results. Since long-memory is an asymptotic notion, we should analyze the correlogram at high lags. However, when for example $H = 0.6$ it is quite hard to distinguish it from short-memory. To avoid this issue, a more suitable plot will be this of $\log \hat{\rho}(N)$ against $\log N$. If the asymptotic decay is precisely hyperbolic, then for large lags the points should be scattered around a straight line with negative slope equal to $2H - 2$ and the data will have long-memory. On the other hand when the plot diverges to $-\infty$ with at least exponential rate, then the memory is short.

Variogram:
The variogram for the lag N is defined as

$$V(N) := \frac{1}{2}\mathbf{E}\left[\left(B_t^H - B_{t-N}^H\right)^2\right].$$

Therefore, it suffices to plot $V(N)$ against N. However, we can see that the interpretation of the variogram is similar to that of the correlogram, since if the process is stationary (which is true for the increments of fractional Brownian motion and all other Hermite processes), then the variogram is asymptotically finite and

$$V(N) = V(\infty)(1 - \rho(N)).$$

In order to determine whether the data exhibit short or long memory this method has the same problems as the correlogram.

The main advantage of these approaches is their simplicity. In addition, due to their non-parametric nature, they can be applied to any long-memory process. However, none of these graphical methods are accurate. Moreover, they can frequently be misleading, indicating existence of long-memory in cases where none exists. For example, when a process has short-memory together with a trend that decays to zero very fast, a correlogram or a variogram could show evidence of long-memory.

In conclusion, a good approach would be to use these methods as a first heuristic analysis to detect the possible presence of long-memory and then use a more rigorous technique, such as those described in the remainder of this section, in order to estimate the long-memory parameter.

2.2. Maximum Likelihood Estimation

The Maximum Likelihood Estimation (*mle*) is the most common technique of parameter estimation in Statistics. In the class of Hermite processes, its use is limited to fBm, since for the other processes we do not have an expression for their distribution function. The *mle* estimation is done in the spectral domain using the spectral density of fBm as follows.

Denote by $X^H = (X_0^H, X_1^H, \ldots, X_N^H)$ the vector of the fractional Gaussian noise (increments of fBm) and by $X^{H\,\prime}$ the transposed (column) vector; this is a Gaussian vector with covariance matrix $\Sigma_N(H) = [\sigma_{ij}(H)]_{i,j=1,\ldots,N}$; we have

$$\sigma_{ij} := Cov\left(X_i^H \; ; \; X_j^H\right) = \frac{1}{2}\left(i^{2H} + j^{2H} - |i-j|^{2H}\right).$$

Then, the *log-likelihood function* has the following expression:

$$\log f(x; H) = -\frac{N}{2}\log 2\pi - \frac{1}{2}\log\left[\det\left(\Sigma_N(H)\right)\right] - \frac{1}{2}X^H\left(\Sigma_N(H)\right)^{-1}X^{H\,\prime}.$$

In order to compute $\hat{H}_{mle}$, the *mle* for H, we need to maximize the log-likelihood equation with respect to H. A detailed derivation can be found in [3] and [9]. The asymptotic behavior of $\hat{H}_{mle}$ is described in the following theorem.

Theorem 2.1. *Define the quantity* $D(H) = \frac{1}{2\pi}\int_{-\pi}^{\pi}\left(\frac{\partial}{\partial H}\log f(x;H)\right)^2 dx$. *Then under certain regularity conditions (that can be found in [9]) the maximum likelihood estimator is weakly consistent and asymptotically normal:*

(i) $\hat{H}_{mle} \to H$, as $N \to \infty$ in probability;
(ii) $\sqrt{N}\sqrt{2\,D(H)}\left(\hat{H}_{mle} - H\right) \to \mathcal{N}(0,1)$ in distribution, as $N \to \infty$.

In order to obtain the *mle* in practice, in almost every step we have to maximize a quantity that involves the computation of the inverse of $\Sigma(H)$, which is not an easy task.

In order to avoid this computational burden, we approximate the likelihood function with the so-called *Whittle* approximate likelihood which can be proved to converge to the true likelihood, [29]. In order to introduce Whittle's approximation we first define the density on the spectral domain.

Definition 2.4. Let X_t be a process with autocovariance function $\gamma(h)$, as in Definition 1.1. The *spectral density function* is defined as the inverse Fourier transform of $\gamma(h)$

$$f(\lambda) := \frac{1}{2\pi} \sum_{h=-\infty}^{\infty} e^{-i\lambda h} \gamma(h).$$

In the fBm case the spectral density can be written as

$$f(\lambda; H) = \frac{1}{2\pi} exp\left\{ -\frac{1}{2\pi} \int_{-\pi}^{\pi} \log f_1 \, d\lambda \right\}, \text{ where}$$

$$f_1(\lambda; H) = \frac{1}{\pi} \Gamma(2H+1) \sin(\pi H)(1 - \cos \lambda) \sum_{j=-\infty}^{\infty} |2\pi j + \lambda|^{-2H-1}$$

The Whittle method approximates each of the terms in the log-likelihood function as follows:

(i) $\lim_{N \to \infty} \log \det(\Sigma_N(H)) = \frac{1}{2\pi} \int_{-\pi}^{\pi} \log f(\lambda; H) d\lambda$.
(ii) The matrix $\Sigma_N^{-1}(H)$ itself is asymptotically equivalent to the matrix $A(H) = [\alpha(j - \ell)]_{j\ell}$, where

$$\alpha(j - \ell) = \frac{1}{(2\pi)^2} \int_{-\pi}^{\pi} \frac{e^{-i(j-\ell)\lambda}}{f(\lambda; H)} d\lambda$$

Combining the approximations above, we now need to minimize the quantity

$$(\log f(\lambda; H))^* = -\frac{N}{2} \log 2\pi - \frac{n}{2} \frac{1}{2\pi} \int_{-\pi}^{\pi} \log f(\lambda; H) d\lambda - \frac{1}{2} X \, A(H) X'.$$

The details in the Whittle *mle* estimation procedure can be found in [3]. For the Whittle *mle* we have the following convergence in distribution result as $N \to \infty$

$$\sqrt{\frac{N}{[2\, D(H)]^{-1}}} \left(\hat{H}_{Wmle} - H \right) \xrightarrow{D} \mathcal{N}(0, 1) \tag{2.1}$$

It can also be shown that the Whittle approximate *mle* remains weakly consistent.

2.3. Wavelet Estimator

Much attention has been devoted to the wavelet decomposition of both fBm and the Rosenblatt process. Following this trend, an estimator for the Hurst parameter based on wavelets has been suggested. The details of the procedure for the constructing this estimator, and the underlying wavelets theory, are beyond the scope of this article. For the proofs and the detailed exposition of the method the reader can refer to [1], [11] and [14]. This section provides a brief exposition.

Let $\psi : \mathbb{R} \to \mathbb{R}$ be a continuous function with support in $[0,1]$. This is also called the *mother wavelet*. $Q \geq 1$ is the number of vanishing moments where

$$\int_{\mathbb{R}} t^p \psi(t) dt = 0, \text{ for } p = 0, 1, \ldots, Q-1,$$

$$\int_{\mathbb{R}} t^Q \psi(t) dt \neq 0.$$

For a "scale" $\alpha \in \mathbb{N}^*$ the corresponding wavelet coefficient is given by

$$d(\alpha, i) = \frac{1}{\sqrt{\alpha}} \int_{-\infty}^{\infty} \psi\left(\frac{t}{\alpha} - i\right) Z_t^H dt,$$

for $i = 1, 2, \ldots, N_\alpha$ with $N_\alpha = \left[\frac{N}{\alpha}\right] - 1$, where N is the sample size. Now, for (α, b) we define the *approximate wavelet coefficient* of $d(\alpha, b)$ as the following Riemann approximation

$$e(\alpha, b) = \frac{1}{\sqrt{\alpha}} \sum_{k=1}^{N} Z_k^H \psi\left(\frac{k}{\alpha} - b\right),$$

where Z^H can be either fBm or Rosenblatt process. Following the analysis by J.-M. Bardet and C.A. Tudor in [1], the suggested estimator can be computed by performing a *log-log regression* of

$$\left(\frac{1}{N_{\alpha_i(N)}} \sum_{j=1}^{N_{\alpha_i(N)}} e^2(\alpha_i(N), j)\right)_{1 \leq i \leq \ell}$$

against $(i\, \alpha_i(N))_{1 \leq i \leq \ell}$, where $\alpha(N)$ is a sequence of integer numbers such that $N\alpha(N)^{-1} \to \infty$ and $\alpha(N) \to \infty$ as $N \to \infty$ and $\alpha_i(N) = i\alpha(N)$. Thus, the obtained estimator, in vectors notation, is the following

$$\hat{H}_{wave} := \left(\frac{1}{2}, 0\right)' \left(Z'_\ell, Z_\ell\right)^{-1} Z_\ell^{-1} \left(\frac{1}{2} \sum_{j=1}^{N_{\alpha_i(N)}} e^2(\alpha_i(N), j)\right)_{1 \leq i \leq \ell} - \frac{1}{2}, \quad (2.2)$$

where $Z_\ell(i, 1) = 1$, $Z_\ell(i, 2) = \log i$ for all $i = 1, \ldots, \ell$, for $\ell \in \mathbb{N} \setminus \{1\}$.

Theorem 2.2. *Let $\alpha(N)$ as above. Assume also that $\psi \in C^m$ with $m \geq 1$ and ψ is supported on $[0,1]$. We have the following convergences in distribution.*

(1) *Let Z^H be a fBm; assume $N\alpha(N)^{-2} \to 0$ as $N \to \infty$ and $m \geq 2$; if $Q \geq 2$, or if $Q = 1$ and $0 < H < 3/4$, then there exists $\gamma^2(H, \ell, \psi) > 0$ such that*

$$\sqrt{\frac{N}{\alpha(N)}} \left(\hat{H}_{wave} - H\right) \xrightarrow{D} \mathcal{N}(0, \gamma^2(H, \ell, \psi)), \text{ as } N \to \infty. \quad (2.3)$$

(2) Let Z^H be a fBm; assume $N\alpha(N)^{-\frac{5-4H}{4-4H}} \to 0$ as N $\alpha(N)^{-\frac{3-2H+m}{3-2H}} \to 0$; if $Q = 1$ and $3/4 < H < 1$, then

$$\left(\frac{N}{\alpha(N)}\right)^{2-2H} \left(\hat{H}_{wave} - H\right) \xrightarrow{D} L, \text{ as } N \to \infty \qquad (2.4)$$

where the distribution law L depends on H, ℓ and ψ.

(3) Let Z^H is be Rosenblatt process; assume $N\alpha(N)^{-\frac{2-2H}{3-2H}} \to 0$ as N $\alpha(N)^{-(1+m)} \to 0$; then

$$\left(\frac{N}{\alpha(N)}\right)^{1-H} \left(\hat{H}_{wave} - H\right) \xrightarrow{D} L, \text{ as } N \to \infty \qquad (2.5)$$

where the distribution law L depends on H, ℓ and ψ.

The limiting distributions L in the theorem above are not explicitly known: they come from a non-trivial non-linear transformation of quantities which are asymptotically normal or Rosenblatt-distributed. A very important advantage of $\hat{H}_{wave}$ over the *mle* for example is that it can be computed in an efficient and fast way. On the other hand, the convergence rate of the estimator depends on the choice of $\alpha(N)$.

3. Multiplication in the Wiener Chaos & Hermite Processes

3.1. *Basic Tools on Multiple Wiener-Itô Integrals*

In this section we describe the basic framework that we need in order to describe and prove the asymptotic properties of the estimator based on the discrete variations of the process. We denote by $\{W_t : t \in [0,1]\}$ a classical Wiener process on a standard Wiener space $(\Omega, \mathcal{F}, P)$. Let $\{B_t^H; t \in [0,1]\}$ be a fractional Brownian motion with Hurst parameter $H \in (0,1)$ and covariance function

$$\langle \mathbf{1}_{[0,s]}, \mathbf{1}_{[0,t]} \rangle = R_H(t,s) := \frac{1}{2}\left(t^{2H} + s^{2H} - |t-s|^{2H}\right). \qquad (3.1)$$

We denote by $\mathcal{H}$ its canonical Hilbert space. When $H = \frac{1}{2}$, then $B^{\frac{1}{2}}$ is the standard Brownian motion on $L^2([0,1])$. Otherwise, $\mathcal{H}$ is a Hilbert space which contains functions on $[0,1]$ under the inner product that extends the rule $\langle \mathbf{1}_{[0,s]}, \mathbf{1}_{[0,t]} \rangle$. Nualart's textbook (Chapter 5, [19]) can be consulted for full details.

We will use the representation of the fractional Brownian motion B^H with respect to the standard Brownian motion W: there exists a Wiener process W and a deterministic kernel $K^H(t,s)$ for $0 \le s \le t$ such that

$$B^H(t) = \int_0^1 K^H(t,s) dW_s = I_1\left(K^H(\cdot, t)\right), \qquad (3.2)$$

where I_1 is the Wiener-Itô integral with respect to W. Now, let $I_n(f)$ be the multiple Wiener-Itô integral, where $f \in L^2([0,1]^n)$ is a symmetric function. One can construct the multiple integral starting from simple functions of the form $f := \sum_{i_1,\ldots,i_n} c_{i_1,\ldots,i_n} 1_{A_{i_1} \times \ldots \times A_{i_n}}$ where the coefficient $c_{i_1,\ldots,i_n}$ is zero if two indices are equal and the sets A_{i_j} are disjoint intervals by

$$I_n(f) := \sum_{i_1,\ldots,i_n} c_{i_1,\ldots,i_n} W(A_{i_1}) \ldots W(A_{i_n}),$$

where $W(1_{[a,b]}) = W([a,b]) = W_b - W_a$. Using a density argument the integral can be extended to all symmetric functions in $L^2([0,1]^n)$. The reader can refer to Chapter 1 [19] for its detailed construction. Here, it is interesting to observe that this construction coincides with the iterated Itô stochastic integral

$$I_n(f) = n! \int_0^1 \int_0^{t_n} \ldots \int_0^{t_2} f(t_1,\ldots,t_n) dW_{t_1} \ldots dW_{t_n}. \qquad (3.3)$$

The application I_n is extended to non-symmetric functions f via

$$I_n(f) = I_n(\tilde{f})$$

where $\tilde{f}$ denotes the symmetrization of f defined by $\tilde{f}(x_1,\ldots,x_N) = \frac{1}{n!} \sum_{\sigma \in S_n} f(x_{\sigma(1)},\ldots,x_{\sigma(n)})$.

I_n is an isometry between the Hilbert space $\mathcal{H}^{\odot n}$ equipped with the scaled norm $\frac{1}{\sqrt{n!}} \|\cdot\|_{\mathcal{H}^{\otimes n}}$. The space of all integrals of order n, $\{I_n(f) : f \in L^2([0,1]^n)\}$, is called n^{th} Wiener chaos. The Wiener chaoses form orthogonal sets in $L^2(\Omega)$:

$$\mathbf{E}(I_n(f)I_m(g)) = n!\langle f, g\rangle_{L^2([0,1]^n)} \quad \text{if } m = n, \qquad (3.4)$$
$$= 0 \quad \text{if } m \neq n.$$

The next multiplication formula will plays a crucial technical role: if $f \in L^2([0,1]^n)$ and $g \in L^2([0,1]^m)$ are symmetric functions, then it holds that

$$I_n(f)I_m(g) = \sum_{\ell=0}^{m \wedge n} \ell! C_m^\ell C_n^\ell I_{m+n-2\ell}(f \otimes_\ell g), \qquad (3.5)$$

where the contraction $f \otimes_\ell g$ belongs to $L^2([0,1]^{m+n-2\ell})$ for $\ell = 0, 1, \ldots, m \wedge n$ and is given by

$$(f \otimes_\ell g)(s_1,\ldots,s_{n-\ell},t_1,\ldots,t_{m-\ell})$$
$$= \int_{[0,1]^\ell} f(s_1,\ldots,s_{n-\ell},u_1,\ldots,u_\ell)g(t_1,\ldots,t_{m-\ell},u_1,\ldots,u_\ell)du_1\ldots du_\ell.$$

Note that the contraction $(f \otimes_\ell g)$ is not necessarily symmetric. We will denote its symmetrization by $(f \tilde{\otimes}_\ell g)$.

We now introduce the Malliavin derivative for random variables in a finite chaos. The derivative operator D is defined on a subset of $L^2(\Omega)$, and takes values in $L^2(\Omega \times [0,1])$. Since it will be used for random variables in a finite chaos, it is sufficient to know that if $f \in L^2([0,1]^n)$ is a symmetric function, $DI_n(f)$ exists and it is given by

$$D_t I_n(f) = n I_{n-1}(f(\cdot,t)), \quad t \in [0,1].$$

D. Nualart and S. Ortiz-Latorre in [21] proved the following characterization of convergence in distribution for any sequence of multiple integrals to the standard normal law.

Proposition 3.1. *Let n be a fixed integer. Let $F_N = I_n(f_N)$ be a sequence of square integrable random variables in the nth Wiener chaos such that $\lim_{N \to \infty} \mathbf{E}\left[F_N^2\right] = 1$. Then the following are equivalent:*

(i) *The sequence $(F_N)_{N \geq 0}$ converges to the normal law $\mathcal{N}(0,1)$.*
(ii) $\|DF_N\|_{L^2[0,1]}^2 = \int_0^1 |D_t I_n(f)|^2 \, dt$ *converges to the constant n in $L^2(\Omega)$ as $N \to \infty$.*

There also exists a multidimensional version of this theorem due to G. Peccati and C. Tudor in [22].

3.2. Main Definitions

The Hermite processes are a family of processes parametrized by the order and the self-similarity parameter with covariance function given by (3.1). They are well-suited to modeling various phenomena that exhibit long-memory and have the self-similarity property, but which are not Gaussian. We denote by $(Z_t^{(q,H)})_{t \in [0,1]}$ the Hermite process of order q with self-similarity parameter $H \in (1/2, 1)$ (here $q \geq 1$ is an integer). The Hermite process can be defined in two ways: as a multiple integral with respect to the standard Wiener process $(W_t)_{t \in [0,1]}$; or as a multiple integral with respect to a fractional Brownian motion with suitable Hurst parameter. We adopt the first approach throughout the paper, which is the one described in the following Definition 3.5.

Definition 3.5. *The Hermite process $(Z_t^{(q,H)})_{t \in [0,1]}$ of order $q \geq 1$ and with self-similarity parameter $H \in (\frac{1}{2}, 1)$ for $t \in [0,1]$ is given by*

$$Z_t^{(q,H)} = d(H) \int_0^t \cdots \int_0^t dW_{y_1} \ldots dW_{y_q} \left(\int_{y_1 \vee \ldots \vee y_q}^t \partial_1 K^{H'}(u, y_1) \ldots \partial_1 K^{H'}(u, y_q) du \right), \tag{3.6}$$

where $K^{H'}$ is the usual kernel of the fractional Brownian motion, $d(H)$ a constant depending on H and

$$H' = 1 + \frac{H-1}{q} \iff (2H' - 2)q = 2H - 2. \tag{3.7}$$

Therefore, the Hermite process of order q is defined as a qth order Wiener-Itô integral of a non-random kernel, i.e.

$$Z_t^{(q,H)} = I_q\left(L(t, \cdot)\right),$$

where $L(t, y_1, \ldots, y_q) = \partial_1 K^{H'}(u, y_1) \ldots \partial_1 K^{H'}(u, y_q) du$.

The basic properties of the Hermite process are listed below:

- the Hermite process $Z^{(q,H)}$ is H-selfsimilar and it has stationary increments;
- the mean square of its increment is given by

$$\mathbf{E}\left[\left|Z_t^{(q,H)} - Z_s^{(q,H)}\right|^2\right] = |t-s|^{2H};$$

as a consequence, it follows from the Kolmogorov continuity criterion that, almost surely, $Z^{(q,H)}$ has Hölder-continuous paths of any order $\delta < H$;
- $Z^{(q,H)}$ exhibits long-range dependence in the sense of Definition 1.2. In fact, the autocorrelation function $\rho(n)$ *of its increments* of length 1 is asymptotically equal to $H(2H-1)n^{2H-2}$. This property is identical to that of fBm since the processes share the same covariance structure, and the property is well-known for fBm with $H > 1/2$. In particular for Hermite processes, the self-similarity and long-memory parameter coincide.

In the sequel, we will also use the *filtered* process to construct an estimator for H.

Definition 3.6. A filter α of length $\ell \in \mathbb{N}$ and order $p \in \mathbb{N} \setminus 0$ is an $(\ell+1)$-dimensional vector $\alpha = \{\alpha_0, \alpha_1, \ldots, \alpha_\ell\}$ such that

$$\sum_{q=0}^{\ell} \alpha_q q^r = 0, \quad \text{for } 0 \le r \le p-1, \ r \in \mathbb{Z}$$

$$\sum_{q=0}^{\ell} \alpha_q q^p \ne 0$$

with the convention $0^0 = 1$.

We assume that we observe the process in discrete times $\{0, \frac{1}{N}, \ldots, \frac{N-1}{N}, 1\}$. The filtered process $Z^{(q,H)}(\alpha)$ is the convolution of the process with the filter, according to the following scheme:

$$Z(\alpha) := \sum_{q=0}^{\ell} \alpha_q Z^{(q,H)}\left(\frac{i-q}{N}\right), \quad \text{for } i = \ell, \ldots, N-1 \qquad (3.8)$$

Some examples are the following:

(1) For $\alpha = \{1, -1\}$

$$Z^{(q,H)}(\alpha) = Z^{(q,H)}\left(\frac{i}{N}\right) - Z^{(q,H)}\left(\frac{i-1}{N}\right).$$

This is a filter of length 1 and order 1.

(2) For $\alpha = \{1, -2, 1\}$

$$Z^{(q,H)}(\alpha) = Z^{(q,H)}\left(\frac{i}{N}\right) - 2Z^{(q,H)}\left(\frac{i-1}{N}\right) + Z^{(q,H)}\left(\frac{i-2}{N}\right).$$

This is a filter of length 2 and order 2.

(3) More generally, longer filters produced by finite-differencing are such that the coefficients of the filter α are the binomial coefficients with alternating signs. Borrowing the notation ∇ from time series analysis, $\nabla Z^{(q,H)}(i/N) = Z^{(q,H)}(i/N) - Z^{(q,H)}((i-1)/N)$, we define $\nabla^j = \nabla \nabla^{j-1}$ and we may write the jth-order finite-difference-filtered process as follows

$$Z^{(q,H)}(\alpha) := \left(\nabla^j Z^{(q,H)}\right)\left(\frac{i}{N}\right).$$

4. Hurst Parameter Estimator Based on Discrete Variations

The estimator based on the discrete variations of the process is described by Coeurjolly in [8] for fractional Brownian motion. Using previous results by Breuer and Major, [5], he was able to prove consistency and derive the asymptotic distribution for the suggested estimator in the case of filter of order 1 for $H < 3/4$ and for all H in the case of a longer filter.

Herein we see how the results by Coeurjolly are generalized: we construct consistent estimators for the self-similarity parameter of a Hermite process of order q based on the discrete observations of the underlying process. In order to determine the corresponding asymptotic behavior we use properties of the Wiener-Itô integrals as well as Malliavin calculus techniques. The estimation procedure is the same irrespective of the specific order of the Hermite process, thus in the sequel we denote the process by $Z := Z^{(q,H)}$.

4.1. *Estimator Construction*

Filter of order 1: $\alpha = \{-1, +1\}$.

We present first the estimation procedure for a filter of order 1, i.e. using the increments of the process. The quadratic variation of Z is

$$S_N(\alpha) = \frac{1}{N}\sum_{i=1}^{N}\left(Z\left(\frac{i}{N}\right) - Z\left(\frac{i-1}{N}\right)\right)^2. \qquad (4.1)$$

We know that the expectation of $S_N(\alpha)$ is $\mathbf{E}[S_N(\alpha)] = N^{-2H}$; thus, given good concentration properties for $S_N(\alpha)$, we may attempt to estimate $S_N(\alpha)$'s expectation by its actual value, i.e. $\mathbf{E}[S_N(\alpha)]$ by $S_N(\alpha)$; suggesting the following estimator for H:

$$\hat{H}_N = -\frac{\log S_N(\alpha)}{2 \log N}. \qquad (4.2)$$

Filter of order p:

In this case we use the filtered process in order to construct the estimator for H. Let α be a filter (as defined in (3.6)) and the corresponding filtered process $Z(\alpha)$ as in (3.8): First we start by computing the quadratic variation of the filtered process

$$S_N(\alpha) = \frac{1}{N} \sum_{i=\ell}^{N} \left(\sum_{q=0}^{\ell} \alpha_q Z\left(\frac{i-q}{N}\right) \right)^2. \qquad (4.3)$$

Similarly as before, in order to construct the estimator, we estimate S_N by its expectation, which computes as $\mathbf{E}[S_N] = -\frac{N^{-2H}}{2} \sum_{q,r=0}^{\ell} \alpha_q \alpha_r |q-r|^{2H}$. Thus, we can obtain $\hat{H}_N$ by solving the following non-linear equation with respect to H

$$S_N = -\frac{N^{-2H}}{2} \sum_{q,r=0}^{\ell} \alpha_q \alpha_r |q-r|^{2H}. \qquad (4.4)$$

We write that $\hat{H}_N = g^{-1}(S_N)$, where $g(x) = -\frac{N^{-2x}}{2} \sum_{q,r=0}^{\ell} \alpha_q \alpha_r |q-r|^{2x}$. In this case, it is not possible to compute an analytical expression for the estimator. However, we can show that there exists a unique solution for $H \in [\frac{1}{2}, 1]$ as long as

$$N > \max_{H \in [\frac{1}{2},1]} \exp\left\{ \frac{\sum_{q,r=0}^{\ell} \alpha_q \alpha_r \log|q-r| \, |q-r|^{2H}}{\sum_{q,r=0}^{\ell} \alpha_q \alpha_r |q-r|^{2H}} \right\}.$$

This restriction is typically satisfied, since we work with relatively large sample sizes.

4.2. Asymptotic Properties of $\hat{H}_N$

The first question is whether the suggested estimator is consistent. This is indeed true: if sampled sufficiently often (i.e. as $N \to \infty$), the estimator converges to the true value of H almost surely, for any order of the filter.

Theorem 4.1. *Let $H \in (\frac{1}{2}, 1)$. Assume we observe the Hermite process Z of order q with Hurst parameter H. Then $\hat{H}_N$ is strongly consistent, i.e.*

$$\lim_{N \to \infty} \hat{H}_N = H \text{ a.s.}$$

In fact, we have more precisely that $\lim_{N \to \infty} \left(H - \hat{H}_N\right) \log N = 0$ a.s.

Remark 4.2. If we look at the above theorem more carefully, we observe that this is a slightly different notion of consistency than the usual one. In the case of the mle, for example, we let N tend to infinity which means that the horizon from which we sample goes to infinity. Here, we do have a fixed horizon $[0,1]$ and by letting $N \to \infty$ we sample infinitely often. If we had convergence in distribution this would not be an issue, since we could rescale the process appropriately by taking advantage of the self-similarity property, but in terms of almost sure or convergence in probability it is not exactly equivalent.

The next step is to determine the asymptotic distribution of $\hat{H}_N$. Obviously, it should depend on the distribution of the underlying process, and in our case, on q and H. We consider the following three cases separately: fBm (Hermite process of order $q = 1$), Rosenblatt process ($q = 2$), and Hermite processes of higher order $q > 2$. We summarize the results in the following theorem, where the limit notation $X_n \xrightarrow{L^2(\Omega)} X$ denotes convergence in the mean square $\lim_{N\to\infty} \mathbf{E}\left[(X_N - X)^2\right] = 0$, and $\xrightarrow{D}$ continues to denote convergence in distribution.

Theorem 4.2.

(1) Let $H \in (0,1)$ and B^H be a fractional Brownian motion with Hurst parameter H.

  (a) Assume that we use the filter of order 1.
   i. If $H \in (0, \frac{3}{4})$, then as $N \to \infty$
$$\sqrt{N} \log N \frac{2}{\sqrt{c_{1,H}}} \left(\hat{H}_N - H\right) \xrightarrow{D} \mathcal{N}(0,1), \tag{4.5}$$
   where $c_{1,H} := 2 + \sum_{k=1}^{\infty} \left(2k^{2H} - (k-1)^{2H} - (k+1)^{2H}\right)^2$.
   ii. If $H \in (\frac{3}{4}, 1)$, then as $N \to \infty$
$$N^{1-H} \log N \frac{2}{\sqrt{c_{2,H}}} \left(\hat{H}_N - H\right) \xrightarrow{L^2(\Omega)} Z^{(2,H)}, \tag{4.6}$$
   where $c_{2,H} := \frac{2H^2(2H-1)}{4H-3}$.
   iii. If $H = \frac{3}{4}$, then as $N \to \infty$
$$\sqrt{N \log N} \frac{2}{\sqrt{c_{3,H}}} \left(\hat{H}_N - H\right) \xrightarrow{D} \mathcal{N}(0,1), \tag{4.7}$$
   where $c_{3,H} := \frac{9}{16}$.

  (b) Now, let α be of any order $p \geq 2$. Then,
$$\sqrt{N} \log N \frac{1}{c_{6,H}} \left(\hat{H}_N - H\right) \xrightarrow{D} \mathcal{N}(0,1), \tag{4.8}$$
  where $c_{6,H} = \frac{1}{2} \sum_{i \in \mathbb{Z}} \rho_H^\alpha(i)^2$.

(2) Suppose that $H > \frac{1}{2}$ and the observed process $Z^{2,H}$ is a Rosenblatt process with Hurst parameter H.

  (a) If α is a filter of order 1, then
$$N^{1-H} \log N \frac{1}{2c_{4,H}} \left(\hat{H}_N - H\right) \xrightarrow{L^2(\Omega)} Z^{(2,H)}, \tag{4.9}$$
  where $c_{4,H} := 16d(H)^2$.

(b) If α is a filter of order $p > 1$, then
$$2c_{7,H}^{-1/2} N^{1-H} \log N \left(\hat{H}_N - H\right) \stackrel{L^2(\Omega)}{\to} Z^{(2,H)} \qquad (4.10)$$

where
$$c_{7,H} = \frac{64}{c(H)^2} \left(\frac{2H-1}{H(H+1)^2}\right) \times$$
$$\left\{\sum_{q,r=0}^{\ell} b_q b_r \left[|1+q-r|^{2H'} + |1-q+r|^{2H'} - 2|q-r|^{2H'}\right]\right\}^2$$

with $b_q = \sum_{r=0}^{q} a_r$. Here ℓ is the length of the filter, which is related to the order p, see Definition 3.6 and examples following.

(3) Let $H \in (\frac{1}{2}, 1)$ and $q \in \mathbb{N} \setminus \{0\}$, $q \geq 2$. Let $Z^{(q,H)}$ be a Hermite process of order q and self-similarity parameter H. Then, for $H' = 1 + \frac{H-1}{q}$ and a filter of order 1,
$$N^{2-2H'} \log N \frac{2}{c_{5,H}} \left(\hat{H}_N(\alpha) - H\right) \stackrel{L^2(\Omega)}{\to} Z^{(2,2H'-1)}, \qquad (4.11)$$

where $c_{5,H} := \frac{4q! d(H)^4 (H'(2H'-1))^{2q-2}}{(4H'-3)(4H'-2)}$.

Remark 4.3. In the notation above, $Z^{(2,K)}$ denotes a standard Rosenblatt random variable, which means a random variable that has the same distribution as the Hermite process of order 2 and parameter K at $t=1$.

Before continuing with a sketch of proof of the theorem, it is important to discuss the theorem's results.

(1) In most of the cases above, we observe that the order of convergence of the estimator depends on H, which is the parameter that we try to estimate. This is not a problem, because it has already been proved, in [6], [7], [27] and [28], that the theorem's convergences still hold when we replace H by $\hat{H}_N$ in the rate of convergence.
(2) The effect of the use of a longer filter is very significant. In the case of fBm, when we use a longer filter, we no longer have the threshold of 3/4 and the suggested estimator is always asymptotically normal. This is important for the following reason: when we start the estimation procedure, we do not know beforehand the value of H and such a threshold would create confusion in choosing the correct rate of convergence in order to scale $\hat{H}_N$ appropriately. Finally, the fact that we have asymptotic normality for all H allows us to construct confidence intervals and perform hypothesis testing and model validation.
(3) Even in the Rosenblatt case the effect of the filter is significant. This is not obvious here, but we will discuss it later in detail. What actually happens is that by filtering the process asymptotic standard error is reduced, i.e. the longer the filter the smaller the standard error.

(4) Finally, one might wonder if the only reason to chose to work with the quadratic variation of the process, instead of a higher order variation (powers higher than 2 in (4.1) and (4.3)), is for the simplicity of calculations. It turns out that there are other, better reasons to do so: Coeurjolly ([8]) proved that the use of higher order variations would lead to higher asymptotic constants and thus to larger standard errors in the case of the fBm. He actually proved that the optimal variation respect to the standard error is the second (*quadratic*).

Proof. [Sketch of proof of Theorem 4.2] We present the key ideas for proving the consistency and asymptotic distribution results. We use a Hermite process of order q and a filter of order 1. However, wherever it is necessary we focus on either fBm or Rosenblatt in order to point out the corresponding differences. The ideas for the proof are very similar in the case of longer filters so the reader should refer to [7] for the details in this approach.

It is convenient to work with the centered normalized quadratic variation defined as

$$V_N := -1 + \frac{1}{N} \sum_{i=0}^{N-1} \frac{\left(Z^{(q,H)}_{\frac{i+1}{N}} - Z^{(q,H)}_{\frac{i}{N}}\right)^2}{N^{-2H}}. \tag{4.12}$$

It is easy to observe that for S_N defined in (4.1),

$$S_N = N^{-2H}\left(1 + V_N\right).$$

Using this relation we can see that $\log\left(1 + V_N\right) = 2\left(\hat{H}_N - H\right)\log N$, therefore in order to prove consistency it suffices to show that V_N converges to 0 as $N \to \infty$ and the asymptotic distribution of $\hat{H}_N$ depends on the asymptotic behavior of V_N.

By the definition of the Hermite process in Definition 3.5, we have that

$$Z^{(q,H)}_{\frac{i+1}{N}} - Z^{(q,H)}_{\frac{i}{N}} = I_q\left(f_{i,N}\right)$$

where we denoted

$$f_{i,N}(y_1,\ldots,y_q) = 1_{[0,\frac{i+1}{N}]}(y_1 \vee \ldots \vee y_q) d(H) \int_{y_1 \vee \ldots \vee y_q}^{\frac{i+1}{N}} \partial_1 K^{H'}(u,y_1)\ldots\partial_1 K^{H'}(u,y_q)du$$

$$- 1_{[0,\frac{i}{N}]}(y_1 \vee \ldots \vee y_q) d(H) \int_{y_1 \vee \ldots \vee y_q}^{\frac{i}{N}} \partial_1 K^{H'}(u,y_1)\ldots\partial_1 K^{H'}(u,y_q)du.$$

Now, using the multiplication property (3.5) of multiple Wiener-Itô integrals we can derive a Wiener chaos decomposition of V_N as follows:

$$V_N = T_{2q} + c_{2q-2}T_{2q-2} + \ldots + c_4 T_4 + c_2 T_2 \tag{4.13}$$

where $c_{2q-2k} := k!\binom{q}{k}^2$ are the combinatorial constants from the product formula for $0 \leq k \leq q-1$, and

$$T_{2q-2k} := N^{2H-1} I_{2q-2k}\left(\sum_{i=0}^{N-1} f_{i,N} \otimes_k f_{i,N}\right),$$

where $f_{i,N} \otimes_k f_{i,N}$ is the kth contraction of $f_{i,N}$ with itself which is a function of $2q - 2k$ parameters.

To determine the magnitude of this Wiener chaos decomposition V_N, we study each of the terms appearing in the decomposition separately. If we compute the L^2 norm of each term, we have

$$\mathbf{E}\left[T_{2q-2k}^2\right] = N^{4H-2}(2q-2k)! \left\| \left(\sum_{i=0}^{N-1} f_{i,N} \otimes_k f_{i,N}\right)^s \right\|_{L^2([0,1]^{2q-2k})}^2$$

$$= N^{4H-2}(2q-2k)! \sum_{i,j=0}^{N-1} \langle f_{i,N} \tilde{\otimes}_k f_{i,N}, f_{j,N} \tilde{\otimes}_k f_{j,N} \rangle_{L^2([0,1]^{2q-2k})}$$

Using properties of the multiple integrals we have the following results

- For $k = q - 1$, $\mathbf{E}\left[T_2^2\right] \sim \frac{4d(H)^4(H'(2H'-1))^{2q-2}}{(4H'-3)(4H'-2)} N^{2(2H'-2)}$
- For $k = 0, \ldots, q - 2$

$$\mathbf{E}\left[N^{2(2-2H')}T_{2q-2k}^2\right] = O\left(N^{-2(2-2H')2(q-k-1)}\right).$$

Thus, we observe that the term T_2 is the dominant term in the decomposition of the variation statistic V_N. Therefore, with

$$c_{1,1,H} = \frac{4d(H)^4(H'(2H'-1))^{2q-2}}{(4H'-3)(4H'-2)},$$

it holds that

$$\lim_{N \to \infty} \mathbf{E}\left[c_{1,1,H}^{-1} N^{(2-2H')2} c_2^{-2} V_N^2\right] = 1.$$

Based on these results we can easily prove that V_N converges to 0 a.s. and then conclude that $\hat{H}_N$ is strongly consistent.

Now, in order to understand the asymptotic behavior of the renormalized sequence V_N it suffices to study the limit of the dominant term

$$I_2\left(N^{2H-1}N^{(2-2H')} \sum_{i=0}^{N-1} f_{i,N} \otimes_{q-1} f_{i,N}\right)$$

When $q = 1$ (the fBm case), we can use the Nualart–Ortiz-Latorre criterion (Proposition 3.1) in order to prove convergence to Normal distribution. However, in the general case for $q > 1$, using the same criterion, we can see that convergence to a Normal law is no longer true. Instead, a direct method can be employed to determine the asymptotic distribution of the above quantity. Let $N^{2H-1}N^{(2-2H')} \sum_{i=0}^{N-1} f_{i,N} \otimes_{q-1} f_{i,N} = f_2^N + r_2$, where r_2 is a remainder term and

$$f_2^N(y,z) := N^{2H-1}N^{(2-2H')}d(H)^2 a(H')^{q-1}$$

$$\sum_{i=0}^{N-1} \mathbf{1}_{[0,\frac{i}{N}]}(y \vee z) \int_{I_i} \int_{I_i} dv du \, \partial_1 K(u,y) \partial_1 K(v,z) |u-v|^{(2H'-2)(q-1)}.$$

It can be shown that the term r_2 converges to 0 in $L^2([0,1]^2)$, while f_2^N converges in $L^2([0,1]^2)$ to the kernel of the Rosenblatt process at time 1, which is by definition

$$(H'(2H'-1))^{(q-1)} d(H)^2 N^{2H-1} N^{2-2H'}$$

$$\times \sum_{i=0}^{N-1} \int_{I_i} \int_{I_i} |u-v|^{(2H'-2)(q-1)} \partial_1 K^{H'}(u,y) \partial_1 K^{H'}(v,z).$$

This implies, by the isometry property (3.4) between double integrals and $L^2([0,1]^2)$, that the dominant term in V_N, i.e. the second-chaos term T_2, converges in $L^2(\Omega)$ to the Rosenblatt process at time 1. The reader can consult [6], [7], [27] and [28] for all details of the proof. □

5. Comparison & Conclusions

In this section, we compare the estimators described in Sections 2 and 4. The performance measure that we adopt is the *asymptotic relative efficiency*, which we now define according to [24]:

Definition 5.7. Let T_n be an estimator of θ for all n and $\{\alpha_n\}$ a sequence of positive numbers such that $\alpha_n \to +\infty$ or $\alpha_n \to \alpha > 0$. Suppose that for some probability law Y with positive and finite second moment,

$$\alpha_n (T_n - \theta) \xrightarrow{D} Y,$$

(i) The *asymptotic mean square error* of T_n ($amse_{T_n}(\theta)$) is defined to be the asymptotic expectation of $(T_n - \theta)^2$, i.e.

$$amse_{T_n}(\theta) = \frac{EY^2}{\alpha_n}.$$

(ii) Let T'_n be another estimator of θ. The *asymptotic relative efficiency* of T'_n with respect to T_n is defined to be

$$e_{T_n, T'_n}(\theta) = \frac{amse_{T_n}(\theta)}{amse_{T'_n}(\theta)}. \tag{5.1}$$

(iii) T_n is said to be *asymptotically more efficient* than T'_n if and only if

$$\limsup_n e_{T_n, T'_n}(\theta) \leq 1, \quad \text{for all } \theta \text{ and}$$

$$\limsup_n e_{T_n, T'_n}(\theta) < 1, \quad \text{for some } \theta.$$

Remark 5.4. These definitions are in the most general setup: indeed (i) they are not restricted by the usual assumption that the estimators converge to a Normal distribution; moreover, (ii) the asymptotic distributions of the estimators do not have to be the same. This will be important in our comparison later.

Our comparative analysis focuses on fBm and the Rosenblatt process, since the maximum likelihood and the wavelets methods cannot be applied to higher order Hermite processes.

5.1. *Variations Estimator vs. mle*

We start with the case of a filter of order 1 for fBm. Since the asymptotic behavior of the variations estimator depends on the true value of H we consider three different cases:

- If $H \in (0, 3/4)$, then

$$e_{\hat{H}_N(\alpha), \hat{H}_{mle}}(H) = \frac{\frac{2+\sum_{k=1}^{\infty}(2k^{2H}-(k-1)^{2H}-(k+1)^{2H})^2}{2\sqrt{N}\log N}}{\frac{[2\,D(H)]^{-1}}{\sqrt{N}}} \approx \frac{1}{\log N}.$$

This implies that

$$\limsup e_{\hat{H}_N(\alpha), \hat{H}_{mle}}(H) = 0,$$

meaning that $\hat{H}_N(\alpha)$ is asymptotically more efficient than $\hat{H}_{mle}$.

- If $H \in (3/4, 1)$, then

$$e_{\hat{H}_N(\alpha), \hat{H}_{mle}}(H) = \frac{\frac{\frac{2H^2(2H-1)}{4H-3}}{4\,N^{1-H}\log N}}{\frac{[2\,D(H)]^{-1}}{\sqrt{N}}} \approx \frac{N^{H-1/2}}{\log N}.$$

This implies that

$$\limsup e_{\hat{H}_N(\alpha),\,\hat{H}_{mle}}(H) = \infty,$$

meaning that $\hat{H}_{mle}$ is asymptotically more efficient than $\hat{H}_N(\alpha)$.

- If $H = 3/4$, then

$$e_{\hat{H}_N(\alpha),\,\hat{H}_N^{mle}}(H) = \frac{\frac{16/9}{4\sqrt{N}\log N}}{\frac{[2\,D(H)]^{-1}}{\sqrt{N}}} \approx \frac{1}{\sqrt{\log N}}.$$

Similarly, as in the first scenario the variations estimator is asymptotically more efficient than the *mle*.

Remark 5.5. By $\hat{H}_{mle}$ we mean either the exact *mle* or the Whittle approximate *mle*, since both have the same asymptotic distribution.

Before discussing the above results let us recall the Cramér-Rao Lower Bound theory (see [24]). Let $X = (X_1, \ldots, X_N)$ be a *sample* (i.e. identically distributed random variables) with common distribution P_H and corresponding density function f_H. If T is an estimator of H such that $E(T) = H$, then

$$Var(T) \geq [I(H)]^{-1} \tag{5.2}$$

where $I(H)$ is the *Fisher information* defined by

$$I(H) := \mathbf{E}\left\{\left[\frac{\partial}{\partial H}\log f_H(X)\right]^2\right\}. \tag{5.3}$$

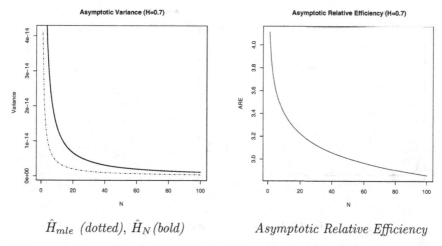

$\hat{H}_{mle}$ (dotted), $\hat{H}_N$ (bold)   Asymptotic Relative Efficiency

Figure 6.3. Comparison of the variations' estimator and mle for a filter of order $p = 1$.

The inverse of the Fisher information is called *Cramér-Rao Lower Bound*.

Remark 5.6. It has been proved by Dahlhaus, [9], that the asymptotic variance of both the approximate and exact *mle* converges to the Cramér-Rao Lower Bound and consequently both estimators are asymptotically efficient according to the Fisher criterion. Thus how can the variations estimator be more efficient in some cases?

The variations-based estimator is computed using data coming from a fixed time horizon and more specifically $[0, 1]$, i.e. data such as $X_a = (X_0, X_{\frac{1}{N}}, \ldots, X_1)$, while the *mle* is computed using data of the form $X_b = (X_0, X_1, \ldots, X_N)$. The time-scaling makes a big difference since the vectors X_a and X_b **do not have the same distribution**. The construction of the Fisher information (and accordingly the asymptotic Cramér-Rao Lower Bound) depends on the underlying distribution of the sample and it is going to be different for X_a and X_b. This implies that the Cramér-Rao Lower Bound attained by the *mle* using X_b is not the same as the Cramér-Rao Lower Bound attained by the *mle* using X_a. By the self-similarity property we can derive that $X_a =^{\mathcal{D}} N^H X_b$, which indicates that if we want to compute the information matrix for the rescaled data, the scaling contains the parameter H and this will alter the information matrix and its rate of convergence.

We begin by observing what happens in practice for a filter of order 1. In the following graphs, we compare the corresponding variations estimator with the mle, in the case of a simulated fractional Brownian motion with $H = 0.65$, by plotting the asymptotic variance against the sample size N.

As we observe in Figure 6.3, the *mle* performs better than the estimator based on the variations of the process with filter order $p = 1$. It has a smaller asymptotic variance and the asymptotic relative efficiency seems to converge to zero extremely slowly, even for a very large sample size N. This is because $\hat{H}_N$ is faster only by

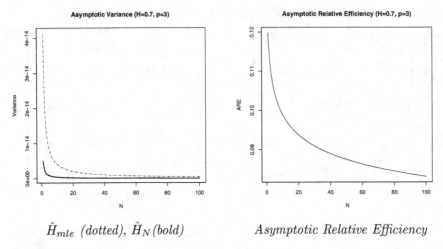

$\hat{H}_{mle}$ (dotted), $\hat{H}_N$ (bold)   Asymptotic Relative Efficiency

Figure 6.4. Comparison of the variations' estimator and *mle* for a filter of order $p = 10$.

the factor $\log N$ (which is quite slow) and the constants in the case of $\hat{H}_N$ are quite large.

Let us consider now the case of longer filters ($p \geq 2$). Using the results proved in the previous sections (esp. Theorem 4.2 part (1.b)), we have that for all $H \in (0, 1)$

$$e_{\hat{H}_N(\alpha),\, \hat{H}_N^{mle}}(H) \approx \frac{1}{\log N}$$

and from this we conclude that the variations estimator is always asymptotically more efficient than the *mle*. If we do the same plots as before we can see (Figure 6.4) that the constant is now significantly smaller.

5.2. Variations' vs. Wavelet Estimator

In this subsection we compare the variations and the wavelets estimators for the both the fBm and the Rosenblatt process.

fBm:

(1) Let $0 < H < 3/4$, then for a filter of order $p \geq 1$ in the variations estimator, and for any $Q \geq 1$ in the wavelets estimator, we have

$$e_{\hat{H}_N(\alpha),\, \hat{H}_{wave}}(H) \approx \frac{1}{\sqrt{\alpha(N)} \log N}.$$

Based on the properties of $\alpha(N)$ as stated before (Theorem 2.2), we conclude that

$$\lim_{N \to 0} e_{\hat{H}_N(\alpha),\, \hat{H}_{mle}}(H) = 0,$$

which implies that the variations estimator is asymptotically more efficient than the wavelets estimator.

(2) When $3/4 < H < 1$, then for a filter of order $p = 1$ in the variations estimator, and $Q = 1$ for the wavelets estimator, we have

$$e_{\hat{H}_N(\alpha),\, \hat{H}_{wave}}(H) \approx \frac{N^{2-2H}}{\alpha(N)^{2-2H} \log N}$$

If we choose $\alpha(N)$ to be the *optimal* as suggested by Bardet and Tudor in [1], i.e. $\alpha(N) = N^{1/2+\delta}$ for δ small, then $e_{\hat{H}_N(\alpha),\, \hat{H}_{wave}}(H) \approx \frac{N^{(1-H)(1-2\delta)}}{\log N}$, which implies that the wavelet estimator performs better.

(3) When $3/4 < H < 1$, then for a filter of order $p \geq 2$ in the variations estimator and $Q = 1$ for the wavelets estimator, using again the optimal choice of $\alpha(N)$ as proposed in [1], we have

$$e_{\hat{H}_N(\alpha),\, \hat{H}_{wave}}(H) \approx \frac{N^{(\frac{1}{2}-H)-2\delta(1-H)}}{\log N},$$

so the variations estimator is asymptotically more efficient than the wavelets one.

Rosenblatt process:

Suppose that $1/2 < H < 1$, then for any filter of any order $p \geq 1$ in the variations estimator, and any $Q \geq 1$ for the wavelets based estimator, we have

$$e_{\hat{H}_N(\alpha),\, \hat{H}_{wave}}(H) \approx \frac{1}{\alpha(N)^{1-H} \log N}.$$

Again, with the behavior of $\alpha(N)$ as stated in Theorem 2.2, we conclude that the variations estimator is asymptotically more efficient than the wavelet estimator.

Overall, it appears that the estimator based on the discrete variations of the process is asymptotically more efficient than the estimator based on wavelets, in most cases. The wavelets estimator does not have the problems of computational time which plague the *mle*: using efficient techniques, such as Mallat's algorithm, the wavelets estimator takes seconds to compute on a standard PC platform. However, the estimator based on variations is much simpler, since it can be constructed by simple transformation of the data.

Summing up, the conclusion is that the heuristic approaches (R/S, variograms, correlograms) are useful for a preliminary analysis to determine whether long memory may be present, due to their simplicity and universality. However, in order to estimate the Hurst parameter it would be preferable to use any of the other techniques. Overall, the estimator based on the discrete variations is asymptotically more efficient than the estimator based on wavelets or the *mle*. Moreover, it can be applied not only when the data come from a fractional Brownian motion, but also when they come from any other non-Gaussian Hermite process of higher order.

Finally, when we apply a longer filter in the estimation procedure, we are able to reduce the asymptotic variance and consequently the standard error significantly.

The benefits of using longer filters needs to be investigated further. It would be interesting to study the choice of different types of filters, such as wavelet-type filters versus finite difference filters. Specifically, the complexity introduced by the construction of the estimator based on a longer filter, which is not as straightforward as in the case of filter of order 1, is something that will be investigated in a subsequent article.

References

1. J-M. Bardet and C.A. Tudor (2008): *A wavelet analysis of the Rosenblatt process: chaos expansion and estimation of the self-similarity parameter.* Preprint.
2. J.B. Bassingthwaighte and G.M. Raymond (1994): Evaluating rescaled range analysis for time series, *Annals of Biomedical Engineering*, **22**, 432-444.
3. J. Beran (1994): *Statistics for Long-Memory Processes.* Chapman and Hall.
4. J.-C. Breton and I. Nourdin (2008): Error bounds on the non-normal approximation of Hermite power variations of fractional Brownian motion. *Electronic Communications in Probability*, **13**, 482-493.
5. P. Breuer and P. Major (1983): Central limit theorems for nonlinear functionals of Gaussian fields. *J. Multivariate Analysis*, **13** (3), 425-441.
6. A. Chronopoulou, C.A. Tudor and F. Viens (2009): Application of Malliavin calculus to long-memory parameter estimation for non-Gaussian processes. *Comptes rendus - Mathematique* **347**, 663-666.
7. A. Chronopoulou, C.A. Tudor and F. Viens (2009): *Variations and Hurst index estimation for a Rosenblatt process using longer filters.* Preprint.
8. J.F. Coeurjolly (2001): Estimating the parameters of a fractional Brownian motion by discrete variations of its sample paths. *Statistical Inference for Stochastic Processes*, **4**, 199-227.
9. R. Dahlhaus (1989): Efficient parameter estimation for self-similar processes. *Annals of Statistics*, **17**, 1749-1766.
10. R.L. Dobrushin and P. Major (1979): Non-central limit theorems for non-linear functionals of Gaussian fields. *Z. Wahrscheinlichkeitstheorie verw. Gebiete*, **50**, 27-52.
11. P. Flandrin (1993): Fractional Brownian motion and wavelets. *Wavelets, Fractals and Fourier transforms.* Clarendon Press, Oxford, 109-122.
12. R. Fox, M. Taqqu (1985): Non-central limit theorems for quadratic forms in random variables having long-range dependence. *Probab. Th. Rel. Fields*, **13**, 428-446.
13. Hurst, H. (1951): Long Term Storage Capacity of Reservoirs, *Transactions of the American Society of Civil Engineers*, **116**, 770-799.
14. A.K. Louis, P. Maass, A. Rieder (1997): *Wavelets: Theory and applications* Pure & Applied Mathematics. Wiley-Interscience series of texts, monographs & tracts.
15. M. Maejima and C.A. Tudor (2007): Wiener integrals and a Non-Central Limit Theorem for Hermite processes, *Stochastic Analysis and Applications*, **25** (5), 1043-1056.
16. B.B. Mandelbrot (1975): Limit theorems of the self-normalized range for weakly and strongly dependent processes. *Z. Wahrscheinlichkeitstheorie verw. Gebiete*, **31**, 271-285.
17. I. Nourdin, D. Nualart and C.A Tudor (2007): *Central and Non-Central Limit Theorems for weighted power variations of the fractional Brownian motion.* Preprint.

18. I. Nourdin, G. Peccati and A. Réveillac (2008): Multivariate normal approximation using Stein's method and Malliavin calculus. *Ann. Inst. H. Poincaré Probab. Statist.*, 18 pages, to appear.
19. D. Nualart (2006): *Malliavin Calculus and Related Topics.* Second Edition. Springer.
20. D. Nualart and G. Peccati (2005): Central limit theorems for sequences of multiple stochastic integrals. *The Annals of Probability*, **33**, 173-193.
21. D. Nualart and S. Ortiz-Latorre (2008): Central limit theorems for multiple stochastic integrals and Malliavin calculus. *Stochastic Processes and their Applications*, **118**, 614-628.
22. G. Peccati and C.A. Tudor (2004): Gaussian limits for vector-valued multiple stochastic integrals. *Séminaire de Probabilités*, **XXXIV**, 247-262.
23. G. Samorodnitsky and M. Taqqu (1994): *Stable Non-Gaussian random variables.* Chapman and Hall, London.
24. J. Shao (2007): *Mathematical Statistics.* Springer.
25. M. Taqqu (1975): Weak convergence to the fractional Brownian motion and to the Rosenblatt process. *Z. Wahrscheinlichkeitstheorie verw. Gebiete*, **31**, 287-302.
26. C.A. Tudor (2008): Analysis of the Rosenblatt process. *ESAIM Probability and Statistics*, **12**, 230-257.
27. C.A. Tudor and F. Viens (2008): Variations and estimators for the selfsimilarity order through Malliavin calculus. *Annals of Probability*, 34 pages, to appear.
28. C.A. Tudor and F. Viens (2008): Variations of the fractional Brownian motion via Malliavin calculus. *Australian Journal of Mathematics*, 13 pages, to appear.
29. P. Whittle (1953): Estimation and information in stationary time series. *Ark. Mat.*, **2**, 423-434.

Chapter 7

Modeling Colored Noise by Fractional Brownian Motion

Jinqiao Duan[1,2], Chujin Li[2] and Xiangjun Wang[2] *

1. Department of Applied Mathematics
Illinois Institute of Technology
Chicago, IL 60616, USA
E-mail: duan@iit.edu

2. School of Mathematics and Statistics
Huazhong University of Science and Technology
Wuhan 430074, China
E-mail: licjhust@sina.com; x.j.wang@163.com

Complex systems are usually under the influences of noises. Appropriately modeling these noises requires knowledge about generalized time derivatives and generalized stochastic processes. To this end, a brief introduction to generalized functions theory is provided. Then this theory is applied to fractional Brownian motion and its derivative, both regarded as generalized stochastic processes, and it is demonstrated that the "time derivative of fractional Brownian motion" is correlated and thus is a mathematical model for colored noise. In particular, the "time derivative of the usual Brownian motion" is uncorrelated and hence is an appropriate model for white noise.

Keywords: White noise, colored noise, stochastic differential equations (SDEs); generalized time derivative, generalized stochastic processes, stationary processes

2000 AMS Subject Classification: 60G20, 60H40, 60H10

Contents

1	What is noise	120
2	Generalized time derivative and generalized stochastic processes	121
3	White noise	122
4	Colored noises and fractional Brownian motion	125
	References	126

---

*The authors gratefully acknowledge the support by the Cheung Kong Scholars Program, the K. C. Wong Education Foundation, Hong Kong, and the NSF grant 0620539.

1. What is noise

Complex systems in science and engineering are often subject to random fluctuations. Although these fluctuations arise from various situations, they appear to share some common features.[16,21,26,29-31] They are generally regarded as stationary stochastic processes, with zero mean and with special correlations at different time instants.

We assume that stochastic processes are defined for all real time $t \in (-\infty, \infty)$. Let $X_t(\omega)$ be a real-valued stochastic process defined in a probability space $(\Omega, \mathcal{F}, \mathbb{P})$. We say that X_t is an stationary stochastic process if, for any integer k and any real numbers $t_1 < t_2 < \cdots < t_k$, the distribution of $(X_{t_1+t}, X_{t_2+t}, \cdots, X_{t_k+t})$ does not depend on t, i.e.

$$\mathbb{P}\left(\{\omega : X_{t_1+t}(\omega), \cdots, X_{t_k+t}(\omega)) \in A\}\right) = \mathbb{P}\left(\{\omega : (X_{t_1}(\omega), \cdots, X_{t_k}(\omega)) \in A\}\right),$$

for every open interval A and all t.

Moreover, we say that X_t has stationary increments if, for any integer k and any real numbers $t_0 < t_1 < \cdots < t_k$, the distribution of $X_{t_j} - X_{t_{j-1}}$ depends on t_j and t_{j-1} only through the difference $t_j - t_{j-1}$ where $j = 1, ..., k$. It means that if $t_j - t_{j-1} = t_i - t_{i-1}$ for some $i, j \in \{1, \cdots, k\}$, then $(X_{t_j} - X_{t_{j-1}}) =^d (X_{t_i} - X_{t_{i-1}})$, i.e., the both sides have the same distributions. Such a stationary process X_t is also called a strongly stationary process or a first order stationary process. It is called a weakly stationary process or the second order stationary process if, for any integer k and any real numbers $t_1 < t_2 < \cdots < t_k$, the mean and covariance matrix of $(X_{t_1+t}, X_{t_2+t}, \cdots, X_{t_k+t})$ does not depend on t.

Noise is a special stationary stochastic process $\eta_t(\omega)$. Its mean $\mathbb{E}\eta_t = 0$ and its covariance $\mathbb{E}(\eta_t \eta_s) = K\, c(t-s)$ for all t and s, with some constant K and a function $c(\cdot)$. When $c(t-s)$ is the Dirac Delta function $\delta(t-s)$, the noise η_t is called a white noise, otherwise it is a colored noise.

For example, Gaussian white noise may be molded in terms of "time derivative" of Brownian motion. Let us first discuss this formally.[21,26] Recall that a scalar Brownian motion B_t is a Gaussian process with stationary (and also independent) increments, together with mean zero $\mathbb{E}B_t = 0$ and covariance $\mathbb{E}(B_t B_s) = t \wedge s = \min\{t, s\}$. By the formal formula $\mathbb{E}(\dot{X}_t \dot{X}_s) = \partial^2 \mathbb{E}(X_t X_s)/\partial t \partial s$, we see that

$$\mathbb{E}(\dot{B}_t \dot{B}_s) = \partial^2 \mathbb{E}(B_t B_s)/\partial t \partial s = \partial^2 (t \wedge s)/\partial t \partial s = \delta(t-s).$$

So the spectral density function for $\dot{B}_t$, i.e., the Fourier transform $\mathcal{F}$ for its covariance function $\mathbb{E}(\dot{B}_t \dot{B}_s)$, is constant

$$\mathcal{F}(\mathbb{E}(\dot{B}_t \dot{B}_s)) = \mathcal{F}(\delta(t-s)) = \frac{1}{2\pi}.$$

Moreover, the increments like $B_{t+\Delta t} - B_t \approx \dot{B}_t$ are stationary, and formally, $\mathbb{E}\dot{B}_t \approx \mathbb{E}\frac{B_{t+\Delta t} - B_t}{\Delta t} = \frac{0}{\Delta t} = 0$. Thus $\eta_t = \dot{B}_t$ is taken as a mathematical model for white noise. Note that Brownian motion does not have usual time derivative. It is necessary to interpret $\dot{B}_t$ as a generalized time derivative and make the above argument rigorous.

This chapter is organized as follows. In §2, we discuss generalized time derivatives for stochastic processes. We then consider white noise and colored noise in §3 and §4, respectively.

2. Generalized time derivative and generalized stochastic processes

We first consider the Heaviside function

$$H(t) = \begin{cases} 1 & \text{if } t \geq 0, \\ 0 & \text{otherwise}. \end{cases} \quad (2.1)$$

This function is certainly not $C^1(\mathbb{R})$, but if it had a (generalized) derivative $H'(t)$, then by formal integration by parts, $H'(t)$ should satisfy

$$\int_{-\infty}^{+\infty} H'(t)v(t)dt = -\int_{-\infty}^{+\infty} H(t)v'(t)dt = -\int_{0}^{+\infty} v'(t)dx = v(0) - v(+\infty) = v(0)$$

for every test function $v \in C_0^1(\mathbb{R})$ (i.e., smooth functions with compact support on $\mathbb{R}$).

Since $H'(t) \equiv 0$ for $t \neq 0$, this suggests that "$H'(0) = \infty$" in such a way that $\int_{-\infty}^{+\infty} H'(t)v(t)dt = v(0)$. Of course, no true function behaves like this, so we call $H'(t)$ a generalized function. Notice that this process enables us to take more derivatives of $H(t)$:

$$\int_{-\infty}^{+\infty} H''(t)v(t)dt = \int_{-\infty}^{+\infty} H(t)v''(t) = \int_{0}^{+\infty} v''(t)dt = -v'(0),$$

$$\int_{-\infty}^{+\infty} H'''(t)v(t)dt = -\int_{-\infty}^{+\infty} H(t)v'''(t) = \int_{0}^{+\infty} v'''(t)dt = -v''(0),$$

provided v is sufficiently smooth and has compact support, say $v \in C_0^\infty(\mathbb{R})$. The above computation of the derivatives is via formal integration by parts against a function in $C_0^\infty(\mathbb{R})$. This is called generalized differentiation, and the function in $C_0^\infty(\mathbb{R})$ are called test functions.

We define the delta function, as a generalized function, to be the object $\delta(t)$ so that formally

$$\int_{-\infty}^{+\infty} \delta(t)v(t)dt = v(0), \quad (2.2)$$

for every test function $v \in C_0^\infty(\mathbb{R})$. Hence we find $H'(t) = \delta(t)$.

Also note that we can view the generalized function $\delta(t)$ as a linear functional on the test space $C_0^\infty(\mathbb{R})$, which assigns each v a real value:

$$F(v) := \int_{-\infty}^{+\infty} \delta(t)v(t)dt = v(0).$$

Notice that we can take any order of generalized derivatives of $\delta(t)$. In fact, for any positive integer m, $D^m\delta(t)$ is the object satisfying $\int D^m\delta(t)v(t)dt = (-1)^m \int \delta(t)D^m v(t)dt = (-1)^m D^m v(0)$. Also, we can translate the singularity in $\delta(t)$ to any point μ by letting $\delta_\mu(t) = \delta(t-\mu)$ so that a change of variables $y = t - \mu$ yields

$$\int_{-\infty}^{+\infty} \delta_\mu(t)v(t)dt = \int_{-\infty}^{+\infty} \delta(t-\mu)v(t)dt = \int_{-\infty}^{+\infty} \delta(y)v(y+\mu)dy = v(\mu).$$

We have thus defined a special generalized function, the delta function, and its generalized derivatives through their actions on test functions.

In fact, this procedure applies to other generalized functions as well.

A generalized function F is a linear functional on the test space, i.e., a linear mapping $F : C_0^\infty(\mathbb{R}) \to \mathbb{R}$ such that $F(v_j) \to 0$ for every sequence $v_j \subset C_0^\infty(\mathbb{R})$ with support in a fixed compact set $K \subset \mathbb{R}$ and whose derivatives $D^m v_j \to 0$ uniformly in K, as $j \to \infty$. If F and F_j are generalized functions in $\mathbb{R}$, then $F_j \to F$ as generalized functions provided that $F_j(v) \to F(v)$ for every $v \in C_0^\infty(\mathbb{R})$. The support of a generalized function F in $\mathbb{R}$ is the smallest closed set $K \subset \mathbb{R}$ such that $F(v) = 0$ whenever $v \equiv 0$ in a neighborhood of K.

For example, $F(v) = \int_\mathbb{R} v(t)f(t)dt$ with $f(t)$ given and $v \in C_0^\infty(\mathbb{R})$, is a generalized function. There are other generalized functions not defined by such integrals; see[28] for more information.

A generalized stochastic process $\eta_t(\omega)$ is a generalized function in time t, almost surely.

In the next two sections, we will consider generalized stochastic processes as mathematical models for white noise and colored noise.

3. White noise

In engineering, white noise is generally understood as a stationary process ξ_t, with zero mean $E\xi_t = 0$ and constant spectral density $f(\lambda)$ on the entire real axis. See[21,31] for discussions of white noise in engineering and see[7,8] for more applications. If $E\xi_s\xi_{t+s} = C(t)$ is the covariance function of ξ_t, then the spectral density is defined as Fourier transform of the covariance function $C(t)$

$$f(\lambda) = \frac{1}{2\pi} \int_{-\infty}^{+\infty} e^{-i\lambda t}C(t)dt = \frac{K}{2\pi}$$

where K is a positive constant. This relation holds for a (generalized) stochastic process ξ_t with covariance function $C(t) = \delta(t)$, the Dirac delta function. This says that white noise is a stationary stochastic process that has zero mean and is uncorrelated at different time instants.

We will see that $\xi_t = \dot{B}_t$ is such a stochastic process.

If white noise ξ_t's covariance $Cov(\xi_t, \xi_s)$, and $\dot{B}_t$'s covariance $Cov(\dot{B}_t, \dot{B}_s)$ are the same, then we can take $\dot{B}_t$ as a mathematical model for white noise ξ_t. This is indeed the case, but we have to verify this in the context of generalized functions, because $\frac{dB_t}{dt}$ has no meaning in the sense of ordinary functions.

In fact, white noise was first correctly described in connection with the theory of generalized functions.[2] From the last section, we know that

$$\Phi_f(\varphi) = \int_{-\infty}^{+\infty} \varphi(t) f(t) dt, \tag{3.1}$$

defines a generalized function for a given f. The function Φ_f depends linearly and continuously on test functions φ. It is the generalized function corresponding to f. With this representation we regard f as a generalized function. In fact, we may identify f with this linear functional Φ_f.

In particular, the generalized function defined by

$$\Phi(\varphi(t)) = \varphi(t_0),$$

with a fixed t_0, for $\varphi \in C_0^\infty(\mathbb{R})$, is called the Dirac delta function, and is also symbolically denoted by $\delta(t-t_0)$. In contrast with classical functions, generalized functions always have derivatives of every order, which again are generalized functions. By the derivative $\dot{\Phi}$ of Φ, we mean the generalized function defined by

$$\dot{\Phi}(\varphi) = -\Phi(\dot{\varphi}).$$

A generalized stochastic process is now simply a random generalized function in the following sense: for every test function φ, a random variable $\Phi(\varphi)$ is assigned such that the functional Φ is linear and continuous.

A generalized stochastic process is said to be Gaussian if, for arbitrary linearly independent functions $\varphi_1, \cdots \varphi_n \in K$, the random variable $(\Phi(\varphi_1), \cdots \Phi(\varphi_n))$ is normally distributed. Just as in the classical case, a generalized Gaussian process is uniquely defined by the continuous linear mean functional

$$\mathbb{E}\Phi(\varphi) = m(\varphi)$$

and the continuous bilinear positive-definite covariance functional

$$\mathbb{E}(\Phi(\varphi) - m(\varphi))(\Phi(\psi) - m(\psi)) = C(\varphi, \psi).$$

One of the important advantages of a generalized stochastic process is the fact that its derivative always exists and is itself a generalized stochastic process. In fact, the derivative $\dot{\Phi}$ of Φ is the process defined by setting

$$\dot{\Phi}(\varphi) = -\Phi(\dot{\varphi}).$$

The derivative of a generalized Gaussian process with mean $m(\varphi)$ and covariance $C(\varphi, \psi)$ is again a generalized Gaussian process and it has mean $\dot{m}(\varphi) = -m(\dot{\varphi})$ and covariance $C(\dot{\varphi}, \dot{\psi})$.

Now let us look at Brownian motion B_t and its derivative. The generalized stochastic process corresponding to B_t is the following linear functional

$$\Phi(\varphi) = \int_{-\infty}^{+\infty} \varphi(t) B_t dt,$$

for $\varphi \in C_0^\infty(\mathbb{R})$. With this representation we regard B_t as a generalized stochastic process. In fact, we may identify B_t with this linear functional Φ. We conclude that the mean functional

$$m(\varphi) \equiv 0$$

and the covariance functional

$$C(\varphi, \psi) = \int_0^\infty \int_0^\infty \min(t, s) \, \varphi(t) \psi(s) dt ds.$$

After some elementary manipulations and integration by parts, we get

$$C(\varphi, \psi) = \int_0^\infty (\hat{\varphi}(t) - \hat{\varphi}(\infty))(\hat{\psi}(t) - \hat{\psi}(\infty)) dt,$$

where

$$\hat{\varphi}(t) = \int_0^t \varphi(s) ds \text{ and } \hat{\psi}(t) = \int_0^\infty \int_0^t \psi(s) ds.$$

The derivative of B_t, i.e., derivative of $\Phi(\varphi) = \int_{-\infty}^{+\infty} \varphi(t) B_t dt$, is also a generalized Gaussian process with mean $\dot{m}(\varphi) = 0 = -m(\dot{\varphi})$ and covariance

$$\dot{C}(\varphi, \psi) = C(\dot{\varphi}, \dot{\psi}) = \int_0^\infty \varphi(t) \psi(s) dt.$$

This formula can be put in the form

$$\dot{C}(\varphi, \psi) = \int_0^\infty \int_0^\infty \delta(t - s) \varphi(t) \psi(s) dt ds.$$

Therefore, the covariance of the derivative of Brownian motion B_t is the generalized stochastic process with mean zero and covariance

$$\dot{C}(s, t) = \delta(t - s).$$

But this is the covariance for white noise ξ_t. Thus, $\dot{B}_t$ is taken as a mathematical model for white noise ξ_t. This justifies the notation

$$\xi_t = \dot{B}_t$$

frequently used in engineering literature and occasionally in stochastic differential equations. We may also write

$$B_t = \int_0^t \xi_s ds.$$

We may say that a Gaussian white noise ξ_t is a generalized Gaussian stochastic process Φ_ξ with mean zero and covariance functional

$$C_\xi(\varphi, \psi) = \int_{-\infty}^{+\infty} \varphi(t)\psi(t)dt.$$

4. Colored noises and fractional Brownian motion

Random fluctuations in complex systems may not be uncorrelated (i.e., may not be white noise). In fact, most fluctuations ξ_t are correlated[5,13,14] and thus the covariance $\mathbb{E}(\xi_t \xi_s) = c(t-s)$ is usually not a delta function. In this case we call the stationary process ξ_t a colored noise. We usually take its mean to be zero, otherwise we define a new stationary process by subtracting the mean. Since covariance function $c(t-s)$ may be arbitrary, there are many colored noises in principle. For example, the Ornstein-Uhlenbeck process, as the solution of a linear Langevin equation, is often used as a model for colored noise.

We here discuss a model of colored noise in terms of fractional Brownian motion (fBM). It has been known that such colored noise arise in mathematical modeling of missing mechanisms or unresolved scales in various complex systems.[4-6] The fractional Brownian motion $B^H(t)$, indexed by a so called Hurst parameter $H \in (0,1)$, is a generalization of the more well-known process of the usual Brownian motion $B(t)$. It is a zero-mean Gaussian process with stationary increments. However, the increments of the fractional Brownian motion are not independent, except in the usual Brownian motion case ($H = \frac{1}{2}$). For more details, see.[10,22-24]

Definition of fractional Brownian motion: For $H \in (0,1)$, a Gaussian process $B^H(t)$ is a fractional Brownian motion if it starts at zero $B^H(0) = 0$, a.s., has mean zero $\mathbb{E}[B^H(t)] = 0$, and has covariance $\mathbb{E}[B^H(t)B^H(s)] = \frac{1}{2}(|t|^{2H} + |s|^{2H} - |t-s|^{2H})$ for all t and s. The standard Brownian motion is a fractional Brownian motion with Hurst parameter $H = \frac{1}{2}$.

Some properties of fractional Brownian motion: A fractional Brownian motion $B^H(t)$ has the following properties:
(i) It has stationary increments;
(ii) When $H = 1/2$, it has independent increments;
(iii) When $H \neq 1/2$, it is neither Markovian, nor a semimartingale.

As for the covariance for the generalized derivative of fractional Brownian noise, $\dot{B}_t^H$, it is complicated due to its non-Markovian nature. Recall that a function f in

$C^\infty(\mathbb{R})$ is a Schwartz function if it goes to zero as $|x| \to \infty$ faster than any inverse power of x, as do all its derivatives. Since $\dot{B}_t^H = M_-^H B_t$ (here and below M_-^H and M_+^H are operators defined in[23]), for any Schwartz functions f and g, we have

$$\langle \mathbb{E}\dot{B}_t^H \dot{B}_s^H, f \otimes g \rangle = \mathbb{E} \int_\mathbb{R} f(t) M_-^H \dot{B}_t dt \int_\mathbb{R} g(s) M_-^H \dot{B}_s ds$$
$$= \mathbb{E} \int_\mathbb{R} M_+^H f(t) dB_t \int_\mathbb{R} M_+^H g(s) dB_s$$
$$= \int_\mathbb{R} M_+^H f(t) M_+^H g(t) dt$$
$$= \langle \delta(t-s), M_+^H f \otimes M_+^H g \rangle$$
$$= \langle M_-^H \otimes M_-^H \delta(t-s), f \otimes g \rangle.$$

This implies that $\dot{B}_t^H$ is correlated or colored noise.

We use the Weierstrass-Mandelbrot function to approximate the fractional Brownian motion. The basic idea is to simulate fractional Brownian motion by randomizing a representation due to Weierstrass. Given the Hurst parameter H with $0 < H < 1$, we define the function $w(t)$ to approximate the fractional Brownian motion:

$$w(t_i) = \sum_{j=-\infty}^{\infty} C_j r^{jH} \sin(2\pi r^{-j} t_i + d_j)$$

where $r = 0.9$ is a constant, C_j's are normally distributed random variables with mean 0 and standard deviation 1, and the d_j's are uniformly distributed random variables in the interval $0 \le d_j < 2\pi$. The underlying theoretical foundation for this approximation can be found in.[27] Three figures here show a few sample paths of the fractional Brownian motion with Hurst parameters $H = 0.25, 0.5, 0.75$, respectively.

References

1. L. Arnold, *Random Dynamical Systems*, Springer, New York, 1998.
2. L. Arnold, *Stochastic DifferentiL Equations*, John Wiley & Sons, New York, 1974.
3. A. Berlinet and C. Thomas-Agnan, *Reproducing Kernel Hilbert Spaces in Probability and Statistics*, Kluwer Academic Publishers, 2004.
4. A. Du and J. Duan. A stochastic approach for parameterizing unresolved scales in a system with memory. *Journal of Algorithms & Computational Technology* **3**(2009), 393-405.
5. J. Duan, Stochastic Modeling of Unresolved Scales in Complex Systems. *Frontiers of Math. in China* **4** (2009), 425-436.
6. J. Duan, Quantifying model uncertainty by correlated noise, *Oberwolfach Reports*, vol. 5, (2008).
7. J. Duan, Predictability in Spatially Extended Systems with Model Uncertainty I & II: *Engineering Simulation* p17–32, No. 2 & p21–35, No. 3, **31** (2009).
8. J. Duan, Predictability in Nonlinear Dynamical Systems with Model Uncertainty. In *Stochastic Physics and Climate Modeling*, T. N. Palmer and P. Williams (eds.), Cambridge Univ. Press, 2009.

9. J. Duan, X. Kan and B. Schmalfuss, Canonical Sample Spaces for Stochastic Dynamical Systems. In "Perspectives in Mathematical Sciences", *Interdisciplinary Math. Sci.* Vol. 9, 2009, pp.53-70.
10. T. E. Duncan, Y. Z. Hu, B. Pasik-Duncan, Stochastic Calculus for Fractional Brownian Motion. I: Theory. *SIAM Journal on Control and Optimization* **38** (2000), 582-612.
11. C. W. Gardiner, *Handbook of Stochastic Methods.* Second Ed., Springer, New York, 1985.
12. J. Garcia-Ojalvo and J. M. Sancho, *Noise in Spatially Extended Systems.* Springer-Verlag, 1999.
13. P. Hanggi, Colored Noise in Dynamical Systems: A Functional Calculus Approach, in: *Noise in Nonlinear Dynamical Systems*, vol. 1, F. Moss and P. V. E. McClintock, eds., chap. 9, pp. 307328, Cambridge University Press,1989.
14. P. Hanggi and P. Jung, Colored Noise in Dynamical Systems. *Advances in Chem. Phys.*, **89**(1995), 239-326.
15. T. Hida. *Brownian Motion.* Springer, New York, 1974.
16. W. Horsthemke and R. Lefever, *Noise-Induced Transitions*, Springer-Verlag, Berlin, 1984.
17. Z. Huang and J. Yan, *Introduction to Infinite Dimensional Stochastic Analysis.* Science Press/Kluwer Academic Pub., Beijing/New York, 1997.
18. M. James, F. Moss and P. Hanggi, Switching in the Presence of Colored Noise: The Decay of an Unstable State, *Phys. Rev. A* 38, 46904695 1988.
19. W. Just, H. Kantz, C. Rodenbeck and M. Helm, Stochastic modelling: replacing fast degrees of freedom by noise. *J. Phys. A: Math. Gen.*, **34** (2001),3199–3213.
20. I. Karatzas and S. E. Shreve, Brownian Motion and Stochastic Calculus 2nd, Springer 1991.
21. V. Krishnan, *Nonlinear Filtering and Smoothing: An Introduction to Martingales, Stochastic Integrals and Estimation.* John Wiley & Sons, New York, 1984.
22. B. Maslowski and B. Schmalfuss, Random dynamical systems and stationary solutions of differential equationsdriven by the fractional Brownian motion. *Stoch. Anal. Appl.*, Volume 22, Issue 6 January 2005, pages 1577 - 1607.
23. Y. S. Mishura, *Stochastic calculus for fractional Brownian motion and related processes*, Springer, New York, 2008.
24. D. Nualart, Stochastic calculus with respect to the fractional Brownian motion and applications. *Contemporary Mathematics* **336**, 3-39, 2003.
25. B. Oksendal, *Stochastic differential equations: An introduction with applications*, Sixth edition, Springer, New York, 2003.
26. A. Papoulis, *Probability, Random Variables, and Stochastic Processes*, McGraw-Hill Companies; 2nd edition, 1984.
27. V. Pipiras and M. S. Taqqu, Convergence of the Wererstrass-Mandelbrot process to fractinal Brownian motion. *Fractals* **Vol. 8, No.4**, (2000), 369-384 .
28. M. Renardy and R. Rogers, *Introduction to Partial Differential Equations*, Springer-Verlag, 1993.
29. N. G. Van Kampen, How do stochastic processes enter into physics? *Lecture Note in Phys.* **1250** (1987) 128–137.
30. N. G. Van Kampen, *Stochastic Processes in Physics and Chemistry.* North-Holland, New York, 1981.
31. E. Wong and B. Hajek. *Stochastic Processes in Engineering Systems.* Spring-Verlag, New York, 1985.

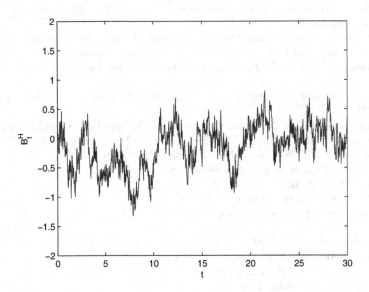

Figure 7.1. A sample path of fractional Brownian motion $B^H(t)$, with $H = 0.25$

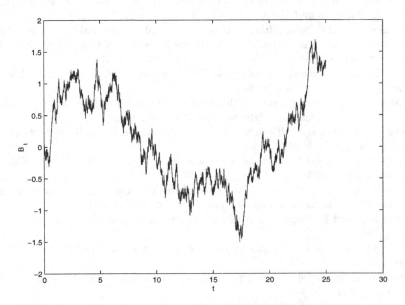

Figure 7.2. A sample path of Brownian motion $B(t)$; namely, fractional Brownian motion with $H = 0.5$

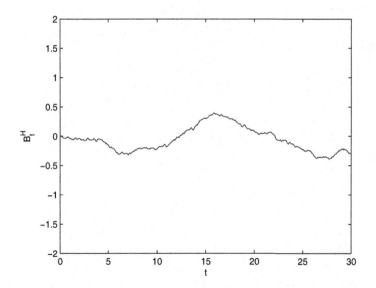

Figure 7.3. A sample path of fractional Brownian motion $B^H(t)$, with $H = 0.75$

Chapter 8

A Sufficient Condition for Non-Explosion for a Class of Stochastic Partial Differential Equations

Hongbo Fu[1], Daomin Cao[2] and Jinqiao Duan[3,1] *

1. School of Mathematics and Statistics,
Huazhong University of Science and Technology
Wuhan 430074, China
E-mail: hbfu_hust@sina.com

2. Institute of Applied Mathematics, Chinese Academy of Sciences
Beijing, 100190, China
E-mail: dmcao@amt.ac.cn

3. Department of Applied Mathematics, Illinois Institute of Technology
Chicago, IL 60616, USA
E-mail: duan@iit.edu

To facilitate random dynamical systems approach for stochastic partial differential equations arising as mathematical models for complex systems under fluctuations in science and engineering, global existence and uniqueness of mild solutions for a class of stochastic partial differential equations with local Lipschitz coefficients are considered. A sufficient condition for non-explosion, or global existence and uniqueness, of mild solutions is provided and a few examples are presented to demonstrate the result.

Keywords : Stochastic partial differential equations (SPDEs); Mild solution; Wellposedness; Local Lipschitz condition

2000 AMS Subject Classification : 60H40

Contents

1	Introduction	132
2	Preliminaries	134
3	Main results	136
4	Examples	140
	References	142

---

*The authors gratefully acknowledge the support by K. C. Wong Education Foundation, Hong Kong, the Science Fund for Creative Research Groups of Natural Science Foundation of China (No.10721101), and the NSF grant 0620539.

1. Introduction

Mathematical models for scientific and engineering systems are often subject to uncertainties, such as fluctuating forcing, random boundary conditions and uncertain parameters. Moreover, some biological, chemical or physical processes are not well understood, not well-observed, and thus are difficult to be represented in the mathematical models. These missing processes or mechanisms may be very small in spatial scale and fast in temporal scale, but their impact on overall system evolution may be delicate or uncertain [1]. Therefore it is important to take such uncertainties into account. Mathematical models for randomly influenced spatio-temporal dynamical systems are usually in the form of stochastic partial differential equations (SPDEs).

To facilitate dynamical systems approach for SPDEs, we need to establish global existence and uniqueness of solutions for SPDEs. One method is the variational approach as presented in [13, 11]. Another approach is the semigroup approach as in [7, 3].

In this chapter, we present the semigroup approach for global existence and uniqueness of mild solutions for SPDEs, through a few examples. We provide a sufficient condition for global existence, the Assumption (**B**) and Theorem 3.2 in §3. This condition is different from those in Da Prato-Zabczyk [7] or Chow [3].

For deterministic partial differential equations, the semigroup approach for well-posedness is presented in, for example, [10, 4, 6].

Let $D \subset \mathbb{R}^d (d \geq 1)$ be a bounded domain with smooth boundary ∂D. We consider the following nonlinear stochastic partial differential equations(SPDEs)

$$\begin{cases} \frac{\partial}{\partial t} u(t,x) = (\kappa \Delta - \alpha) u(t,x) + f(u(t,x)) + \sigma(u(t,x)) \frac{\partial}{\partial t} w(t,x), \\ u|_{\partial D} = 0, u(x,0) = h(x). \end{cases} \quad (1.1)$$

where $t > 0$, $x \in D$, κ and α are positive, $\Delta = \sum_{i=1}^{d} \frac{\partial^2}{\partial x_i^2}$ is the Laplace operator and $h(x)$ is a given function in $L^2(D)$ which be denoted by H with inner product $(\cdot, \cdot)$ and norm $\|\cdot\|$. The coefficients $f, \sigma : \mathbb{R} \to \mathbb{R}$ are given measurable functions and $w(t,x)$ is a H-valued $\mathcal{R}$-Wiener process to be defined below. Such equations model a variety of phenomena in many fields, such as biology, quantum field, neurophysiology and so on, see [9, 14, 8].

Throughout this paper, we shall assume a complete probability space $(\Omega, \mathcal{F}, \{\mathcal{F}_t\}_{t \geq 0}, \mathbb{P})$ equipped with the filtration $\{\mathcal{F}_t\}_{t \geq 0}$, which satisfies the usual conditions, such that $\mathcal{F}_0$ contains all $\mathbb{P}$-null sets.

For our system we present some information on $\mathcal{R}$-Wiener process [3]. Let $\mathcal{R}$ be a linear integral operator defined, for all $\phi \in H$, by

$$\mathcal{R}(\phi)(x) = \int_D r(x,y) \phi(y) dy \quad x \in D,$$

where the integral nuclear $r(x,y) = r(y,x)$ is positive and square integral such that $\int_D \int_D |r(x,y)|^2 dx dy < \infty$. Then the eigenvalues $\{\mu_k\}_{k\geq 1}$ of $\mathcal{R}$ are positive and the normalized eigenfunctions $\{\phi_k\}_{k\geq 1}$ form a complete orthonormal basis for H. In the case that eigenvalues satisfy $\sum_{k=1}^{\infty} \mu_k < \infty$, the $\mathcal{R}$-Wiener process in H has an infinite series representation

$$w(t,x) = \sum_{k=1}^{\infty} \sqrt{\mu_k} w_t^k \phi(x),$$

where $\{w_t^k\}_{k\geq 1}$ is an independent sequence of real-valued Wiener process.

Now let us state the formal problem in Eq. (1.1) in its rigorous meaning. A predictable random field (see [5] for definition) $\{u(t,x), t \geq 0, x \in D\}$ is called a mild solution of the Eq. (1.1) if

$$u(t,x) = \int_D G(t,x,y)h(y)dy + \int_0^t \int_D G(t-s,x,y)f(u(s,y))dyds$$
$$+ \int_0^t \int_D G(t-s,x,y)\sigma(u(s,y))dyw(y,ds), \quad \mathbb{P}-a.s. \quad (1.2)$$

for each $t \geq 0$, where $G(t,x,y)$ stands for the fundamental solution of the heat equation of $\frac{\partial}{\partial t}u(t,x) = (\kappa\Delta - \alpha)u(t,x)$ with the boundary conditions specified before. In fact, $G(t,x,y)$ can be expressed as $G(t,x,y) = \sum_{k=1}^{\infty} e^{-\lambda_k t} e_k(x) e_k(y)$, here $\{e_k(x)\}_{k\geq 1}$ denote the complete orthornormal system of eigenfunctions in H such that, for $k = 1, 2 \cdots$,

$$(-\kappa\Delta + \alpha)e_k = \lambda_k e_k, \quad e_k|_{\partial D} = 0, \quad (1.3)$$

with $\alpha \leq \lambda_1 \leq \lambda_2 \leq \cdots \lambda_k \leq \cdots$.

If we set $(G_t g)(x) = \int_D G(t,x,y)g(y)dy$, $g \in H$. Then G_t is a semigroup on H; see [10] for details. Let us also write $u(t) = u(t,\cdot)$ and $dw_t = w(\cdot, dt)$, then Eq. (1.2) can be written as

$$u(t) = G_t h + \int_0^t G_{t-s} f(u(s))ds + \int_0^t G_{t-s} \sigma(u(s))dw_s, \quad \mathbb{P}-a.s. \quad (1.4)$$

Global existence and uniqueness for mild solution of the SPDE Eq. (1.4) under global Lipschitz condition and linear growth on the coefficients f and σ have be studied in [7]. Here we present global existence and uniqueness results for mild solutions of above mentioned SPDE under local Lipschitz condition, mainly following [3] but with a different sufficient condition to guarantee global existence; see the Assumption (**B**) in §3.

After some preliminaries in §2, we present a global wellposedness result in §3, and discuss a few examples in §4.

2. Preliminaries

Let $T > 0$ be a fixed number and denote by $L^2(\Omega, C([0,T]; H))$ the space of all H-valued $\mathcal{F}_t$-adapt processes $X(t, \omega)$ defined on $[0, T] \times \Omega$ which are continuous in t for a.e. fixed $\omega \in \Omega$ and for which $\|X(\cdot,\cdot)\|_T = \{\mathbb{E} \sup_{0 \leq t \leq T} \|X(t,\omega)\|^2\}^{\frac{1}{2}} < \infty$, then the space $L^2(\Omega, C([0,T]; H))$ is a Banach space with the norm $\|\cdot\|_T$.

Let $L^2(\Omega \times [0,T])$ be the space of all H-valued predictable process $X(t,\omega)$ defined on $[0,T] \times \Omega$ and for which $\|X(\cdot,\cdot)\|_2 = \{\mathbb{E} \int_0^T \|X(t,\omega)\|^2 dt\}^{\frac{1}{2}}$, then the space $L^2(\Omega \times [0,T]; H)$ is also a Banach space with the norm $\|\cdot\|_2$.

Set $A := -\kappa\Delta + \alpha$ and we can use (1.3) to define its fractional power for $\gamma \in (0,1)$ by natural expression (see [12])

$$A^\gamma u = \sum_{i=1}^\infty \lambda_i^\gamma (u, e_i) e_i,$$

when the right hand is well defined. For more information on fractional power we infer to [10] or [4]. Let H^γ denote the domain of A^γ in H, i.e. $H^\gamma = \{u \in H; \sum_{i=1}^\infty \lambda_i^\gamma(u, e_i)e_i \text{ convergences in } H\}$, then H^γ is a Banach space endowed with norm $\|u\|_\gamma := \{\sum_{j=1}^\infty \lambda_j^{2\gamma}(u, e_j)^2\}^{\frac{1}{2}}$. It is clear that $H^{\gamma_1} \subset H^{\gamma_2}$ continuously for any $0 < \gamma_2 < \gamma_1 < 1$.

Set $\|\sigma\|_\mathcal{R} := (\int_D r(x,x)\sigma^2(x)dx)^{\frac{1}{2}}$ and assume that $r(x,y) \leq r_0$ for some positive number r_0, where $r(x,y)$ is the integral nuclear of integral operator $\mathcal{R}$.

Before starting to prove our main theorem, for the reader's convenience, we shall formulate some foundational inequalities with $\gamma \in (0, \frac{1}{2}]$, which will be used in the proofs (See [3]).

Lemma 2.1. *Suppose $h \in H$ and $v(t, \cdot)$ is a predictable random field in H such that $\mathbb{E} \int_0^T \|v(t,\cdot)\|^2 dt < \infty$, then the following inequalities hold:*

$$\sup_{0 \leq t \leq T} \|G_t h\| \leq \|h\|, \tag{2.1}$$

$$\mathbb{E} \sup_{0 \leq t \leq T} \|\int_0^t G_{t-r} v(r,\cdot) dr\|^2 \leq T\mathbb{E} \int_0^T \|v(t,\cdot)\|^2 dt, \tag{2.2}$$

Lemma 2.2. *Suppose that $\sigma(t,\cdot)$ is a predictable random field in H such that $\mathbb{E} \int_0^T \int_D r(x,x)\sigma^2(t,x)dxdt = \mathbb{E} \int_0^T \|\sigma(t)\|_\mathcal{R}^2 dt < \infty$, then we have*

$$\mathbb{E} \sup_{0 \leq t \leq T} \|\int_0^t G_{t-s} \sigma(s,\cdot) dw_s\|^2 \leq 16 \mathbb{E} \int_0^T \|\sigma(t)\|_\mathcal{R}^2 dt. \tag{2.3}$$

By the same techniques to proof of Lemma 2.1 and Lemma 2.2, we have the following two Lemmas.

Lemma 2.3. *Suppose $h \in H$ and $v(t,\cdot)$ is a predictable random field in H such that $\mathbb{E} \int_0^T \|v(t,\cdot)\|^2 dt < \infty$, then $G_t h \in H^{\frac{1}{2}}$, $\int_0^t G_{t-s} v(s,\cdot) ds$ is a $H^{\frac{1}{2}}$-valued random* field and the following inequalities hold:

$$\|G_t h\|_\gamma^2 \le \frac{1}{\alpha^{1-2\gamma}} \|h\|^2, \tag{2.4}$$

$$\int_0^T \|G_t h\|_\gamma^2 dt \le \frac{1}{2\alpha^{1-2\gamma}} \|h\|^2, \tag{2.5}$$

$$\mathbb{E} \int_0^T \| \int_0^t G_{t-s} v(s,\cdot) ds \|_\gamma^2 dt \le \frac{T}{2\alpha^{1-2\gamma}} \mathbb{E} \int_0^T \|v(t,\cdot)\|^2 dt. \tag{2.6}$$

$$\mathbb{E} \sup_{0 \le t \le T} \| \int_0^t G_{t-s} v(s,\cdot) ds \|_\gamma^2 \le \frac{1}{2} \mathbb{E} \int_0^T \|v(t,\cdot)\|^2 dt. \tag{2.7}$$

Lemma 2.4. *Suppose that $\sigma(t,\cdot)$ is a predictable random field in H such that $\mathbb{E} \int_0^T \int_D r(x,x) \sigma^2(t,x) dx dt = \mathbb{E} \int_0^T \|\sigma(t)\|_{\mathcal{R}}^2 dt < \infty$, then $\int_0^t G_{t-s}\sigma(s,\cdot) dw_s$ is a $H^{\frac{1}{2}}$-valued process and we have*

$$\mathbb{E} \int_0^T \| \int_0^t G_{t-s}\sigma(s,\cdot) dw_s \|_\gamma^2 dt \le \frac{1}{2\alpha^{1-2\gamma}} \mathbb{E} \int_0^T \|\sigma(t)\|_{\mathcal{R}}^2 dt. \tag{2.8}$$

Moreover, if $\sigma(t,\cdot)$ is a H^γ-valued process, then we have

$$\mathbb{E} \| \int_0^t G_{t-s}\sigma(s,\cdot) dw_s \|_\gamma^2 \le r_0 \mathbb{E} \int_0^t \|\sigma(s)\|_\gamma^2 ds, \quad 0 \le t \le T. \tag{2.9}$$

We point out that the proofs for Lemma 2.3 and Lemma 2.4 are straightforward. For instance, consider the inequality (2.8) as follows:

$$\mathbb{E} \int_0^T \| \int_0^t G_{t-s}\sigma(s,\cdot) dw_s \|_\gamma^2 dt = \mathbb{E} \int_0^T \sum_{j=1}^\infty \lambda_j^{2\gamma} (\int_0^t G_{t-s}\sigma(s,\cdot) dw_s, e_j)^2 dt$$

$$= \int_0^T \left(\sum_{j=1}^\infty \lambda_j^{2\gamma} \mathbb{E} \left| \int_0^t e^{-\lambda_j(t-s)} (e_j, \sigma(s,\cdot)) dw_s \right|^2 \right) dt$$

$$= \int_0^T \left(\sum_{j=1}^\infty \lambda_j^{2\gamma-1} \mathbb{E} \int_0^t \lambda_j e^{-2\lambda_j(t-s)} (Q_s e_j, e_j) ds \right) dt$$

$$\le \frac{1}{2\alpha^{1-2\gamma}} \mathbb{E} \int_0^T \sum_{j=i}^\infty (Q_t e_j, e_j) dt$$

$$= \frac{1}{2\alpha^{1-2\gamma}} \mathbb{E} \int_0^T \|\sigma(t)\|_{\mathcal{R}}^2 dt.$$

Here Q_s denotes the local characteristic operator (see [3] Chapter 3, Section 2) for H-valued martingale $\int_0^t \sigma(s) dw_s$, which is defined by $(Q_t g, h) = \int_D \int_D r(x,y) \sigma(t,x) \sigma(t,y) g(x) h(y) dx dy$, for any $g, h \in H$.

3. Main results

We consider the SPDE (1.1) or its equivalent mild form (1.2), in two separate cases: **global Lipschitz** coefficients and **local Lipschitz** coefficients.

First we consider the global Lipschitz case. We make following global Lipschitz assumption on coefficients f and σ in Eq. (1.2):

(H) Global Lipschitz condition: $f(r)$ and $\sigma(r)$ are real-valued, measurable functions defined on $\mathbb{R}$, and there exist positive constants β, r_1 and r_1 such that for any $u, v \in H^\gamma$

$$\|f(u) - f(v)\|^2 \leq \beta\|u - v\|^2 + r_1\|u - v\|_\gamma^2,$$

$$\|\sigma(u) - \sigma(v)\|_\mathcal{R}^2 \leq \beta\|u - v\|^2 + r_2\|u - v\|_\gamma^2.$$

In fact from **(H)** it is clear that there exits a constant C such that for any $u \in H^\gamma$, the following **sublinear growth** condition holds:

$$\|f(u)\|^2 + \|\sigma(u)\|_\mathcal{R}^2 \leq C(1 + \|u\|^2 + \|u\|_\gamma^2). \tag{3.1}$$

Remark 3.1. Note that for non-autonomous SPDEs, i.e., when f and σ in SPDE (1.1) explicitly depend on time t, the global Lipschitz condition **(H)** does not usually imply the sublinear growth condition. In that case, we need to impose the additional sublinear growth condition as in [3, 7].

Remark 3.2. Since $H^\gamma \subset H$ continuously, (3.1) and the inequalities in **(H)** can be rewritten to concise forms, which the norm $\|\cdot\|$ is dominated by the norm $\|\cdot\|_\gamma$.

Remark 3.3. In case of $\gamma = \frac{1}{2}$, it is proved in [3] that the assumption **(H)** on f and σ ensure the global existence and uniqueness of the solution of Eq. (1.2) if $r_1 + r_2 < 1$ holds.

From now, we shall restrict that $0 < \gamma < \frac{1}{2}$, then we have the following theorem when we discuss Eq. (1.2) on a finite time interval $[0,T]$ for any fixed $T > 0$. The following result is essentially in [7, 3], but since we consider more regular mild solutions, the proof is thus modified.

Theorem 3.1. *[Wellposedness under global Lipschitz condition]*
*Assume that the global Lipschitz condition **(H)** holds and consider the SPDE (1.2) with initial data $h \in H^\gamma$ for $0 < \gamma < \frac{1}{2}$. Then there exists a unique solution u, as an adapt, continuous process in H. Moreover for any $T > 0$, u belongs to $L^2(\Omega; C([0,T]; H)) \bigcap L^2(\Omega \times [0,T]; H^\gamma)$ such that*

$$\mathbb{E}\{\sup_{0 \leq t \leq T} \|u(t)\|^2 + \int_0^T \|u(t)\|_\gamma^2 dt\} < \infty.$$

Proof. We choose some sufficiently small $T_0 < T$ and denote by Y_{T_0} the set of predictable random field $\{u(t)\}_{0 < t \leq T}$ belong to $L^2(\Omega; C([0, T_0]; H)) \cap L^2(\Omega \times [0, T_0]; H^\gamma)$ and for which

$$\|u\|_{T_0} = \{\mathbb{E}(\sup_{0 \leq t \leq T_0} \|u(t)\|^2 + \int_0^{T_0} \|u(t)\|_\gamma^2 dt)\}^{\frac{1}{2}} < \infty,$$

then Y_{T_0} is a Banach space with norm $\|\cdot\|_{T_0}$.

Let Γ denote a mapping in Y_{T_0} defined by

$$\Gamma_t u = G_t h + \int_0^t G_{t-s} f(u(s)) ds + \int_0^t G_{t-s} \sigma(u(s)) dw_s, \quad t \in [0, T_0].$$

We first verify that $\Gamma : Y_{T_0} \to Y_{T_0}$ is well-defined and bounded.

In the following, C' will denotes a positive constant whose values might change from line to line.

It follows form (2.1) and (2.5) that

$$\mathbb{E} \sup_{0 \leq t \leq T_0} \|G_t h\|^2 + \mathbb{E} \int_0^{T_0} \|G_t h\|_\gamma^2 dt \leq C' \|h\|^2. \tag{3.2}$$

Let $\{u(s)\}_{t \in [0, T_0]} \in Y_{T_0}$ and use (2.2), (2.6) firstly, then (3.1). We obtain

$$\mathbb{E} \sup_{0 \leq t \leq T_0} \|\int_0^t G_{t-s} f(u(s)) ds\|^2 + \mathbb{E} \int_0^{T_0} \|\int_0^t G_{t-s} f(u(s)) ds\|_\gamma^2 dt$$

$$\leq C' \mathbb{E} \int_0^{T_0} \|f(u(s))\|^2 ds$$

$$\leq C' \mathbb{E} \int_0^{T_0} (1 + \|u(s)\|^2 + \|u(s)\|_\gamma^2) ds$$

$$\leq C'(1 + \|u(s)\|_{T_0}^2). \tag{3.3}$$

Similarly, by making use of (2.3), (2.8) and (3.1), it is easy to check

$$\mathbb{E} \sup_{0 \leq t \leq T_0} \|\int_0^t G_{t-s} \sigma(u(s)) dw_s\|^2 + \int_0^{T_0} \|\int_0^t G_{t-s} \sigma(u(s)) dw_s\|_\gamma^2 dt \leq C'(1 + \|u(s)\|_{T_0}^2). \tag{3.4}$$

From (3.2), (3.3) and (3.4), it follows that $\Gamma : Y_{T_0} \to Y_{T_0}$ is well-defined and bounded. To show Γ is a contraction operator in Y_{T_0}, we introduce another equivalent norm $\|\cdot\|_{\mu, Y_{T_0}}$ in Y_{T_0} as follow:

$$\|u\|_{\mu, T_0} = \{\mathbb{E}(\sup_{0 \leq t \leq T_0} \|u(t)\|^2 + \mu \int_0^{T_0} \|u(t)\|_\gamma^2 dt)\}^{\frac{1}{2}},$$

where μ is a parameter, then for $u, v \in Y_{T_0}$,

$$\|\Gamma u - \Gamma v\|_{\mu, T_0}^2 = \mathbb{E}\{\sup_{0 \leq t \leq T_0} \|\Gamma_t u - \Gamma_t v\|^2 + \mu \int_0^{T_0} \|\Gamma_t u - \Gamma_t v\|_\gamma^2 dt\}.$$

By making use of (2.2), (2.3) and simple inequality $(a+b)^2 \leq C_\epsilon a^2 + (1+\epsilon)b^2$ with $C_\epsilon = \frac{1+\epsilon}{\epsilon}$ for any $\epsilon > 0$, we get

$$\mathbb{E} \sup_{0 \leq t \leq T_0} \|\Gamma_t(u) - \Gamma_t(v)\|^2 = \mathbb{E} \sup_{0 \leq t \leq T_0} \{\|\int_0^t G_{t-s}(f(u(s)) - f(v(s)))ds$$

$$+ \int_0^t G_{t-s}(\sigma(u(s)) - \sigma(v(s)))dw_s\|^2\}$$

$$\leq C_\epsilon T_0 \mathbb{E} \int_0^{T_0} \|f(u(s)) - f(v(s))\|^2 dt$$

$$+ 16(1+\epsilon) \mathbb{E} \int_0^{T_0} \|\sigma(u(s)) - \sigma(v(s))\|_\mathcal{R}^2 dt.$$

Similarly, by making use of (2.6), (2.8) and the simple inequality mentioned above, we obtain

$$\mathbb{E} \int_0^{T_0} \|\Gamma_t u - \Gamma_t v\|_\gamma^2 dt \leq \frac{T_0 C_\epsilon}{2\alpha^{1-2\gamma}} \mathbb{E} \int_0^{T_0} \|f(u(s)) - f(v(s))\|^2 dt$$

$$+ \frac{1+\epsilon}{2\alpha^{1-2\gamma}} \mathbb{E} \int_0^{T_0} \|\sigma(u(s)) - \sigma(v(s))\|_\mathcal{R}^2 dt.$$

Hence, by assumption (**H**), we get

$$\|\Gamma(u) - \Gamma(v)\|_{\mu,T_0}^2 \leq (C_\epsilon T_0 + \frac{C_\epsilon T_0 \mu}{2\alpha^{1-2\gamma}}) \mathbb{E} \int_0^{T_0} (\beta\|u(s) - v(s)\|^2 + r_1\|u(s) - v(s)\|_\gamma^2) ds$$

$$+ (16(1+\epsilon) + \frac{(1+\epsilon)\mu}{2\alpha^{1-2\gamma}}) \mathbb{E} \int_0^{T_0} (\beta\|u(s) - v(s)\|^2 + r_2\|u(s) - v(s)\|_\gamma^2) ds$$

$$\leq \rho_1 \mathbb{E} \sup_{0 \leq t \leq T_0} \|u(t) - v(t)\|^2 + \rho_2 \mu \mathbb{E} \int_0^{T_0} \|u(s) - v(s)\|_\gamma^2 ds,$$

here

$$\rho_1 = \beta(1+\epsilon)T_0(16 + \frac{T_0}{\epsilon} + \frac{T_0 \mu'}{2\epsilon} + \frac{\mu'}{2}),$$

$$\rho_2 = (1+\epsilon)(\frac{1}{2} + \frac{T_0}{\epsilon \mu'} + \frac{T_0}{2\epsilon} + \frac{16}{\mu'})(\frac{r_1+r_2}{\alpha^{1-2\gamma}})$$

with $\mu' = \frac{\mu}{\alpha^{1-2\gamma}}$.

Note that we can always assume that $\frac{r_1+r_2}{\alpha^{1-2\gamma}} < 1$. If not that, choose $M > 0$ such that $\frac{r_1+r_2}{(\alpha+M)^{1-2\gamma}} < 1$, and rewrite Equation (1.1) as

$$\frac{\partial}{\partial t} u(t,x) = [\kappa \Delta - (\alpha + M)]u(t,x) + [f(u(t,x)) + M \cdot u(t,x)] + \sigma(u(t,x))\frac{\partial}{\partial t} w(t,x),$$

it is clear that (**H**) holds with β, replaced by $\beta + M^2$. So it is possible to choose μ' sufficiently large and ϵ, T_0 sufficiently small such that $\rho = \rho_1 \vee \rho_2 < 1$, which implies that Γ is a contraction operator in Y_{T_0}. This means that there exists a unique local solution of the Eq. (1.2) over $[0, T_0]$, the solution can be extend over the finite interval $[0, T]$ by standard arguments.

The proof of the theorem is completed. □

Remark 3.4. To show that it is possible to choose the parameters μ', ϵ and T_0 to make $\rho < 1$, we first let $\frac{T_0}{\epsilon} < \frac{1}{64}$ and $\mu' = 64^2$, then we have $\rho_2 \leq (1+\epsilon)(\frac{1}{2} + \frac{1}{64^3} + \frac{1}{128} + \frac{16}{64^3})$, this yields $\rho_2 < \frac{3}{4}$ if we choose $\epsilon = \frac{1}{64}$. It is clear that ρ_1 can be make less than $\frac{3}{4}$ by taking T_0 to be sufficiently small.

In **Theorem 3.1**, if the global Lipschitz condition on coefficients is relaxed to hold locally, then we only obtain a local solution which may blow up (or explode or have explosion) in finite time. To get global solution, we impose the following conditions:

(Hn) Local Lipschitz condition: $f(r)$ and $\sigma(r)$ are real-valued, measurable functions defined on $\mathbb{R}$ and there exit constants $r_n > 0$ such that

$$\|f(u) - f(v)\|^2 \vee \|\sigma(u) - \sigma(v)\|^2 \leq r_n \|u - v\|_\gamma^2,$$

for all $u, v \in H^\gamma$ with $\|u\|_\gamma \vee \|v\|_\gamma \leq n, n = 1, 2, \cdots$. Here we use the notation $a \vee b = \max(a, b)$.

(B) A priori estimate: For the solution $u(t)$, $\|u(t)\|_\gamma$ is continuous a.s for all $t > 0$ and satisfies a priori bound

$$\mathbb{E}\|u(t)\|_\gamma^2 \leq K(t),\ 0 \leq t < \infty,$$

where $K(t)$ is defined and finite for all $t > 0$.

Theorem 3.2. *[Wellposedness under local Lipschitz condition]*
*Assume that the local Lipschitz condition **(Hn)** and the finite time a priori estimate **(B)** hold and consider the Eq. (1.2) with initial data $h \in H^\gamma$ for $0 < \gamma < \frac{1}{2}$. Then there exists a unique solution u as an adapt, continuous process in H. Moreover, for any $T > 0$, u belongs to $L^2(\Omega; C([0,T]; H)) \bigcap L^2(\Omega \times [0,T]; H^\gamma)$ such that*

$$\mathbb{E}\{\sup_{0 \leq t \leq T} \|u(t)\|^2 + \int_0^T \|u(t)\|_\gamma^2 dt\} < \infty.$$

Proof. For any integer $n \geq 1$, let $\eta_n : [0, \infty) \to [0, 1]$ is a C^∞-function such that

$$\eta_n(r) = \begin{cases} 1, & 0 \leq r \leq n, \\ 0, & r \geq 2n. \end{cases}$$

We will consider the truncated systems

$$\begin{cases} \frac{\partial}{\partial t} u(t, x) = (\kappa \Delta - \alpha) u(t, x) + f_n(u(t, x)) + \sigma_n(u(t, x)) \frac{\partial}{\partial t} w(t, x), \\ u|_{\partial D} = 0, u(x, 0) = h(x), \end{cases} \quad (3.5)$$

where $f_n(u) = f(\eta_n(\|u\|_\gamma) \cdot u)$, $\sigma_n(u) = \sigma(\eta_n(\|u\|_\gamma \cdot u))$. Then the assumption (**Hn**) implies that f_n and σ_n satisfy the global conditions (**H**). Hence, by **Theorem 3.1** the system (3.5) has a unique solution $u^n(t) \in L^2(\Omega; C([0,T]; H)) \cap L^2(\Omega \times [0,T]; H^\gamma)$. Define a increasing sequence of stopping time $\{\tau_n\}_{n \geq 1}$ by

$$\tau_n = \inf\{t > 0;\ \|u^n(t)\|_\gamma > n\}$$

if it exits, and $\tau_n = \infty$ otherwise. Let $\tau_\infty = \lim_{n \to \infty} \tau_n$ a.s. and set $u^{\tau_n}(t) = u^n(t \wedge \tau_n)$, then $u^{\tau_n}(t)$ is a local solution of Equation (1.2). By assumption (**B**), we have, for any $T > 0$,

$$\mathbb{E}\|u^{\tau_n}(T)\|_\gamma^2 \leq K(T). \tag{3.6}$$

Since

$$\begin{aligned}\mathbb{E}\|u^n(T \wedge \tau_n)\|_\gamma^2 &= \mathbb{E}\|u^{\tau_n}(T)\|_\gamma^2 \\ &\geq \mathbb{E}\{1_{\{\tau_n \leq T\}}\|u^n(T \wedge \tau_n)\|_\gamma^2\} \\ &\geq \mathbb{P}\{\tau_n \leq T\}n^2. \end{aligned} \tag{3.7}$$

In view of (3.6) and (3.7), we get $\mathbb{P}\{\tau_n \leq T\} \leq \frac{K}{n^2}$, which, by invoking the Borel-Cantelli Lemma,

$$\mathbb{P}\{\tau_\infty > T\} = 1$$

is obtained. Hence $u(t) := \lim_{n \to \infty} u^n(t)$ is a global solution.

The proof of the theorem is completed. $\square$

Remark 3.5. Our framework is mainly adapted from works by P. L. Chow [3], which also deals with, in chapter 6, the wellposedness of strong solutions for stochastic evolution equations by Galerkin approximate under local Lipschitz condition, coercivity condition and monotonicity condition. In addition, we point out that the prior continuity of $\|u(t)\|_\gamma$ in assumption (**B**) is not easy to check, thus we will not discuss it in this article.

4. Examples

Let us look at a couple of examples. Let $D \subset \mathbb{R}^2$ be a bounded domain with smooth boundary ∂D and denote $H = L^2(D)$ as before.

Example 4.1. Global Lipschitz case.
Consider the following SPDE on D for $t > 0$:

$$\begin{cases} \frac{\partial}{\partial t}u(t) = (\Delta - 1)u(t) + \sin u(t) + \cos u(t)\frac{\partial}{\partial t}w(t), \\ u|_{\partial D} = 0,\ u(x,0) = h(x). \end{cases} \tag{4.1}$$

It is easy to check that global Lipschitz condition (**H**) holds for the SPDE (4.1). Therefore, by **Theorem 3.1**, the SPDE (4.1) has a unique mild solution $\{u(t,x)\}_{t\geq 0}$ with given $h \in H^\gamma$ for $\gamma \in (0, \frac{1}{2})$.

Example 4.2. Local Lipschitz case.
Now consider the following SPDE on D for $t > 0$:

$$\begin{cases} \frac{\partial}{\partial t}u(t) = \Delta u(t) + u(t) - u^3(t) + u(t)\frac{\partial}{\partial t}w(t), \\ u|_{\partial D} = 0, u(x,0) = h(x). \end{cases} \quad (4.2)$$

We choose $\gamma = \frac{7}{16}$ for simplicity. To apply **Theorem 3.2**, we can set $\alpha = \kappa = 1$, $f(u) = 2u - u^3$ and $\sigma(u) = u$. Note that since $H^{\frac{7}{16}}$ is continuously imbedded in $L^8(D)$ (see [4]), there exists a constant $L_{\frac{7}{16}}$ such that for any $u \in H^{\frac{7}{16}}$,

$$\|u\|_{L^8}^2 + \|u\|_{L^6}^2 + \|u\|_{L^4}^2 \leq L_{\frac{7}{16}} \|u\|_{\frac{7}{16}}^2, \quad (4.3)$$

where $\|\cdot\|_{L^p}$ denotes the usual $L^p(D)$ norm. Then, using the Hölder inequality and (4.3), for any $u,v \in H^{\frac{7}{16}}$ we have

$$\|f(u) - f(v)\|^2 + \|\sigma(u) - \sigma(v)\|_{\mathcal{R}}^2 \leq 4\|u-v\|_{L^4}^2 \|u^2 + v^2 + 1\|_{L^4}^2 + r_0\|u-v\|_{\frac{7}{16}}^2$$

$$\leq C(\|u\|_{\frac{7}{16}}, \|v\|_{\frac{7}{16}})\|u-v\|_{\frac{7}{16}}^2,$$

where $C = C(\|u\|_{\frac{7}{16}}, \|v\|_{\frac{7}{16}})$ depending on $\|u\|_{\frac{7}{16}}$ and $\|v\|_{\frac{7}{16}}$. Thus, condtion (**Hn**) holds for Eq. (4.2).

To check condition (**B**), suppose a H-valued process $\{u(t)\}_{t\geq 0}$ is the unique solution for Eq. (4.2) such that

$$u(t) = G_t h + \int_0^t G_{t-s}(2u(s) - u^3(s))ds + \int_0^t G_{t-s}u(s)dw_s, \quad t \geq 0.$$

With the aid of (2.4), (2.7), (2.9), and (4.3) we can get, for any fixed $T > 0$,

$$\mathbb{E}\|u(t)\|_{\frac{7}{16}}^2 \leq M(T) + 3r_0 \mathbb{E}\int_0^t \|u(s)\|_{\frac{7}{16}}^2 ds, \quad 0 \leq t \leq T,$$

where $M = M(T)$ depending on T. It then follows from the Gronwall inequality that

$$\mathbb{E}\|u(t)\|_{\frac{7}{16}}^2 \leq M(T)\exp\{3r_0 T\} < \infty, \ 0 \leq t \leq T,$$

which means priori bound is satisfied. If we can check the prior continuity of $\|u(t)\|_\gamma$, then we can conclude the Eq. (4.2) has a unique mild solution $\{u(t,x)\}_{t\geq 0}$ with given $h(x) \in H^{\frac{7}{16}}$.

Acknowledgements

We would like to thank Hongjun Gao, Jicheng Liu and Wei Wang for helpful discussions and comments.

References

1. L. Arnold. *Random Dynamical Systems*. Springer, New York, 1998.
2. D. Barbu, Local and global existence for mild soltions of stochastic differential equations, *Port. Math.* **55** (4) (1998) 411-424.
3. P. L. Chow, *Stochastic Partial Differential Equations*. Chapman & Hall/CRC, New York, 2007.
4. D. Henry, *Geometric Theory of Semilinear Parabolic Equations,*. Springer-Verlag, Berlin, 1981.
5. H. Kunita, *Stochastic Flows and Stochastic Differential Equations*, Cambridge University Press, Cambridge, UK, 1990.
6. R.C McOwen. *Partial Differential Equantions*. Pearson Education, New Jetsy, 2003.
7. G. Da Prato, J. Zabczyk, *Stochastic Equations in Infinte Dimensions*. Cambridge University Press, Cambridge, UK, 1992.
8. R. Marcus, stochastic diffusion on an unbounded domain, *Pacific J. Math.* **84** (4) (1979) 143-153.
9. E. Pardoux, Stochastic partial differential equations, a review, *Bull. Sci. Math.* **117** (1) (1993) 29-47.
10. A. Pazy, *Semigroups of Linear Operators and Applications to Partial Differential Equations*, Springer, Berlin, 1985.
11. C. Prevot and M. Rockner, *A Concise Course on Stochastic Partial Differential Equations*, Lecture Notes in Mathematics, Vol. 1905. Springer, New York, 2007.
12. J. C. Robinson, *Infinite Dimensional Dynamical Systems*, Cambridge University Press, Cambridge, UK, 2001, pp.83-84.
13. B. L. Rozovskii, *Stochastic Evolution Equations*. Kluwer Academic Publishers, Boston, 1990.
14. T. Shiga, Two contrasting properties of solutions for one-dimension stochastic partial differntial equations, *Canad. J. Math.* **46** (2) (1994) 415-437.

Chapter 9

The Influence of Transaction Costs on Optimal Control for an Insurance Company with a New Value Function

Lin He, Zongxia Liang* and Fei Xing[†]

Zhou Pei-Yuan Center for Applied Mathematics, Tsinghua University, Beijing 100084, China. Email: hel04@mails.tsinghua.edu.cn

In this chapter, we consider the optimal proportional reinsurance policy for an insurance company and focus on the case that the reinsurer asks for positive transaction costs from the insurance company. That is, the reserve of the insurance company $\{R_t\}$ is governed by a SDE $dR_t^\pi = (\mu - (1-a_\pi(t))\lambda)dt + a_\pi(t)\sigma dW_t$, where W_t is a standard Brownian motion, π denotes the admissible proportional reinsurance policy, μ, λ and σ are positive constants with $\mu \leq \lambda$. The aim of this paper is to find a policy that maximizes the return function $J(x,\pi) = x + \mathbf{E}\{\int_0^{\tau_\pi} e^{-ct}dR_t^\pi\}$, where $c > 0$, τ_π is the time of bankruptcy and x represents the initial reserve. We find the optimal policy and the optimal return function for the insurance company via stochastic control theory. We also give transparent economic interpretation of the return function $J(x,\pi)$ and show that it could be a better interpretation than the traditional one. Finally, we make some numerical calculations and give some economic analysis on the influence of the transaction costs.

Contents

1 Introduction 143
2 Stochastic Control Model 144
3 Comparisons of the Two Models 146
4 HJB Equation and Verification Theorem 147
5 Construction of Solution to the HJB Equation 150
6 Economic Analysis 157
References 159

1. Introduction

In this chapter, we consider the optimal risk control policy for an insurance company which uses proportional reinsurance policy to reduce its risk. In our model, the reinsurance company asks for some extra 'transaction costs' from the company.

---
*Department of Mathematical Sciences, Tsinghua University, Beijing 100084, China. Email: zliang@math.tsinghua.edu.cn
[†]Department of Mathematical Sciences, Tsinghua University, Beijing 100084, China. Email: firefreezing@gmail.com

That is, the reinsurer uses a safety loading not smaller than the insurer. Our target is to find the appropriate reinsurance rate to maximize the total value of the company. We will work out the optimal control policy via HJB methods.

The liquid reserves of the insurance company is governed by a Brownian motion with constant drift and diffusion coefficient. One common control policy for the management of the company is the reinsurance policy. They use it to reduce risk by simultaneously reducing drift and the diffusion coefficient. We refer the readers to Refs. 1, 4 and 9. The objective of the management is to maximize the total value of the company. The traditional value function of the insurance company is often interpreted as $\int_0^\tau e^{-ct} R_t dt$, where τ is the bankruptcy time, c is the discount rate and R_t is the liquid reserve of the company. This kind of model is solved in Ref. 5. Unfortunately, the model is opaquely defined. It has some flaws both interpreted as the return of shareholders and the value of the company. We propose a new value model in Ref. 2 and we consider it to be a better interpretation of company value. Indeed, $\int_0^\tau e^{-ct} dR_t$ stands for the total discounted liquid reserve changes during the life time of the company. It could also be explained as the total discounted net incomes of the company. We attribute $R_0 + \int_0^\tau e^{-ct} dR_t$ as a better value function of the insurance company. Our work are all based on this model.

Based on the above model, we consider the case that the reinsurer asks more risk premium from the insurance company than what the insurer asks from their insured. Our objective is to find the optimal control policy to maximize the value of the insurance company. Højgaard, Taksar have solved this kind of optimal stochastic control problem under the traditional value function, one can refer to Ref. 3. In this paper, we discuss the influence of transaction costs on the insurance company based on the new value function.

With the help of the HJB methods, we solve the problem effectively and get some results which are quite helpful for the companies to make their choices.

The paper is organized as follows: In section 2, we introduce the stochastic control model of the insurance company. The transparent economic interpretation of the new model is given in section 3. In section 4, we construct the HJB equation for this stochastic control model and prove the validation theorem. The most important results are given in section 5. We solve the HJB equation and give the analytical solutions of the optimal control policy and the optimal value function. We give some economic explanation and numerical calculations in the last section.

2. Stochastic Control Model

In this chapter, we consider the situation that an insurance company uses proportional reinsurance policy to control its risk while the reinsurance company chooses a safety loading not smaller than the insurer, that is, the total premium that the insurer pays to the reinsurer is not smaller than the premium he gets from the insured.

In this model, if there is no dividends payments and we suppose that only the proportional reinsurance policy is used to control the risk, then the liquid reserve of the insurance company is modeled by the following stochastic differential equation,

$$dR_t = (\mu - (1 - a(t))\lambda) dt + \sigma a(t) dW_t,$$

where W_t is a standard Brownian Motion, $a(t) \in [0, 1]$ is the proportional reinsurance rate and the constants μ and λ are regarded as the safety loadings of the insurer and the reinsurer, respectively.

To give a rigor mathematical foundation of this optimization problem, we fix a filtered probability space $(\Omega, \mathcal{F}, \{\mathcal{F}_t\}_{t\geq 0}, P)$, $\{W_t\}$ is a standard Brownian motion on this probability space, $\{\mathcal{F}_t\}_{t\geq 0}$ is a family of sub-σ- algebras of $\mathcal{F}$ satisfying the usual conditions. $\mathcal{F}_t$ represents the information available at time t and all decisions are made based on this information.

A control policy π is described by $\pi = \{a_\pi(t)\}$. Given a control policy π, we assume the liquid reserve of the insurance company is modeled by the following stochastic differential equations,

$$dR_t^\pi = (\mu - (1 - a_\pi(t))\lambda) dt + \sigma a_\pi(t) dW_t, \quad R_0^\pi = x, \qquad (2.1)$$

where x is the initial reserve of the company. The time of bankruptcy or the ruin time is defined by

$$\tau_\pi = \inf\{t : R_t^\pi = 0\}. \qquad (2.2)$$

The objective of the insurance company is to maximize the company value by choosing control policy π, i.e., we want to find the optimal value function $V(x)$ and the optimal policy π^* such that

$$V(x) = J(x, \pi^*)$$

where $V(x)$ is defined by

$$V(x) = \sup_\pi \{J(x, \pi)\}$$

and

$$J(x, \pi) = x + \mathbf{E}\left\{ \int_0^{\tau_\pi} e^{-ct} dR_t^\pi \right\} \qquad (2.3)$$

where x is the initial reserve, c denotes the discount rate. We explain the right hand side of Eq.(2.3) as the value of the company. The first term is the initial reserve. While the second term is the total discounted liquid reserve changes during the life time of the company. Another way of explanation is the total discounted net incomes until bankruptcy. Therefore, it could be a better interpretation of the company value than the traditional one. We will give the transparent economic interpretation of the new model in next section.

3. Comparisons of the Two Models

In this section, we compare the new model with the traditional models. In order to simplify the problem, we only consider the differences between the two models under the cheap reinsurance condition, i.e., $\lambda = \mu$. Suppose the liquid reserves of the company is modeled by

$$dR_t^\pi = \mu a_\pi(t)dt + \sigma a_\pi(t)dW_t, \quad R_0^\pi = x. \tag{3.1}$$

In the traditional model, the value of the company is defined as

$$J_{old}(x, \pi) = \mathbf{E}\left\{ \int_0^{T_\pi} e^{-ct} R_t dt \right\}, \tag{3.2}$$

we call it model 1. In the new model, the value of the company is defined as

$$J_{new}(x, \pi) = x + \mathbf{E}\left\{ \int_0^{T_\pi} e^{-ct} dR_t \right\}, \tag{3.3}$$

we call it model 2. We think it's a better interpretation of value of the company than model 1 and we will explain the reasons as follows.

Model 1 is widely used as the value of the company. Unfortunately, its economic interpretation is opaquely defined. We find an explanation in Ref. 10. They supposed the dividends are paid off at the rate proportional to the current surplus. Then (3.2) stands for the total discounted return of the shareholders. Unfortunately, the reserve process can not be defined as in (3.1) and it should be rewritten as

$$dR_t^\pi = \mu a_\pi(t)dt + \sigma a_\pi(t)dW_t - bR_t^\pi dt, \quad R_0^\pi = x, \tag{3.4}$$

b is the constant dividends payout rate. Obviously, this will lead to a totally different solution.

In fact, the company seldom uses the proportional dividends payout strategy as mentioned in Ref. 10. Since the strategy will lead to fluctuated dividends payout, this seems to be a bad management of the company.

Another interpretation of model 1 is that it stands for the value of the company. But, there are also some flaws in the model. In fact, it's the integral of value with respect to time t and it has no meanings. In order to compare the two models clearly, we consider the following extreme cases.

Firstly, we set up the discrete forms of the two different value functions. Suppose that $0 = t_0 < t_1 < ... < t_n < ...,$

$$J_{old}(x, \pi) = \mathbf{E}\left\{ \lim_{\|\Delta\|\to 0} \sum_{i=0}^{\infty} e^{-ct_i \wedge T_\pi} R_{t_i \wedge T_\pi}(t_{i+1} \wedge T_\pi - t_i \wedge T_\pi) \right\} \tag{3.5}$$

$$J(x, \pi) = x + \mathbf{E}\left\{ \lim_{\|\Delta\|\to 0} \sum_{i=0}^{\infty} e^{-ct_i \wedge T_\pi}(R_{t_{i+1} \wedge T_\pi} - R_{t_i \wedge T_\pi}) \right\}, \tag{3.6}$$

where $\|\Delta\| = \max\{t_i - t_{i-1}\}$.

Clearly, model 1 stands for the total discounted future liquid reserves and model 2 stands for the total discounted liquid reserve changes before bankruptcy. If the time value is not taken into consideration, i.e., $c = 0$, the value function $J_{old}(x, \pi) > 0$ and $J(x, \pi) = 0$. In fact, at the time of bankruptcy, all the profit equals to the loss and the company worths nothing neglecting the time value. In our case, bankruptcy is not the optional choice but the compulsory one which only happens when $R_t^\pi = 0$. So, $J(x, \pi) \equiv 0$ as $c = 0$. On the contrary, $J_{old}(x, \pi) > 0$, we believe that model 1 gives too much privilege for the former performance of the company.

In order to make the economic interpretation more clear, we consider another extreme case. Suppose the management of the company always chooses the strategy $\pi = \{a \equiv 0\}$. Then the company will survive for infinite time and the reserve of the company is x determinately. Using the integral in (3.5) and (3.6), we get $J_{old}(x, \pi) = \frac{x}{c}$ and $J(x, \pi) = x$. Since the company chooses such a risk free strategy, the discount rate c should be zero. Then $J_{old}(x, \pi) \to \infty$ in this case. In model 1, they treat this case as a good one and its value could be huge. In fact, this is merely passive management, the company worths nothing except for the initial value. So, model 1 is not effective in this extreme case. According to the above extreme examples, we can see that model 1 gives too much privilege of the former performance of the company and it is not a good representation of the company value. In the meanwhile, model 2 calculates the total liquid reserve changes of the company before bankruptcy and it is perfectly reasonable in the extreme cases.

Secondly, model 2 can be interpreted as the total discounted net incomes during the lifetime of the insurance company. $R_{t_{i+1}} - R_{t_i}$ is the net income happened in the interval $[t_i, t_{i+1}]$. According to the valuation of the company, the company value is the total discounted net incomes before bankruptcy. So, the new model is a good interpretation of the company value.

4. HJB Equation and Verification Theorem

Theorem 4.1. *Assume $f \in C^2$ is a solution of the following HJB equation:*

$$\max_{a \in [0,1]} \{\frac{\sigma^2 a^2}{2} f''(x) + (\mu - (1-a)\lambda)f'(x) - cf(x)$$

$$+ \mu - (1-a)\lambda\} = 0 \quad (4.1)$$

with boundary condition

$$f(0) = 0, \quad (4.2)$$

and

$$\limsup_{x \to \infty} \frac{|f(x)|}{x} < \infty. \quad (4.3)$$

Then for any admissible control π,

$$g(x) = f(x) + x \geq J(x, \pi).$$

Proof. Fix a policy π. Choose $\varepsilon > 0$ and let $\tau_\pi^\varepsilon = \inf\{t : R_t^\pi \leq \varepsilon\}$. Then, by Itô's formula

$$e^{-c(t \wedge \tau_\pi^\varepsilon)} f(R_{t \wedge \tau_\pi^\varepsilon}^\pi)$$
$$= f(x) + \int_0^{t \wedge \tau_\pi^\varepsilon} e^{-cs}(\mu - (1-a)\lambda) f'(R_s^\pi) ds$$
$$+ \frac{1}{2} \int_0^{t \wedge \tau_\pi^\varepsilon} e^{-cs} a^2 \sigma^2 f''(R_s^\pi) ds - c \int_0^{t \wedge \tau_\pi^\varepsilon} e^{-cs} f(R_s^\pi) ds$$
$$+ \int_0^{t \wedge \tau_\pi^\varepsilon} e^{-cs} a\sigma f'(R_s^\pi) dW_s.$$

Since

$$\int_0^{t \wedge \tau_\pi^\varepsilon} e^{-cs} dR_s^\pi = \int_0^{t \wedge \tau_\pi^\varepsilon} e^{-cs}(\mu - (1-a)\lambda) ds + \int_0^{t \wedge \tau_\pi^\varepsilon} e^{-cs} a\sigma dW_s,$$

we have

$$e^{-c(t \wedge \tau_\pi^\varepsilon)} f(R_{t \wedge \tau_\pi^\varepsilon}^\pi) + \int_0^{t \wedge \tau_\pi^\varepsilon} e^{-cs} dR_s^\pi$$
$$= f(x) + \int_0^{t \wedge \tau_\pi^\varepsilon} e^{-cs}(\mu - (1-a)\lambda) f'(R_s^\pi) ds$$
$$+ \frac{1}{2} \int_0^{t \wedge \tau_\pi^\varepsilon} e^{-cs} a^2 \sigma^2 f''(R_s^\pi) ds - c \int_0^{t \wedge \tau_\pi^\varepsilon} e^{-cs} f(R_s^\pi) ds$$
$$+ \int_0^{t \wedge \tau_\pi^\varepsilon} e^{-cs} a\sigma f'(R_s^\pi) dW_s + \int_0^{t \wedge \tau_\pi^\varepsilon} e^{-cs}(\mu - (1-a)\lambda) ds$$
$$+ \int_0^{t \wedge \tau_\pi^\varepsilon} e^{-cs} a\sigma dW_s.$$

Taking expectations at both sides of the last equations, we have

$$\mathbf{E} \int_0^{t \wedge \tau_\pi^\varepsilon} e^{-cs} dR_s^\pi + \mathbf{E} e^{-c(t \wedge \tau_\pi^\varepsilon)} f(R_{t \wedge \tau_\pi^\varepsilon}^\pi) \leq f(x)$$
$$+\mathbf{E} \int_0^{t \wedge \tau_\pi^\varepsilon} e^{-cs} a\sigma f'(R_s^\pi) dW_s + \mathbf{E} \int_0^{t \wedge \tau_\pi^\varepsilon} e^{-cs} a\sigma dW_s \qquad (4.4)$$

due to the fact

$$\frac{\sigma^2 a^2}{2} f''(x) + (\mu - (1-a)\lambda) f'(x) - cf(x) + \mu - (1-a)\lambda \leq 0$$

for all $a \in [0, 1]$.

By concavity of f, $0 \leq f'(R_s^\pi) \leq f'(\varepsilon)$ on $(0, \tau_\pi)$, the second term in the right hand side of (4.4) is a zero-mean square integrable martingale, i.e.,

$$\mathbf{E}\{\int_0^{t \wedge \tau_\pi^\varepsilon} e^{-cs} a\sigma f'(R_s^\pi) dW_s\} = 0.$$

Similarly,
$$\mathbf{E}\{\int_0^{t \wedge \tau_n^\varepsilon} e^{-cs} a\sigma dW_s\} = 0.$$

So (4.4) becomes
$$\mathbf{E}\int_0^{t \wedge \tau_n^\varepsilon} e^{-cs} dR_s^\pi + \mathbf{E}\{e^{-c(t \wedge \tau_n^\varepsilon)} f(R_{t \wedge \tau_n^\varepsilon}^\pi)\} \leq f(x). \tag{4.5}$$

Letting $\varepsilon \to 0$ and noticing that
$$\lim_{t \to \infty} \mathbf{E}\{e^{-c(t \wedge \tau_\pi)} f(R_{t \wedge \tau_\pi}^\pi) I_{\{\tau_\pi < \infty\}}\}$$
$$= \mathbf{E}\{e^{-c\tau_\pi} f(R_{\tau_\pi}^\pi) I_{\{\tau_\pi < \infty\}}\}$$
$$= \mathbf{E}\{e^{-c\tau_\pi} f(0) I_{\{\tau_\pi < \infty\}}\}$$
$$= 0 \tag{4.6}$$

and
$$\lim_{t \to \infty} \mathbf{E}\{e^{-c(t \wedge \tau_\pi)} f(R_{t \wedge \tau_\pi}^\pi) I_{\{\tau_\pi = \infty\}}\}$$
$$= \lim_{t \to \infty} \mathbf{E}\{e^{-ct} f(R_t) I_{\{\tau_\pi = \infty\}}\}$$
$$\leq \lim_{t \to \infty} k e^{-ct} \mathbf{E}\{R_t I_{\{\tau_\pi = \infty\}}\}$$
$$\leq \lim_{t \to \infty} \frac{k(x + \lambda t)}{e^{ct}} = 0, \tag{4.7}$$

where $k := \limsup_{x \to \infty} \frac{|f(x)|}{x} < \infty$, we obtain
$$\lim_{t \to \infty} \mathbf{E}\{e^{-c(t \wedge \tau_\pi)} f(R_{t \wedge \tau_\pi}^\pi)\} = 0. \tag{4.8}$$

Putting (4.5) and (4.8) together implies that
$$J(x, \tau_\pi) - x = \mathbf{E}\{\int_0^{t \wedge \tau_\pi} e^{-cs} dR_s^\pi\} \leq f(x). \tag{4.9}$$

Thus we obtain
$$g(x) = f(x) + x \geq J(x, \pi) \tag{4.10}$$
for any admissible control π. □

Theorem 4.2. *Let $\pi^* = a^*(x)$ be the maximizer of the left hand side of (4.1) and $R_t^{\pi^*}$ is a solution to the following stochastic differential equation*
$$dR_t^{\pi^*} = (\mu - (1 - a^*(R_t^{\pi^*}))\lambda)dt + a^*(R_t^{\pi^*})\sigma dW_t,$$
with the boundary condition $R_0^{\pi^*} = x$. Then
$$V(x) = J(x, \pi^*) = g(x).$$

Proof. We follow the same procedures in Theorem 4.1, and all the inequities in (4.4)(4.5)(4.9)(4.10) could become equalities. It is easy to get $g(x) = J(x, \pi^*)$, i.e., $g(x) \leq V(x)$. Combined with $g(x) \geq V(x)$, we get the conclusion $V(x) = J(x, \pi^*) = g(x)$. □

5. Construction of Solution to the HJB Equation

According to Theorem 4.1, we only need to find a C^2 function f satisfying the HJB equation (4.1) with the boundary conditions (4.2) and (4.3).

The case $\lambda = \mu$

This case is also known as cheap reinsurance and has already been solved in Lin He and Zongxia Liang (2007), where a solution f to (4.1) can be found as follows:

$$f(x) = \begin{cases} k_1 Q(G^{-1}(\frac{x}{k_1})), & 0 \leq x < x_0, \\ k_2 e^{d(x-x_0)} + \frac{\mu}{c}, & x \geq x_0, \end{cases}$$

where k_1, k_2, x_0, d are constants determined via exogenous parameters of the problem and G, Q are special functions given by

$$G(x) = \int^x e^{\frac{\gamma}{1+y}} y^\gamma (1+y)^{-2-\gamma} dy,$$

$$Q(x) = \int^x e^{\frac{\gamma}{1+y}} y^{\gamma-1} (1+y)^{-2-\gamma} dy.$$

The maximizing function is

$$a(x) = \begin{cases} \frac{k_1 \mu}{\sigma^2}(G^{-1}(\frac{x}{k_1}) + (G^{-1}(\frac{x}{k_1}))^2) g(G^{-1}(\frac{x}{k_1})), & 0 \leq x < x_0, \\ 1, & x \geq x_0, \end{cases}$$

where g is the integral kernel of G, that is, $g(x) = e^{\frac{\gamma}{1+x}} x^\gamma (1+x)^{-2-\gamma}$.

The case $\lambda > \mu$

First we guess that $a(x) = 1$ for all x, where (4.1) becomes

$$\frac{\sigma^2}{2} f''(x) + \mu f'(x) - cf(x) + \mu = 0. \tag{5.1}$$

Using the concavity of f as well as $f(0) = 0$, we get the solution of (5.1),

$$f(x) = -\frac{\mu}{c} e^{dx} + \frac{\mu}{c}, \tag{5.2}$$

where

$$d = \frac{-\mu - \sqrt{\mu^2 + 2\sigma^2 c}}{\sigma^2} \tag{5.3}$$

is a negative solution to the following equation

$$\frac{\sigma^2}{2}y^2 + \mu y - c = 0. \tag{5.4}$$

Then we can get the following important result.

Proposition 5.1. *Let f be defined by (5.2). Then f is a solution to (4.1) if and only if $\lambda \geq 2\mu$.*

Proof. First we prove the sufficiency. To show that f is a concave solution to (4.1), we only need to show that for any $a \in [0,1]$

$$F(a,x) = \frac{(1-a^2)\sigma^2}{2}f''(x) + (1-a)\lambda f'(x) + (1-a)\lambda \geq 0. \tag{5.5}$$

In fact, we have for any $a \in [0,1], x \in [0, +\infty)$

$$F(a,x) = \frac{(1-a^2)\sigma^2}{2}(-\frac{\mu d^2}{c}e^{dx}) + (1-a)\lambda(-\frac{\mu d}{c}e^{dx} + 1)$$

$$= (1-a)e^{dx}[\frac{(1+a)\sigma^2}{2}(-\frac{\mu d^2}{c}) + \lambda(-\frac{\mu d}{c} + e^{-dx})]$$

$$\geq (1-a)e^{dx}[\frac{(1+a)\sigma^2}{2}(-\frac{\mu d^2}{c}) + \lambda(-\frac{\mu d}{c} + 1)] \tag{5.6}$$

$$= (1-a)e^{dx}[-\frac{\mu}{c}(1+a)(c-\mu d) + \frac{\lambda}{c}(-\mu d + c)] \tag{5.7}$$

$$= (1-a)e^{dx}\frac{c-\mu d}{c}[\lambda - \mu(1+a)]$$

$$\geq 0 \tag{5.8}$$

where (5.6) satisfies because of $d < 0$ and $x \geq 0$, which implies that $e^{-dx} \geq 1$; (5.7) satisfies since $\frac{1}{2}\sigma^2 d^2 + \mu d - c = 0$ and (5.8) satisfies since $c - \mu d > 0$ and $\lambda \geq 2\mu$.

So we have proved sufficiency. As to the necessity, assume $\mu < \lambda < 2\mu$. Then there exists a $a_0 \in (0,1)$ such that $m := \lambda - \mu(1+a_0) < 0$. Let $x_0 = -\frac{1}{d}\log(1 - \frac{m(c-\mu d)}{2c\lambda})$. Clearly, $x_0 \in (0, \infty)$. Then we can compute

$$F(a_0, x_0) = \frac{c-\mu d}{2c}m(1-a_0)e^{dx_0} < 0,$$

which contradicts the assumption. Therefore we obtain the necessity. □

Now assume $\mu < \lambda < 2\mu$.

Suppose that the concave function f is found and there exists x_1 such that $a(x)$ satisfies $0 < a(x) < 1$ for all $0 < x < x_1$. When $x < x_1$, we can find $a(x)$ by differentiating the expression $\Phi(a) \equiv \frac{\sigma^2 a^2}{2}f''(x) + (\mu - (1-a)\lambda)f'(x) - cf(x) + \mu -$

$(1-a)\lambda$ in bracket $\{\cdot\}$ of Eq.(4.1) with respect to a and letting $\Phi'(a) = 0$. Here x_1 is an unknown variable to be specified later and $a(x)$ is given as

$$a(x) = -\frac{\lambda(1+f')}{\sigma^2 f''}. \tag{5.9}$$

Substituting (5.9) into (4.1), f satisfies

$$-\frac{\lambda^2(f'+1)^2}{2\sigma^2 f''} + (\mu - \lambda)f' - cf + \mu - \lambda = 0, \text{ for } 0 \le x < x_1. \tag{5.10}$$

By our assumption, $f(x)$ is concave for $0 \le x < x_1$, so there exists a function $X(z) : \mathcal{R} \longrightarrow [0, +\infty)$ such that $-\ln(f'(X(z)) + 1) = z$. Here $z \le 0$. In addition, the following two equations are valid,

$$f'(X(z)) = e^{-z} - 1, \tag{5.11}$$

$$f''(X(z)) = \frac{-e^{-z}}{X'(z)}.$$

Define M such that $X(-M) = 0$, i.e., $f'(0) = e^M - 1$. Then $X : [-M, 0) \to [0, \infty)$. Putting $x = X(z)$ into (5.10), we obtain that

$$\frac{\lambda^2}{2\sigma^2}X'(z)e^{-z} - cf(X(z)) + (\mu - \lambda)e^{-z} = 0. \tag{5.12}$$

Differentiating (5.12) w.r.t z and using (5.11), it is easy to get

$$\frac{\lambda^2}{2\sigma^2}X''(z)e^{-z} - (\frac{\lambda^2}{2\sigma^2} + c - ce^z)X'(z)e^{-z} - (\mu - \lambda)e^{-z} = 0. \tag{5.13}$$

Defining $\gamma = \frac{2\sigma^2}{\lambda^2}$, (5.13) can be rewritten as

$$X''(z) - (1 + c\gamma - c\gamma e^z)X'(z) = (\mu - \lambda)\gamma.$$

So

$$X'(z) = \left[\gamma(\mu - \lambda)\int_{-M}^{z} \exp(-(1+c\gamma)y + c\gamma e^y)dy \right.$$
$$\left. + k_1\right] \exp((1+c\gamma)z - c\gamma e^z),$$

where k_1 is a free parameter.

Let g be the p.d.f of Gamma distribution with parameters $(c\gamma + 1, 1/c\gamma)$, we obtain

$$X'(z) = \gamma(\mu - \lambda)e^z g(e^z)\int_{-M}^{z} \frac{1}{e^y g(e^y)}dy + k_1 e^z g(e^z)$$
$$= \gamma(\mu - \lambda)e^z g(e^z)H(e^z) + k_1 e^z g(e^z),$$

where

$$H(z) = \int_{e^{-M}}^{z} \frac{1}{y^2 g(y)}dy, \quad \forall z \ge e^{-M}. \tag{5.14}$$

Since $X(-M) = 0$, we have

$$X(z) = \int_{-M}^{z} \left(\gamma(\mu - \lambda)e^y g(e^y) H(e^y) + k_1 e^y g(e^y)\right) dy$$

$$= \int_{e^{-M}}^{e^z} \left(\gamma(\mu - \lambda) H(y) g(y) + k_1 g(y)\right) dy$$

$$= K(e^z), \qquad (5.15)$$

where

$$K(z) := \int_{e^{-M}}^{z} \left(\gamma(\mu - \lambda) H(y) g(y) + k_1 g(y)\right) dy, \quad \forall z \geq e^{-M}. \qquad (5.16)$$

Because $X(z)$ and e^z is monotone increasing in z, $K(z)$ is also monotone increasing in z. Therefore we can calculate the inverse function of K:

$$K^{-1}(X(z)) = \frac{1}{f'(X(z)) + 1},$$

that is,

$$f'(x) = \frac{1}{K^{-1}(x)} - 1. \qquad (5.17)$$

Consequently,

$$f(x) = \int_0^x \frac{1}{K^{-1}(y)} dy - x, \qquad \forall x \in [0, x_1]. \qquad (5.18)$$

Thus

$$a(x) = -\frac{\lambda(f'(x) + 1)}{\sigma^2 f''(x)} = \frac{\lambda}{\sigma^2} K^{-1}(x) k(K^{-1}(x)), \qquad (5.19)$$

where

$$k(y) := \big(K(y)\big)' = \gamma(\mu - \lambda) H(y) g(y) + k_1 g(y). \qquad (5.20)$$

Now for $x > x_1$ we have $a(x) = 1$. Letting $a(x) = 1$ in (4.1) and then using the concavity, we obtain the following solution:

$$f(x) = k_2 e^{dx} + \frac{\mu}{c}, \qquad \forall x \geq x_1$$

with d given in (5.3). Therefore we obtain the following solution:

$$f(x) = \begin{cases} \int_0^x \frac{1}{K^{-1}(y)} dy - x, & 0 \leq x < x_1, \\ k_2 e^{dx} + \frac{\mu}{c}, & x \geq x_1, \end{cases} \qquad (5.21)$$

where k_1, k_2, x_1, M are constants to be determined. To ensure f to be twice continuously differentiable at x_1 we have

$$f'(x_1-) = k_2 d e^{dx_1},$$
$$f''(x_1-) = k_2 d^2 e^{dx_1},$$

that is,

$$\frac{1}{K^{-1}(x_1)} - 1 = k_2 d e^{dx_1}, \qquad (5.22)$$

$$-\frac{\lambda}{\sigma^2} \frac{1}{K^{-1}(x_1)} = k_2 d^2 e^{dx_1}, \qquad (5.23)$$

where the left hand side of (5.23) is obtained via using

$$a(x_1) = -\frac{\lambda(f'(x_1)+1)}{\sigma^2 f''(x_1)} = 1,$$

which means that

$$f''(x_1) = -\frac{\lambda}{\sigma^2}(f'(x_1)+1) = -\frac{\lambda}{\sigma^2} \frac{1}{K^{-1}(x_1)}.$$

Let

$$\alpha = K^{-1}(x_1), \qquad \beta = k_2 e^{dx_1}.$$

We see from (5.22) and (5.23) that

$$\frac{1}{\alpha} - 1 = \beta d, \qquad -\frac{\lambda}{\sigma^2} \frac{1}{\alpha} = \beta d^2.$$

These equations have a solution

$$(\alpha, \beta) = \left(\frac{\lambda}{d\sigma^2} + 1, \frac{\sigma^2}{\lambda + d\sigma^2} - \frac{1}{d}\right). \qquad (5.24)$$

By (5.3),(5.4) and the assumption $\lambda < 2\mu$, we have $\beta < 0$ and $0 < \alpha < 1$. Using (5.19) and the condition $a(x_1) = 1$, we have

$$\frac{\lambda \alpha}{\sigma^2}\left(\gamma(\mu - \lambda)H(\alpha)g(\alpha) + k_1 g(\alpha)\right) = 1.$$

Therefore,

$$(x_1, k_1) = \left(K(\alpha), \frac{\sigma^2}{\lambda \alpha g(\alpha)} + \gamma(\lambda - \mu)H(\alpha)\right). \qquad (5.25)$$

So $f(x)$ must be the following form,

$$f(x) = \begin{cases} \int_0^x \frac{1}{K^{-1}(y)} dy - x, & 0 \le x < x_1, \\ \beta e^{d(x-x_1)} + \frac{\mu}{c}, & x \ge x_1, \end{cases} \qquad (5.26)$$

and the maximizing function has the following form,

$$a(x) = \begin{cases} \frac{\lambda}{\sigma^2} K^{-1}(x) k(K^{-1}(x)), & 0 \le x < x_1, \\ 1, & x \ge x_1, \end{cases} \qquad (5.27)$$

where K is given by (5.16). The constant x_1, k_1 and β are given by (5.24) and (5.25), and d is given by (5.3). Now we only need to determine the constant M

such that $f(0+)$ is a solution to (5.10) in the limit. Letting $x \to 0$ in (5.10) we get the equation

$$\frac{\lambda a(0+)}{2}(f'(0+)+1) + (\mu - \lambda)f'(0+) + (\mu - \lambda) = 0.$$

Due to the fact that $f(0+) \geq 0$, this is satisfied if and only if

$$a(0+) = \frac{2(\lambda - \mu)}{\lambda} := B \tag{5.28}$$

Since $\mu < \lambda < 2\mu$, $B \in (0,1)$. By (5.27) and the fact that $K(e^{-M}) = X(-M) = 0$, we obtain

$$a(0+) = \frac{\lambda}{\sigma^2}e^{-M}g(e^{-M})\left(\gamma(\lambda - \mu)(H(\alpha) - H(e^{-M})) + \frac{\sigma^2}{\lambda a g(\alpha)}\right)$$

$$= \frac{e^{-M}g(e^{-M})}{ag(\alpha)} + Be^{-M}g(e^{-M})H(\alpha). \tag{5.29}$$

By (5.28) and (5.29), we have

$$\frac{e^{-M}g(e^{-M})}{ag(\alpha)} + Be^{-M}g(e^{-M})H(\alpha) = B. \tag{5.30}$$

Denote

$$F(y) := \frac{yg(y)}{ag(\alpha)} + Byg(y)\int_y^\alpha \frac{1}{z^2 g(z)}dz, \qquad y \in [0,\alpha]. \tag{5.31}$$

To ensure the existence of M in (5.31), we need only to prove there exists $y_0 \in [0, \alpha]$, such that $F(y_0) = B$. Then $M = -\ln y_0$ is what we need. □

Lemma 5.1. *Suppose $f(x) \in C^2[a,b]$ and $f(a) < f(b)$. If for any $x \in [a,b]$, $f'(x) = 0$ leads to $f''(x) < 0$, then for any $y_0 \in (f(a), f(b))$, there exists a unique $x_0 \in [a,b]$ such that $f(x_0) = y_0$.*

Proof. Suppose there exist $x_1, x_2 \in [a,b], x_1 < x_2$, such that $y_0 := f(x_1) = f(x_2) \in (f(a), f(b))$. Then we claim that $y_1 := \min_{x \in [x_1, x_2]} f(x) \geq y_0$. Since if it is not the case, then we can find $c \in (x_1, x_2)$ such that $f(c) = y_1$, which implies that $f'(c) = 0$ and $f''(c) \geq 0$. Contradicting to the assumption of this lemma. Using the same method we obtain $y_2 := \min_{x \in [x_2, b]} f(x) \geq y_0$. Then we obtain that $y_0 = \min_{x \in [x_1, b]} f(x)$, which implies that $f'(x_2) = 0$ and $f''(x_2) \geq 0$. Contradicting to the assumption of this lemma. Therefore, we have finished the proof. □

Proposition 5.2. *$F(x) = B$ has a unique solution in $[0, \alpha]$.*

Proof. First, $F(\alpha) = 1 > B$. By L'Hospital rule and (5.31) it can be seen that $F(x) \to B/(1 + c\gamma) < B$ when $x \to 0$.

Next, notice that

$$(xg(x))' = g(x)(c\gamma - c\gamma x + 1). \tag{5.32}$$

Using this we can obtain

$$F'(x) = \frac{1}{x}(F(x)(c\gamma - c\gamma x + 1) - B)$$

and

$$F''(x) = \frac{c\gamma}{x} F'(x)(1-x) - \frac{c\gamma F(x)}{x}.$$

Assume there exists $\hat{x} \in (0, \alpha)$ satisfying $F'(\hat{x}) = 0$, we have

$$F''(\hat{x}) = -\frac{c\gamma F(\hat{x})}{\hat{x}} < 0. \tag{5.33}$$

By Lemma 5.1 we obtain the result. □

Recall that we assume $a(x) < 1$ for all $x < x_1$ in the beginning of this section. Now we are going to check this case. First we need to prove a lemma.

Lemma 5.2. *Suppose $f(x) \in C^2[a,b]$, $f(a) < f(b)$ and $f'(b) > 0$. If for any $x \in [a,b]$, $f'(x) = 0$ leads to $f''(x) < 0$, then for any $x \in (a,b)$ we have $f(x) < f(b)$.*

Proof. Suppose there exists $x_0 \in (a,b)$ satisfying $x_0 \geq f(b)$. Since $f'(b-) > 0$, then there exists $x_1 \in (x_0, b)$ such that $f(x_1) < f(b) \leq f(x_0)$. Therefore, we can find $c \in (x_0, b)$ such that $f(c) = \min_{x \in [x_0,b]} f(x)$. Since $f(x) \in C^2[a,b]$, then we have $f'(c) = 0$ and $f''(c) \geq 0$, which contradicts the assumption. □

Proposition 5.3. *Let $a(x)$ be defined by (5.27). Then $a(x) < 1$ for all $x < x_1$.*

Proof. We only need to consider

$$a(y) = \frac{\lambda}{\sigma^2} y k(y), \qquad e^{-M} < y < \alpha.$$

Using (5.32) again it follows that

$$a'(y) = \frac{1}{y}(a(y)(1 + c\gamma - c\gamma y) + B) \tag{5.34}$$

and

$$a''(y) = c\gamma a'(y)(1 - \frac{1}{y}) - \frac{c\gamma}{y} a(y). \tag{5.35}$$

By (5.34) and (5.35) if $a'(y) = 0$, we have $a''(y) < 0$. Since $a(\alpha) = 1$, it follows from (5.34) that $a'(\alpha -) = \frac{1}{\alpha}(1 + c\gamma - c\gamma y + B) > 0$, then by Lemma 5.2 we complete the proof. □

Theorem 5.1. *Let f be defined by (5.26) where e^{-M} is the unique solution to (5.31). Then $f \in C^2$ and is a concave solution to (4.1) for $\mu < \lambda < 2\mu$.*

Proof. By the construction of f we obtain that f is concave and $f \in C^2$, so we only need to prove that f satisfies (4.1). By Proposition 5.3, we know that the maximizing function $a(x)$ satisfies $a(x) < 1$ for all $x < x_1$, then it satisfies (4.1) for all $x < x_1$ from the construction of f. Now for $x \geq x_1$, equality holds with $a(x) = 1$. So we need only to prove that for all $0 \leq a \leq 1$

$$F(a,x) = \frac{(1-a^2)\sigma^2}{2} f''(x) + (1-a)\lambda f'(x) + (1-a)\lambda \geq 0.$$

In fact, we can compute that

$$\begin{aligned}
F(a,x) &= \frac{(1-a^2)\sigma^2}{2} f''(x) + (1-a)\lambda f'(x) + (1-a)\lambda \geq 0 \\
&= \frac{1}{2}\sigma^2 \beta d^2 (1-a)^2 e^{d(x-x_1)} + (1-a)\lambda \beta d e^{d(x-x_1)} + (1-a)\lambda \\
&= (1-a) e^{d(x-x_1)} \Big[\frac{1}{2}(1+a)\sigma^2 \beta d^2 + \lambda \beta d + \lambda e^{-d(x-x_1)}\Big] \\
&\geq (1-a) e^{d(x-x_1)} \Big[\frac{1}{2}(1+a)\sigma^2 \beta d^2 + \lambda \beta d + \lambda\Big], \quad (5.36)
\end{aligned}$$

where (5.36) satisfies since $d < 0$ and $x \geq x_1$. Notice that $(1-a)e^{d(x-x_1)} \geq 0$ for all $a \leq 0$, so it suffices to show that $\frac{1}{2}(1+a)\sigma^2 \beta d^2 + \lambda \beta d + \lambda \geq 0$. Since $\sigma^2 \beta d^2 < 0$, using (5.24) we have

$$\begin{aligned}
&\frac{1}{2}(1+a)\sigma^2 \beta d^2 + \lambda \beta d + \lambda \\
&\geq \sigma^2 \beta d^2 + \lambda \beta d + \lambda \\
&= (\sigma^2 d + \lambda)\Big(\frac{\sigma^2 d}{\lambda + \sigma^2 d} - 1\Big) + \lambda \\
&\geq 0,
\end{aligned}$$

so $F(a,x) \geq 0$ for all $a \leq 1$. Thus we complete the proof. □

6. Economic Analysis

In this section, we calculate the optimal return function $V(x)$, the optimal control strategy a^* and the switch point x_1 for $\mu = 1$, $\sigma = 1$, $c = 0.1$ and different values of λ between 1 and 2.

From Figure 1, we see that the valuation function $V(x)$ is a monotone decreasing function with respect to λ. Obviously, higher risk premium charged by the reinsurance company reduces the profit of the insurance company. In the meanwhile, we find that the optimal value is about ten percent of the optimal value obtained via the traditional model. We think the new model is more realistic. (See Ref. 3 for details)

From Figure 2, we see that when λ is more than two folds of μ, the management of the insurance company would like to take full responsibility of the risk. This means huge risk premium charged by the reinsurance company leads to no business

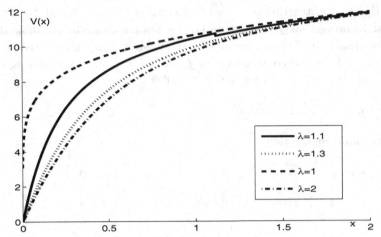

Figure 1. The optimal return function $V(x)$ for $\mu = 1$, $\sigma = 1$, $c = 0.1$.

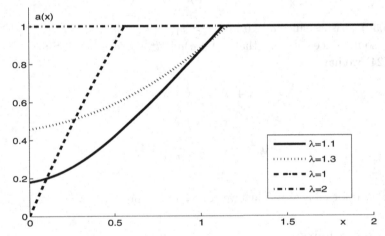

Figure 2. The optimal control policy a^* for $\mu = 1$, $\sigma = 1$, $c = 0.1$.

at all. The other interesting point is that when the liquid reserves approaches 0, the reinsurance rate is 1 in the cheap reinsurance case. So the company never goes bankruptcy in the case of $\lambda = \mu$. (See Ref. 2 for details) In the non-cheap reinsurance case, the management of the insurance company chooses to take some risk when the reserves approaches 0. When the reserves is small, the management of the company can't afford the transaction costs generated from the reinsurance procedure and have to take some risk.

From Figure 3, we see that $x(1)$ is a concave function of λ. First, x_1 increases with respect to λ, then decreases in λ. This is quite interesting. x_1 stands for the switch point that the management of the company would like to take full responsibility of the risk. The company reduces its risk appetite when the λ becomes larger

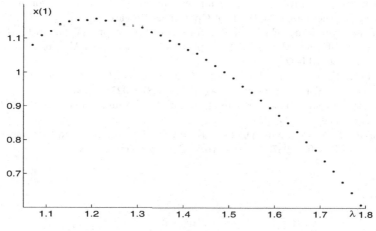

Figure 3. The switch point x_1 of λ.

at first. The company enlarges its responsibility of the risk slowly with respect to the liquid reserve level. This may be induced by the balance between the cost reduction and the risk aversion. The company would like to pay some cost to reduce the possibility of bankruptcy when the cost is small. When λ is large enough, the management of the company would to take more risk when λ becomes larger. The incentive for cost reduction overcomes the motivation for risk aversion at last.

Acknowledgement

This work is supported by NSFC No. 10771114, and SRF for ROCS, SEM, and the Korea Foundation for Advanced Studies. We would like to thank for generous financial support.

References

1. Dayananda, P.W.A.: Optimal reinsurance, Journal of Applied Probability 7, 134-156, 1970.
2. He, Lin and Liang, Zongxia: Optimal Proportional Reinsurance Policies for the Insurance Company with a New Value Function, Preprint(2007).
3. Højgaard, B., Taksar: Optimal Proportional Reinsurance Policies for Diffusion Models with Transaction Costs, Insurance: Mathematics and Economics, Vol. 22, 41-51, 1998.
4. Højgaard, B., Taksar, M.: Controlling Risk Exposure and Dividend Payout Schemes: Insurance Company Example, Mathematical Finance, Vol. 9, No. 2, 153-182, 1999.
5. Højgaard, B., Taksar, M.: Optimal Proportional Reinsurance Policies for Diffusion Models, Insurance: Mathematics and Economics, Volume 23, 303, 1998.
6. Øksendal, B.: Stochastic Differential Equations, Berlin: Springer Verlag, 1985
7. Samuel, k., Taylor, H. M.: A Second Course in Stochastic Processes, ISBN 0123986508, New York: Academic Press, 1981.
8. Sethi, S.: Optimal Consumption and Investment with Bankruptcy, Kluwer, 1997.

9. Taskar M.: Optimal Risk and Dividend Distribution Control Models for an Insurance Company, Mathematical Methods of Operations Research, Vol. 51, 1-42, 2000.
10. Taksar, M., Hunderup, C., L.: The Influence of Bankruptcy Value on Optimal Risk Control for Diffusion Models with Proportional Reinsurance, Mathematics and Economics, Vol. 40, 311-321, 2007.
11. Wendell H. Fleming, H. Mete Soner: Controlled Markov Processes and Viscosity Solutions, New York : Springer-Verlag, ISBN 0387979271, 1993.
12. Whittle,P.: Optimization over Time-Dynamic Programming and Stochastic Control, New York: Wiley , 1983.
13. Yong Jiong Min, Zhou Xun Yu: Stochastic Controls: Hamiltonian Systems and HJB Equations, ISBN 0387987231, New York: Springer-Verlag, 1999.

Chapter 10

Limit Theorems for p-Variations of Solutions of SDEs Driven by Additive Stable Lévy Noise and Model Selection for Paleo-Climatic Data

Claudia Hein, Peter Imkeller and Ilya Pavlyukevich

Institut für Mathematik, Humboldt Universität zu Berlin
Unter den Linden 6, 10099 Berlin, Germany
heinc@math.hu-berlin.de (Claudia Hein)
imkeller@math.hu-berlin.de (Peter Imkeller)
pavljuke@math.hu-berlin.de (Ilya Pavlyukevich)

In this chapter asymptotic properties of the p-variations of stochastic processes of the type $X = Y + L$ are studied, where L is an α-stable Lévy process, and Y a perturbation which satisfies some mild Lipschitz continuity assumptions. Local functional limit theorems for the power variation processes of X are established. In case X is a solution of a stochastic differential equation driven by the process L, these limit theorems provide estimators of the stability index α. They are applied to paleo-climatic temperature time series taken from the Greenland ice core, describing the fine structure of temperature variability during the last ice age, in particular exhibiting the intermediate warming periods known as Dansgaard–Oeschger events. Our results provide the best fitting α in a parameter test in which the time series are modeled by stochastic differential equations with α-stable noise component.

Keywords: p-variation; stable Lévy process; tightness; Skorokhod topology; stability index; model selection; estimation; Kolmogorov–Smirnov distance; paleo-climatic time series; Greenland ice core data

2000 AMS Subject Classification: primary 60G52, 60F17; secondary 60H10, 62F10, 62M10, 86A40

Contents

1 Introduction 162
2 Object of study and main results 164
3 Applications to real data 166
4 Functional convergence for $V_p^n(L)$ 169
5 Generalisation to sums of processes 170
    5.1 Equivalence for $p \leq 1$ 170
    5.2 Equivalence for $p > \alpha$ 170
    5.3 Equivalence for $\alpha \in (1, 2)$, $p \in (1, \alpha]$ 171

References 175

1. Introduction

The research leading to this paper was inspired and triggered by two papers by Ditlevsen[1,2] which will be in its focus. In his work, Ditlevsen uses simple stochastic differential equations with additive noise as a model fit for temperature data (yearly averages) obtained from the Greenland ice core describing aspects of the evolution of the Earth's climate during the last Ice Age, which extended over about 100,000 years. This time series features in particular the catastrophic warmings and coolings in the Northern hemisphere, the so-called Dansgaard–Oeschger events.[3] The *dynamical systems component* of the modeling stochastic differential equation describes the evolution of temperature as a function of time, thus lives in one dimensional Euclidean space, and can therefore be chosen to be given by the gradient of a double well potential (climatic quasi-potential), in which the local minima correspond to cold and warm meta-stable climate states. In order to find a good fit for the *noise component*, Ditlevsen performs a histogram analysis for the residuals of the ice core time series, the temperature increments measured between adjacent data points, i.e. years. He conjectures that the noise may contain a strong α–stable component with $\alpha \approx 1.75$, and plots an estimate for the drift term assuming the stationarity of the solution.

Stochastic differential equations have been used for quite a while as meso-scopic descriptions of models for natural phenomena. In their simplest variants, they consist of deterministic differential equations describing dynamical systems perturbed by a noise term. The subclass in which the noise source is Gaussian arises for instance from microscopic models of coupled systems of deterministic differential equations on different time scales, in the limit of infinite speed of the fast scale, and describe the fluctuations of the slow scale component around its averaged version with the fast scale component as a stochastic perturbation. With a view in particular towards the mathematical interpretation of financial time series, the theory for stochastic differential equations the noise term of which is given by more general (discontinuous and non-Gaussian) *semimartingales* has received considerable attention during the recent years.

It is reasonable in a quite general framework to model real data by stochastic differential equations. Usually neither their dynamical systems nor their noise component can be deduced from first principles for instance from microscopic models. However, they may be selected by statistical inference or *model fit* from the time series they are supposed to interpret. The central question of the corresponding *model selection problem* in Ditlevsen's papers motivating the study which led to this paper asks for the best choice of the noise term.

More formally, suppose we wish to model a real time series by the dynamics

$X = (X_t)_{t \geq 0}$ of a real valued process of the type

$$X_t = x + \int_0^t f(s, X_s)\, ds + \eta_t, \quad t \geq 0, \tag{1.1}$$

where the process component η describes the noise perturbation. Then the problem of model fit consists in the choice of a drift term f and a noise term η, so that the solution of (1.1) is in the best possible agreement with the data of the given time series.

We resume Ditlevsen's model selection problem for the fit of the noise component from the perspective of a new testing method details of which have still to be developed. Following Ditlevsen, we work under the model assumption that the noise η has an α-stable Lévy component which may be symmetric or skewed. We search for a test statistics discriminating well between different α, and capable of testing for the right one. We shall show that this job is well done by the *equidistant p-variation* — in the sequel called *p-variation* — of the process X defined for all $p > 0$ as

$$V_p^n(X)_t := \sum_{i=1}^{[nt]} |\Delta_i^n X|^p, \tag{1.2}$$

where $\Delta_i^n X := X_{\frac{i}{n}} - X_{\frac{i-1}{n}}$ for $1 \leq i \leq n$, $n \geq 1$. We first observe that for those values of p relevant for our analysis the main contribution to the p-variation of X comes from the α-stable component of the noise. We next prove local limit theorems which hold under very mild assumptions on the drift term f and allow to determine the stability index α asymptotically. We finally use these limit theorems in Section 3 below to analyze the real data from the Greenland ice core with our methods, and come to an estimate $\alpha \approx 0.7$, surprisingly quite different from the one obtained by Ditlevsen.[1,2] In fact, the diagram of distances between observed and theoretical levels α taken in the Kolmogorov–Smirnov sense, exhibits two local minima which change their shape and position as a scale parameter of the laws is modified. The deeper of the two corresponds to the estimate just quoted, while the shallower one could correspond to Ditlevsen's findings.

The paper is organized as follows. In section 2 we set the stage for stating our main results, which are applied to the Greenland ice core data in section 3. In section 4, the convergence of the finite dimensional laws of renormalized processes of p-variations is shown. By independence of increments, this immediately implies functional convergence, as can be seen by classical results. In section 5, the functional convergence of laws is extended to sums of Lévy processes and processes of finite variation, i.e. to processes typical for the structure of an SDE perturbed by Lévy noise.

In the following, '$\xrightarrow{\mathcal{D}}$' denotes convergence in the Skorokhod topology, '$\xrightarrow{d}$' denotes convergence of finite-dimensional (marginal) distributions, and '$\xrightarrow{u.c.p.}$' stands for uniform convergence on compacts in probability. We denote the indicator function of a set A by $\mathbb{I}_A$, and $\bar{A}$ denotes the complement of the set A.

2. Object of study and main results

Let $(\Omega, \mathcal{F}, (\mathcal{F}_t)_{t\geq 0}, \mathbf{P})$ be a filtered probability space. We assume that the filtration satisfies the usual hypotheses in the sense of the book,[4] i.e. it is right continuous, and $\mathcal{F}_0$ contains all the $\mathbf{P}$-null sets of $\mathcal{F}$.

For $\alpha \in (0, 2]$ let $L = (L_t)_{t\geq 0}$ be an α-stable Lévy process, i.e. a process with right continuous trajectories possessing left side limits (rcll) and stationary independent increments whose marginal laws satisfy

$$\ln \mathbf{E} e^{i\lambda L_t} = \begin{cases} -tC^\alpha |\lambda|^\alpha \left(1 - i\beta \operatorname{sgn}(\lambda) \tan \frac{\pi\alpha}{2}\right), & \alpha \neq 1, \\ -tC|\lambda|\left(1 - i\beta \frac{2}{\pi} \operatorname{sgn}(\lambda) \log |\lambda|\right), & \alpha = 1, \end{cases} \quad t \geq 0, \qquad (2.1)$$

$C > 0$ being the scale parameter and $\beta \in [-1, 1]$ the skewness. We adopt the standard notation from Ref. 5 and write $L_1 \sim S_\alpha(C, \beta, 0)$.

We also make use of the Lévy–Khinchin formula for L which takes the following form in our case (see Ref. ?Chapter XVII.3]Feller-71 for details):

$$\ln \mathbf{E} e^{i\lambda L_1} = \begin{cases} C_F \int_{\mathbb{R}\setminus\{0\}} (e^{i\lambda x} - 1 - i\lambda x \mathbb{I}_{\{|x|<1\}}) \left[\frac{1-\beta}{2}\mathbb{I}_{\{x<0\}} + \frac{1+\beta}{2}\mathbb{I}_{\{x>0\}}\right] \frac{dx}{|x|^{1+\alpha}}, & \alpha \neq 1, \\ C_F \int_{\mathbb{R}\setminus\{0\}} (e^{i\lambda x} - 1 - i\lambda \sin x) \left[\frac{1-\beta}{2}\mathbb{I}_{\{x<0\}} + \frac{1+\beta}{2}\mathbb{I}_{\{x>0\}}\right] \frac{dx}{|x|^{1+\alpha}}, & \alpha = 1, \end{cases}$$
(2.2)

where C_F denotes the scale parameter in Feller's notation and equals

$$C_F = \begin{cases} C^\alpha \left(\cos\left(\frac{\pi\alpha}{2}\right)\Gamma(\alpha)\right)^{-1}, & \alpha \neq 1, \\ C\frac{2}{\pi}, & \alpha = 1. \end{cases} \qquad (2.3)$$

Recall that a totally asymmetric process L with $\beta = 1$ ($\beta = -1$) is called spectrally positive (negative). A spectrally positive α-stable process with $\alpha \in (0, 1)$ has a.s. increasing sample paths and is called *subordinator*.

The main results of this paper are presented in the following three theorems. The first theorem deals with the asymptotic behavior of the p-variation for a stable Lévy process itself. As we will see later, this behavior does not change under perturbations by stochastic processes that satisfy some mild conditions.

Theorem 2.1. *Let $(L_t)_{t\geq 0}$ be an α-stable Lévy process with $L_1 \sim S_\alpha(C, \beta, 0)$. If $p > \alpha/2$ then*

$$\left(V_p^n(L)_t - nt B_n(\alpha, p)\right)_{t\geq 0} \xrightarrow{\mathcal{D}} (L'_t)_{t\geq 0} \quad \text{as } n \to \infty, \qquad (2.4)$$

where $L'_1 \sim S_{\alpha/p}(C', 1, 0)$ with the scale

$$C' = \begin{cases} C^p \left(\dfrac{\cos(\frac{\pi\alpha}{2p})\Gamma(1-\frac{\alpha}{p})}{\cos(\frac{\pi\alpha}{2})\Gamma(1-\alpha)} \right)^{p/\alpha}, & \alpha \neq p, \\ C^p \dfrac{\pi}{2} \dfrac{1}{\cos(\frac{\pi\alpha}{2})\Gamma(1-\alpha)}, & \alpha = p. \end{cases} \quad (2.5)$$

The normalising sequence $(B_n(\alpha, p))_{n \geq 1}$ is deterministic and given by

$$B_n(\alpha, p) = \begin{cases} n^{-p/\alpha} \mathbf{E}|L_1|^p, & p \in (\alpha/2, \alpha), \\ \mathbf{E}\sin(n^{-1}|L_1|^\alpha), & p = \alpha, \\ 0, & p > \alpha. \end{cases} \quad (2.6)$$

We remark that the skewness parameter β of L does not influence the convergence of $V_p^n(L)_t$ and does not appear in the limiting process since the p-variation depends only on the absolute values of the increments of L. Moreover, for $p > \alpha$ the limiting process L' is a subordinator.

We next perturb L by some other process Y. We impose no restrictions on dependence properties of Y and L. The only conditions on Y concern the behavior of its p-variation. We formulate two theorems, the first for $p \in (\alpha/2, 1) \cup (\alpha, \infty)$, and the second for $p \in (1, \alpha]$.

Theorem 2.2. Let $(L_t)_{t \geq 0}$ be an α-stable stochastic process, with $L_1 \sim S_\alpha(C, \beta, 0)$ and $(Y_t)_{t \geq 0}$ be another stochastic process that satisfies

$$V_p^n(Y) \xrightarrow{u.c.p.} 0, \quad n \to \infty, \quad (2.7)$$

for some $p \in (\alpha/2, 1) \cup (\alpha, \infty)$. Then

$$(V_p^n(L+Y)_t - nt\, B_n(\alpha, p))_{t \geq 0} \xrightarrow{D} (L'_t)_{t \geq 0} \quad \text{as } n \to \infty, \quad (2.8)$$

with L' and $(B_n(\alpha, p))_{n \geq 1}$ defined in (2.5) and (2.6).

The methods used to prove Theorem 2.2 do not work for $p \in (1, \alpha]$, and in this case we have to impose stronger conditions on the process Y.

Theorem 2.3. Let $(L_t)_{t \geq 0}$ be an α-stable stochastic process with $L_1 \sim S_\alpha(C, \beta, 0)$, $\alpha \in (1, 2)$ and let $(Y_t)_{t \geq 0}$ be another stochastic process. Let $p \in (1, \alpha]$ and $T > 0$. If Y is such that for every $\delta > 0$ there exists $K(\delta) > 0$ that satisfies

$$\mathbf{P}(|Y_s(\omega) - Y_t(\omega)| \leq K(\delta)|s - t| \text{ for all } s, t \in [0, T]) \geq 1 - \delta, \quad (2.9)$$

the process Y does not contribute to the limit of $V_p^n(L+Y)$, i.e.

$$(V_p^n(L+Y)_t - nt\, B_n(\alpha, p))_{0 \leq t \leq T} \xrightarrow{D} (L'_t)_{0 \leq t \leq T}, \quad n \to \infty \quad (2.10)$$

with L' and $(B_n(\alpha, p))_{n \geq 1}$ defined in (2.5) and (2.6).

To be able to study models of the type (1.1) we formulate the following corollary of Theorems 2.3 and 2.2 which takes into account that Lebesgue integral processes are absolutely continuous w.r.t. the time variable and thus qualify as small process perturbations in the sense of the Theorems.

Corollary 2.1. *Let $(L_t)_{t\geq 0}$ be an α-stable stochastic process, with $L_1 \sim S_\alpha(C,\beta,0)$, and let $f : \mathbb{R}_+ \times \mathbb{R} \to \mathbb{R}$ be a locally bounded function such that for some $x \in \mathbb{R}$ and $T > 0$ the unique solution for*

$$X_t = x + \int_0^t f(s, X_s)\, ds + L_t \tag{2.11}$$

exists on the time interval $[0,T]$. Then for $p > \max\{1, \alpha/2\}$ we have

$$(V_p^n(X)_t - nt\, B_n(\alpha, p))_{0 \leq t \leq T} \xrightarrow{D} (L'_t)_{0 \leq t \leq T} \tag{2.12}$$

as $n \to \infty$, with L' and $(B_n(\alpha, p))_{n \geq 1}$ defined in (2.5) and (2.6).

The functional convergence of power variations of symmetric stable Lévy processes to stable processes was first studied by Greenwood in Ref. 7, where more general *non-equidistant* power variations were considered, and in particular for $p > \alpha$ the convergence to subordinators was proved. Further, more general results on power variations of Lévy processes are obtained by Greenwood and Fristedt.[8] Lepingle[9] proves convergence of power variations for semimartingales in probability. In Refs. 10 and 11, Jacod proves convergence results for p-variations of general Lévy processes and semimartingales. In particular, several laws of large numbers and central limit theorems are established. Our results are different from Jacod's because we consider processes possessing no second moments so that only the generalized central limit theorem can apply. Moreover, we consider in addition convergence of perturbed processes. Corcuera, Nualart and Woerner[12] consider p-variations of a (perturbed) integrated α-stable process of the type $X = Y + \int_0^{\cdot} u_{s-}\, dL_s$ with some cadlag adapted process u. For $u = 1$, our setting results. The paper[12] contains a law of large numbers for $0 < p < \alpha$ and a functional central limit theorem for $0 < p \leq \alpha/2$. However, very restrictive conditions on possible perturbation processes Y are imposed, so that the results are not applicable to processes of the type (1.1).

3. Applications to real data

In this section we illustrate our convergence results and show how they can be applied to the estimation of the stability index α. We emphasize that the conclusions obtained are somewhat heuristic. Additional work has to be done to provide more precise statistical properties of the p-variation processes as estimators for the stability index, and to describe the decision rule of the testing procedure along with its quality features.

We first work with simulated data. Assume they are realizations of the SDE (2.11) where L is a stable process with unknown stability index α and scale $C > 0$. From the data set we extract m samples with n data points each by taking adjacent non-overlapping groups of n consecutive points. This way we get the samples $(X^{(i)})_{1 \le i \le m}$ with $X^{(i)} = (X^{(i)}_{\frac{1}{n}}, X^{(i)}_{\frac{2}{n}}, \ldots, X^{(i)}_{1})$, $1 \le i \le m$. Assume for a moment that $p = 2\alpha$ and $p > 1$. Then Corollary 2.1 can be applied with $B_n(\alpha, p) = 0$. Along each sample we calculate the p-variation

$$V_p^n(X^{(i)})_1 = \sum_{j=1}^{n} |\Delta_j^n X^{(i)}|^p, \quad 1 \le i \le m, \tag{3.1}$$

which is close in law to the stable random variable L'_1, possessing stability index $\alpha/p = 1/2$ and the scale $C' = C'(C)$ connected with the scale parameter C of L by the relation (2.5). The probability distribution function $F_{1/2,C'}$ of the limiting spectrally positive $1/2$-stable random variable L'_1 is known explicitly:

$$F_{1/2,C'}(x) = \sqrt{\frac{C'}{2\pi}} \int_0^x \frac{e^{-C'/2y}}{y^{3/2}} dy, \quad x \ge 0. \tag{3.2}$$

Convergence in law is in turn equivalent to the convergence of the correponding Kolmogorov–Smirnov distance. Thus calculating from the data the empirical distribution function of $(V_p^n(X^{(i)})_1)_{1 \le i \le n}$ given by

$$G_{p,n}(x) := \frac{1}{m} \sum_{i=1}^{m} \mathbb{I}_{(-\infty, x]}(V_p^n(X^{(i)})_1), \quad x \in \mathbb{R}, \tag{3.3}$$

the Corollary 2.1 implies that for n big enough

$$D_n(C, p) = \sup_{x \ge 0} |G_{p,n}(x) - F_{1/2,C'(C)}(x)| \approx 0. \tag{3.4}$$

This argument allows us to estimate the unknown values α and C. Indeed, calculating numerically the empirical distribution functions $G_{p,n}$ for different values $p \in [p_1, p_2]$, $1 < p_1 < p_2$, we minimize the Kolmogorov–Smirnov distance $D_n(C, p)$ over the parameter domain $p \in [p_1, p_2]$ and $C \in [C_1, C_2]$. If $D_n(C, p)$ attains its unique minimum at $C = C^*$ and $p = p^*$ we conclude that L has the stability index $\alpha^* = p^*/2$ and the scale C^*.

To test this method, we simulate $m = 200$ samples of the data from equation (2.11) with $f(\cdot, x) = \cos x$, $x \in \mathbb{R}$, and $L_1 \sim S_{0.75}(6.35, 0, 0)$, $n = 200$. We find that the Kolmogorov–Smirnov distance $D_n(C, p)$ attains a unique global minimum at $C^* \approx 6.35$ and $p^* \approx 1.5$ corresponding to the true values of α and C (see Fig. 10.1).

We next study the real ice-core data, analysed earlier by Ditlevsen.[1,2] The log-calcium signal covers the time period from approximately 90 150 to 10 150 years before present. We divide it into $m = 282$ samples, each containing $n = 282$ data points. Then the Kolmogorov–Smirnov distance is minimized numerically over p and C according to (3.4).

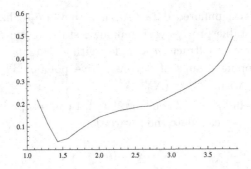

Figure 10.1. $D_n(C^*, p)$ for the simulated data, L is a 0.75-stable Lévy process, $n = m = 200$.

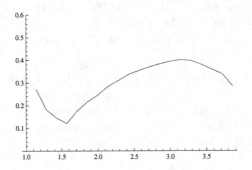

Figure 10.2. $D_n(C^*, p)$ for the ice-core data set, $C^* \approx 7.2$, $\alpha^* \approx 0.75$, $n = m = 282$.

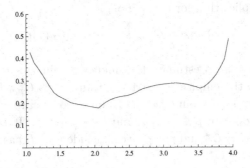

Figure 10.3. $D_n(C, p)$ for the ice-core data set, $C = 3.28$, $n = m = 282$.

It turns out that $D_n(C, p)$ for the real data also exhibits a unique global minimum in the (C, p)-domain, which yields the estimate $\alpha^* \approx 0.75$ for $C^* \approx 7.2$, $D_n(C^*, p^*) \approx 0.1$ (see Fig. 10.2). It is striking that our estimate differs from Ditlevsen's by a quantity very close to 1. This discrepancy can be explained as follows. It turns out that the function $p \mapsto D_n(C, p)$ has two local minima for some values of C different from the optimal value C^*. For example, for $C = 3.28$ there are two local minima corresponding to the stability indices $\alpha_1 \approx 1.02$ and $\alpha_2 \approx 1.75$ with the distances $D_n(C, 2\alpha_1) \approx 0.18$ and $D_n(C, 2\alpha_2) \approx 0.27$. Unfortunately the paper[2] only contains the estimated value of the stability index $\alpha \approx 1.75$ of the (symmetric) forcing L, and not its scale. It is possible that under some *a priori* assumptions on C, Ditlevsen's method provides a locally best fit which is not globally optimal (see Fig. 10.3).

4. Functional convergence for $V_p^n(L)$

To prove the convergence of the marginal distributions we use the following theorem which is a direct result of the well known generalized central limit theorem for i.i.d. random variables with infinite variance (e.g. see Theorem 3 in Feller [6, Chapter XVII.5]).

Prop 10.1. Let $(\eta_i)_{i \geq 1}$ be a sequence of non-negative i.i.d. random variables with a regularly varying tail such that

$$\mathbf{P}(\eta_1 > x) \approx C \frac{2-\alpha}{\alpha} x^{-\alpha/p} \quad \text{as } x \to +\infty. \tag{4.1}$$

for some $\alpha \in (0, 2)$, $p > \alpha/2$ and $C > 0$. Then for any $t > 0$ we have

$$\left(\frac{t}{n}\right)^{p/\alpha} \sum_{i=1}^{n} \eta_i - b_{t,n}(\alpha, p) \xrightarrow{d} t^{p/\alpha} Z, \quad \text{as } n \to \infty, \tag{4.2}$$

with

$$b_{t,n}(\alpha, p) = \begin{cases} n\left(\frac{t}{n}\right)^{p/\alpha} \mathbf{E}\eta_1, & p \in (\alpha/2, \alpha) \\ n\mathbf{E}\sin\left(\frac{t\eta_1}{n}\right), & p = \alpha, \\ 0, & p > \alpha. \end{cases} \tag{4.3}$$

where $Z \sim S_{\alpha/p}(C', 1, 0)$ with C' as defined in (2.5).

Let L be an α-stable Lévy process as defined in (2.1) and let $p > \alpha/2$. To study the finite dimensional distributions of $V_p^n(L)_t$ we note that due to the independence of increments of L it suffices to establish the convergence of marginal laws for a fixed $t > 0$. Further, the stationarity and independence of increments of L and the self-similarity property $L_t \stackrel{d}{=} t^{1/\alpha} L_1$ implies that

$$V_p^n(L)_t = \sum_{i=1}^{[nt]} |\Delta_i^n L|^p \stackrel{d}{=} \sum_{i=1}^{[nt]} \frac{|\Delta_i^1 L|^p}{n^{p/\alpha}}, \tag{4.4}$$

with $\Delta_i^1 L \stackrel{d}{=} L_1 \sim S_\alpha(C, \beta, 0)$ being i.i.d. random variables.

It is easy to see that the random variables $|\Delta_i^1 L|^p$ have a distribution function with a regularly varying tail, namely

$$\mathbf{P}(|\Delta_i^1 L|^p > x) = \mathbf{P}(|L_1| > x^{1/p}) \approx C_F \frac{2-\alpha}{\alpha} x^{-\alpha/p} \quad \text{as } x \to +\infty, \tag{4.5}$$

and thus we can apply Proposition 10.1 to the sum (4.4). Taking into account that $nt/[nt] \to 1$ as $n \to \infty$, we obtain convergence of the finite dimensional laws in Theorem 2.1.

Again as these are sums of i.i.d. random variables, we can combine the results from Theorems VII.2.35 and VII.2.29 in Ref. 13 to obtain functional convergence for the process.

5. Generalisation to sums of processes

We finally discuss the situation of Theorem 2.3, where besides the Lévy process L another process Y is given. To see that $V_p^n(L)$ and $V_p^n(L+Y)$ have equivalent asymptotic behaviour we apply Lemma VI.3.31 in Ref. 13. Under the conditions of Theorems 2.2 and 2.3 it is enough to show that $V_p^n(L+Y) - V_p^n(L) \xrightarrow{\text{u.c.p.}} 0$ as $n \to \infty$.

5.1. *Equivalence for $p \leq 1$*

Let L be an α-stable Lévy process, $p \in (\alpha/2, 1]$ and let Y be such that for $V_p^n(Y) \xrightarrow{\text{u.c.p.}} 0$, as $n \to \infty$. Note that due to the monotonicity properties of $V_p^n(Y)$, the latter convergence condition is equivalent to $V_p^n(Y)_N \xrightarrow{\mathbf{P}} 0$ for any $N \geq 1$. Then a simple application of the triangle inequality yields the proof. In fact, for any $N \geq 1$ we have

$$\sup_{0 \leq t \leq N} |V_p^n(L+Y)_t - V_p^n(L)_t| \leq \sum_{i=1}^{nN} \left||\Delta_i^n(L+Y)|^p - |\Delta_i^n L|^p\right|$$

$$\leq \sum_{i=1}^{nN} |\Delta_i^n(L+Y) - \Delta_i^n L|^p = V_p^n(Y)_N \xrightarrow{\mathbf{P}} 0, \quad n \to \infty. \tag{5.1}$$

5.2. *Equivalence for $p > \alpha$*

Assume again that $V_p^n(Y)_t \xrightarrow{\mathbf{P}} 0$, $n \to \infty$, for any $t \geq 0$. Denote $m := [p]$, so that $p = m + q$ with $q \in [0,1)$. Then for any $N \geq 1$ we have

$$\sup_{t\le N}\left|V_p^n(L+Y)_t - V_p(L)_t\right| = \sup_{t\le N}\left|\sum_{i=1}^{[nt]}|\Delta_i^n(L+Y)|^{m+q} - |\Delta_n L_i|^{m+q}\right|$$

$$= \sup_{t\le N}\left|\sum_{i=1}^{[nt]}|\Delta_i^n(L+Y)|^q \sum_{k=0}^{m}\binom{m}{k}|\Delta_n L_i|^k|\Delta_i^n Y|^{m-k} - |\Delta_i^n L|^{m+q}\right|$$

$$\le \sum_{k=0}^{m-1}\binom{m}{k}\sum_{i=1}^{nN}\left||\Delta_i^n(L+Y)|^q|\Delta_i^n L|^k|\Delta_i^n Y|^{m-k}\right|$$

$$+ \sum_{i=1}^{nN}\left||\Delta_i^n(L+Y)|^q|\Delta_i^n L|^m - |\Delta_i^n L|^{m+q}\right| \qquad (5.2)$$

$$\le \sum_{k=0}^{m-1}\binom{m}{k}\sum_{i=1}^{nN}|\Delta_i^n L|^{k+q}|\Delta_i^n Y|^{m-k} + \sum_{k=0}^{m-1}\binom{m}{k}\sum_{i=1}^{nN}|\Delta_i^n L|^k|\Delta_i^n Y|^{m-k+q}$$

$$+ \sum_{i=1}^{nN}|\Delta_i^n L|^m\left||\Delta_i^n(L+Y)|^q - |\Delta_i^n L|^q\right|$$

$$\le \sum_{k=0}^{m-1}\binom{m}{k}\sum_{i=1}^{nN}|\Delta_i^n L|^{k+q}|\Delta_i^n Y|^{m-k} + \sum_{k=0}^{m-1}\binom{m}{k}\sum_{i=1}^{nN}|\Delta_i^n L|^k|\Delta_i^n Y|^{m-k+q}$$

$$+ \mathbb{I}_{\{q>0\}}\sum_{i=1}^{nN}|\Delta_i^n L|^m|\Delta_i^n Y|^q.$$

The right-hand side of the latter inequality is essentially a sum of $2m+1$ terms of the type $\sum_{i=1}^{nN}|\Delta_i^n L|^k|\Delta_i^n Y|^{p-k}$, where $k\in [0,p]\cap \mathbb{N}$.

Applying Hölder's inequality we get

$$\sum_{i=1}^{[nt]}|\Delta_n L_i|^k|\Delta_n Y_i|^{p-k} \le \left(\sum_{i=1}^{[nt]}|\Delta_n L_i|^p\right)^{k/p}\left(\sum_{i=1}^{[nt]}|\Delta_n Y_i|^p\right)^{(p-k)/p}. \qquad (5.3)$$

The first factor in the latter formula converges in probability to a finite limit, since $p > \alpha$. The second factor converges to 0 in probability due to the assumption on Y, and Theorem 2.2 is proven.

5.3. Equivalence for $\alpha \in (1,2)$, $p \in (1,\alpha]$

The main technical difficulty of this case arises from the fact that the p-variation of L for $p < \alpha$ does not exist. In particular, the events when increments of the stable process L become very large have to be considered carefully. For $T > 0$ and some $c > 0$ which will be specified later define the following sets:

$$\begin{aligned}J_c^n(\omega) &:= \{i\in [0,[nT]]:\ |\Delta_i^n L(\omega)| > c\},\\ A_c^n(j) &:= \{\omega \in \Omega:\ |J_c^n(\omega)| = j\},\quad j = 0,\ldots,[nT].\end{aligned} \qquad (5.4)$$

The set J_c^n contains the time instants (in the scale $\frac{1}{n}$), where the increments of the process L are 'large', i.e. exceed c. The set $A_c^n(j)$ describes the event that the number of large increments equals j.

Let $\delta > 0$, $\varepsilon > 0$. According to the conditions of Theorem (2.3), let $K' := K(\frac{\delta}{2}) > 0$, and set

$$B = \{\omega : |Y_s(\omega) - Y_t(\omega)| \leq K'|s-t| \text{ for all } s,t \in [0,T]\}, \qquad (5.5)$$

so that $\mathbf{P}(B) \geq 1 - \frac{\delta}{2}$.

We estimate

$$\mathbf{P}\Big(\sup_{0 \leq t \leq T} |V_p^n(L+Y)_t - V_p^n(L)_t| > \varepsilon\Big)$$

$$\leq \mathbf{P}\Big(\Big\{\sum_{i=1}^{[nT]} \big||\Delta_i^n(L+Y)|^p - |\Delta_i^n L|^p\big| > \varepsilon\Big\} \cap B\Big) + \mathbf{P}(\bar{B})$$

$$\leq \sum_{j=0}^{[nT]} \mathbf{P}\Big(\Big\{\sum_{i=1}^{[nT]} \big||\Delta_i^n(L+Y)|^p - |\Delta_i^n L|^p\big| > \varepsilon\Big\} \cap A_c^n(j) \cap B\Big) + \frac{\delta}{2} \qquad (5.6)$$

$$\leq \sum_{j=0}^{[nT]} \mathbf{P}\Big(\Big\{\sum_{i \notin J_c^n} \big||\Delta_i^n(L+Y)|^p - |\Delta_i^n L|^p\big| > \frac{\varepsilon}{2}\Big\} \cap A_c^n(j) \cap B\Big)$$

$$+ \sum_{j=0}^{[nT]} \mathbf{P}\Big(\Big\{\sum_{i \in J_c^n} \big||\Delta_i^n(L+Y)|^p - |\Delta_i^n L|^p\big| > \frac{\varepsilon}{2}\Big\} \cap A_c^n(j) \cap B\Big) + \frac{\delta}{2}$$

$$=: D^{(1)}(n,c) + D^{(2)}(n,c) + \frac{\delta}{2}.$$

In the following two steps we show that for appropriately chosen $c = c(\varepsilon) > 0$ and n big enough, $D^{(1)}(n,c) = 0$ and $D^{(2)}(n,c) < \delta/2$. This will finish the proof.

Step 1. To estimate $D^{(1)}(n,c)$, let $\omega \in B$. Using the elementary inequality $|a^p - b^p| \leq \max\{a,b\}^{p-1} p|a-b|$ which holds for $a,b \geq 0$ and $p \geq 1$ we estimate

$$\sum_{i \notin J_c^n(\omega)} \big||\Delta_i^n(L(\omega) + Y(\omega))|^p - |\Delta_i^n L(\omega)|^p\big|$$

$$\leq \sum_{i \notin J_c^n(\omega)} p\Big(c + \frac{K'}{n}\Big)^{p-1} \big||\Delta_i^n(L(\omega) + Y(\omega))| - |\Delta_i^n L(\omega)|\big|$$

$$\leq \sum_{i \notin J_c^n(\omega)} p\Big(c + \frac{K'}{n}\Big)^{p-1} |\Delta_i^n Y| \qquad (5.7)$$

$$\leq \sum_{i \notin J_c^n(\omega)} p\Big(c + \frac{K'}{n}\Big)^{p-1} \frac{K'}{n}$$

$$\leq p\Big(c + \frac{K'}{n}\Big)^{p-1} K'T,$$

where we have used that the 'small' increments of L are bounded by c. So we have

$$D^{(1)}(n,c) = \sum_{j=0}^{[nT]} \mathbf{P}\left(\left\{\left|\sum_{i \notin J_c^n} |\Delta_i^n L + \Delta_i^n Y|^p - |\Delta_i^n L|^p\right| > \frac{\varepsilon}{2}\right\} \cap A_c^n(j) \cap B\right)$$

$$\leq \sum_{j=0}^{[nT]} \mathbf{P}\left(\left\{p\left(c + \frac{K'}{n}\right)^{p-1} K'T > \frac{\varepsilon}{2}\right\} \cap A_c^n(j) \cap B\right) \qquad (5.8)$$

$$\leq \mathbf{P}\left(p\left(c + \frac{K'}{n}\right)^{p-1} K'T > \frac{\varepsilon}{2}\right) = 0$$

for all $n \geq n_1$ with

$$n_1 := \left[2K'\left(\frac{2pK'T}{\varepsilon}\right)^{\frac{1}{p-1}}\right] \quad \text{and} \quad c = c(\varepsilon) := \frac{1}{4}\left(\frac{\varepsilon}{2pK'T}\right)^{\frac{1}{p-1}}. \qquad (5.9)$$

Step 2. To estimate $D^{(2)}(n,c)$, let $n \geq \frac{K'}{c}$ so that for $\omega \in B$ and $i \in J_c^n(\omega)$ we have

$$\left|\frac{\Delta_i^n Y(\omega)}{\Delta_i^n L(\omega)}\right| \leq \frac{K'}{nc} \leq 1. \qquad (5.10)$$

By means of the elementary inequality $(1+|x|)^p - 1 \leq 3|x|$ which holds for $1 \leq p \leq 2$ and $|x| \leq 1$ this implies that

$$\sum_{i \in J_c^n(\omega)} \left||\Delta_i^n(L(\omega) + Y(\omega))|^p - |\Delta_i^n L(\omega)|^p\right|$$

$$\leq \sum_{i \in J_c^n(\omega)} |\Delta_i^n L(\omega)|^p \left[\left(1 + \left|\frac{\Delta_i^n Y(\omega)}{\Delta_i^n L(\omega)}\right|\right)^p - 1\right]$$

$$\leq \sum_{i \in J_c^n(\omega)} |\Delta_i^n L(\omega)|^p \, 3 \left|\frac{\Delta_i^n Y(\omega)}{\Delta_i^n L(\omega)}\right| \qquad (5.11)$$

$$\leq \sum_{i \in J_c^n(\omega)} |\Delta_i^n L(\omega)|^p \frac{3K'}{nc}.$$

This in turn immediately yields

$$D^{(2)}(n,c) \leq \sum_{j=1}^{[nT]} \mathbf{P}\left(A_c^n(j) \cap \left\{\sum_{i \in J_c^n} |\Delta_i^n L|^p \frac{3K'}{nc} > \frac{\varepsilon}{2}\right\}\right)$$

$$\leq \sum_{j=1}^{[nT]} \mathbf{P}\left(A_c^n(j) \cap \bigcup_{i \in J_c^n} \left\{|\Delta_i^n L|^p > \frac{\varepsilon n c}{6K'j}\right\}\right). \qquad (5.12)$$

Since all $\Delta_i^n L$, $i = 1, \ldots, [nT]$, are i.i.d., and only j of them exceed the threshold c, we can estimate the probability for this event precisely. Indeed, denoting

$$p_n := \mathbf{P}(|\Delta_1^n L| > c) = \mathbf{P}(|L_1| > cn^{1/\alpha}), \qquad (5.13)$$

we continue the estimates in (5.12) to get

$$D^{(2)}(n,c) \leq \sum_{j=1}^{[nT]} \binom{[nT]}{j} \mathbf{P}\Big(\{|\Delta_i^n L| > c, i=1,\ldots,j\}$$

$$\cap \{|\Delta_i^n L| \leq c, i=j+1,\ldots,[nT]\} \cap \bigcup_{i=1}^{j}\{|\Delta_i^n L|^p > \frac{\varepsilon nc}{6jK'}\}\Big)$$

$$= \sum_{j=1}^{[nT]} \binom{[nT]}{j} j \mathbf{P}\Big(\{|\Delta_i^n L| > c, i=1,\ldots,j\} \quad (5.14)$$

$$\cap \{|\Delta_i^n L| \leq c, i=j+1,\ldots,[nT]\} \cap \{|\Delta_1^n L|^p > \frac{\varepsilon nc}{6jK'}\}\Big)$$

$$\leq \sum_{j=1}^{[nT]} \binom{[nT]}{j} j p_n^{j-1}(1-p_n)^{[nT]-j} \mathbf{P}\Big(|\Delta_1^n L|^p > \frac{\varepsilon nc}{6jK'}\Big).$$

From (4.5) we know that there is a $\tilde{c}$ such that

$$p_n \leq \frac{\tilde{c}}{c^\alpha n} \quad (5.15)$$

and

$$\mathbf{P}\Big(|\Delta_1^n L|^p > \frac{\varepsilon nc}{6jK'}\Big) = \mathbf{P}\Big(|L_1| > \Big(\frac{\varepsilon nc}{6K'j}\Big)^{1/p} n^{1/\alpha}\Big) \leq \frac{\tilde{c}}{n}\Big(\frac{\varepsilon nc}{6K'j}\Big)^{-\alpha/p} \quad (5.16)$$

holding for some $\tilde{c} > 0$, for all $n \geq n_2 \geq \frac{K'}{c}$ and $1 \leq j \leq [nT]$. Combining (5.14) and (5.16), denoting the constant pre-factor by C, and recalling that $\frac{\alpha}{p} \leq 2$, we obtain for $n \geq n_2$ that

$$D^{(2)}(n,c) \leq \frac{C}{p_n n^{1+\alpha/p}} \sum_{j=1}^{[nT]} \binom{[nT]}{j} j^{1+\alpha/p} p_n^j (1-p_n)^{[nT]-j}$$

$$\leq \frac{C}{p_n n^{1+\alpha/p}} \sum_{j=1}^{[nT]} \binom{[nT]}{j} j^3 p_n^j (1-p_n)^{[nT]-j}. \quad (5.17)$$

The sum in the previous formula represents the third moment of a binomial distribution, and thus can be calculated explicitly. By means of the asymptotic behaviour (4.5) we get

$$D^{(2)}(n,c) \leq C \frac{(nT-2)(nT-1)nTp_n^3 + 3(nT-1)nTp_n^2 + nTp_n}{p_n n^{1+\alpha/p}}$$

$$\leq \frac{CT}{n^{\alpha/p}}((nTp_n)^2 + 3nTp_n + 1). \quad (5.18)$$

Now choose $n \geq n_3 \geq n_2$ big enough to ensure that this expression is smaller than $\frac{\delta}{2}$. This completes the proof of Theorem 2.3.

Acknowledgement

P.I. and I.P. thank DFG SFB 555 *Complex Nonlinear Processes* for financial support. C.H. thanks DFG IRTG *Stochastic Models of Complex Processes* for financial support. C.H. is grateful to R. Schilling for his valuable comments. The authors thank P. Ditlevsen for providing the ice-core data.

References

1. P. D. Ditlevsen, Anomalous jumping in a double-well potential, *Physical Review E.* **60**(1), 172–179, (1999).
2. P. D. Ditlevsen, Observation of α-stable noise induced millenial climate changes from an ice record, *Geophysical Research Letters.* **26**(10), 1441–1444 (May, 1999).
3. W. Dansgaard, S. J. Johnsen, H. B. Clausen, D. Dahl-Jensen, N. S. Gundestrup, C. V. Hammer, C. S. Hvidberg, J. P. Stefensen, A. E. Sveinbjornsdottir, J. Jouzel, and G. Bond, Evidence for general instability of past climate from 250 kyr ice-core record, *Nature.* **364**, 218–220, (1993).
4. P. E. Protter, *Stochastic integration and differential equations.* vol. 21, Applications of Mathematics, (Springer, 2004), second edition.
5. G. Samorodnitsky and M. S. Taqqu, *Stable non-Gaussian random processes.* (Chapman&Hall/CRC, 1994).
6. W. Feller, *An introduction to probability theory and its applications: volume II.* (John Wiley & Sons, 1971).
7. P. E. Greenwood, The variation of a stable path is stable, *Zeitschrift für Wahrscheinlichkeitstheorie und verwandte Gebiete.* **14**(2), 140–148, (1969).
8. P. E. Greenwood and B. Fristedt, Variations of processes with stationary, independent increments, *Zeitschrift für Wahrscheinlichkeitstheorie und verwandte Gebiete.* **23**(3), 171–186, (1972).
9. D. Lepingle, La variation d'ordre p des semi-martingales, *Zeitschrift für Wahrscheinlichkeitstheorie und verwandte Gebiete.* **36**, 295–316, (1976).
10. J. Jacod, Asymptotic properties of power variations of Lévy processes, *ESAIM: Probability and Statistics.* **11**, 173–196, (2007).
11. J. Jacod, Asymptotic properties of realized power variations and related functionals of semimartingales., *Stochastic Processes and their Applications.* **118**(4), 517–559, (2008).
12. J. M. Corcuera, D. Nualart, and J. H. C. Woerner, A functional central limit theorem for the realized power variation of integrated stable processes, *Stochastic Analysis and Applications.* **25**, 169–186, (2007).
13. J. Jacod and A. N. Shiryaev, *Limit theorems for stochastic processes.* vol. 288, Grundlehren der Mathematischen Wissenschaften, (Springer, 2003), second edition.

Chapter 11

Class II Semi-Subgroups of the Infinite Dimensional Rotation Group and Associated Lie Algebra

Takeyuki Hida and Si Si

Professor Emeritus, Nagoya University and Meijo University Nagoya, Japan.

Faculty of Information Science and Technology Aichi Prefectural University, Aichi Prefecture, Japan.

The infinite dimensional rotation group plays important roles in white noise analysis. In particular, the class II subgroup describes significant probabilistic properties. We shall find a local Lie group involving some half whiskers in the class II and discuss its Lie algebra.

Keywords: white noise, infinite dimensional rotation group, half whisker

2000 AMS Subject Classification: 60H40

Contents

1 Introduction 177
2 Class II Subgroups of $O(E)$ 178
3 Half whiskers 180
4 Lie algebra 181
5 Concluding remarks 182
References 182

1. Introduction

There are two main characteristics in white noise analysis.

(1). Generalized white noise functionals.

It has reasonably been developed as a typical theory of infinite dimensional calculus. There should be reminded an important view point for the discussion of finite dimensional approximation, that is, in terms of Volterra-Lévy *le passage du fini à l'infini*.

We emphasize the theory dealing with essentially infinite dimensional calculus, where we have a base involving continuously many (independent) vectors. We, therefore, come naturally to a space of generalized white noise functionals.

(2). Infinite dimensional rotation group $O(E)$.

Following H. Yoshizawa we start with the rotation group $O(E)$ of E ($\subset L^2(R)$). The group consists of homeomorphisms of E which are orthogonal in $L^2(R)$. It is topologized by the compact-open topology.

The collection $O^*(E^*)$ of the adjoint operators g^* of $g \in O(E)$ forms a group which is isomorphic to $O(E)$. The significance of the group $O^*(E^*)$ is that every g^* in $O^*(E^*)$ keeps the white noise measure μ to be invariant:

$$g^*\mu = \mu.$$

By using the group $O(E)$, isomorphic to $O^*(E^*)$, we can carry on, as it were, an infinite dimensional harmonic analysis, which should be a part of the white noise analysis. We now have a brief observation of the group $O(E)$ as a preliminary of the present report. One might think that $O(E)$ is a limit of the finite dimensional rotation groups $SO(n)$ as $n \to \infty$, but not quite. The limit can occupy a very minor part of $O(E)$: of course it is almost impossible to measure the size of the limit occupied in the entire group $O(E)$.

Our idea of investigating the group is as follows:

i) Since $O(E)$ is very big (neither compact nor locally compact), we take subgroups that can be managed. First the entire group is divided into two parts: Class I and Class II. The Class I involves members that can be determined by using a base, say $\{\xi_n\}$ of E and any member of the class II comes from a diffeomorphism of the parameter space R^1.

ii) Each class has subgroups. We are now interested in new subsemigroups in the class II that are isomorphic to local Lie groups and have certain probabilistic meanings.

The present report shall discuss particularly subgroups belonging to the class II. The class I has been discussed extensively, while we know only few of subgroups belonging to the class II. Of course, known subgroups are quite important, however we are sure that there should be some more (certainly quite many) significant subgroups that are important for our white noise analysis; actually we have been in search of finding such subgroups. Now we shall show some examples in the class II.

2. Class II Subgroups of $O(E)$

This section is devoted to a brief review of the known results and find some hints to find new good subgroups.

Each member of this class, say $\{g_t, t : \text{real}\}$, should be defined by a system of parameterized diffeomorphisms $\{\psi_t(u)\}$ of $\bar{R} = R \cup \infty$: one point compactification. Namely,

$$\xi(u) \mapsto (g\xi(\psi_t(u)))\sqrt{|\psi_t'(u)|}, \tag{2.1}$$

where $\psi'_t(u)$ is the derivative of $\psi_t(u)$ in the variable u.

We are interested in a subgroup which can be made to be a local Lie group embedded in $O(E)$. In what follows the basic nuclear space is specified to D_o (see [3]).

More practically, we restrict our attention to the case where $g_t, t \in R$ forms a one-parameter group such that g_t is continuous in t. Assume further that there exists the (infinitesimal) generator $A = \frac{d}{dt} g_t|_{t=0}$ of this group g_t.

There exits, by the assumptions of the group property and continuity, a family $\{\psi_t(u), t \in R\}$ such that $\psi_t(u)$ is measurable in (t, u) and satisfies

$$\psi_t \cdot \psi_t = \psi_{t+s}$$
$$\psi_0(u) = u.$$

Such a one-parameter group is called *whisker*.

Let g_t^* be the adjoint operator to g_t. Then, the system $\{g_t^*\}$ again forms a one-parameter group of μ (the white noise measure) preserving transformations g_t^*. The system is a flow on the white noise space $E^*(\mu)$.

By J. Aezél [1], we have an expression for $\psi_t(u)$:

$$\psi_t(u) = f(f^{-1}(u) + t) \tag{2.2}$$

where f is continuous and strictly monotone. Its (infinitesimal) generator α, if f is differentiable, can now be expressed in the form

$$\alpha = a(u)\frac{d}{du} + \frac{1}{2}a'(u), \tag{2.3}$$

where

$$a(u) = f'(f^{-1}(u)). \tag{2.4}$$

See e.g. [4], [5].

We have already established the results that there exists a three dimensional subgroup of class II with significant probabilistic meanings. The group consists of three one-parameter subgroups, the generators of which are expressed by $a(u) = 1, a(u) = u, a(u) = u^2$, respectively. Namely, we show a list:

$$s = \frac{d}{du},$$
$$\tau = u\frac{d}{du} + \frac{1}{2},$$
$$\kappa = u^2\frac{d}{du} + u.$$

One of the interesting interpretations may be said that they put together describe the projective invariance of Brownian motion.

Those generators form a base of a three dimensional Lie algebra under the Lie product $[\alpha, \beta] = \alpha\beta - \beta\alpha$, satisfying the Jacobi identity.

The algebra given above is isomorphic to $sl(2, R)$. This fact can easily be seen by the commutation relations:

$$[\tau, s] = -s, \quad [\tau, \kappa] = \kappa, \quad [\kappa, s] = 2\tau.$$

There is a remark that the shift with generator s is sitting as a key member of the generators. It corresponds to the *flow of Brownian motion*, significance of which is quite clear.

Also, one can take τ to be another key generator. The τ describes the Ornstein-Uhlenbeck Brownian motion which is Gaussian and simple Markov.

We are now in search of *new* whiskers that show some significant probabilistic properties as above three whiskers under somewhat general setup. There a whisker may be changed to a half-whisker.

3. Half whiskers

First we recall the notes [11] p. 60, section O_∞ 1, where a new whisker with generator

$$\alpha^p = u^p \frac{d}{du} + \frac{p}{2} u^{p-1} \tag{3.1}$$

is suggested to be investigated, where p is not necessarily be integer. (The power p was written as α in [11], but to avoid notational confusion, we write p instead of α.)

Since fractional power p is involved, we tacitly assume that u is positive, We, therefore, take a white noise with time-parameter $[0, \infty)$. The basic nuclear space E is chosen to be D_{00} which is isomorphic to D_0, eventually isomorphic to $C^\infty(S^1)$.

We are now ready to state an answer.

As was remarked in the last section, the power $p = 1$ is the key number and, in fact, it is exceptional. In this case the variable u can run through R, that is, corresponds to a whisker with generator τ. In what follows we escape from the case $p = 1$.

We remind the relationship between f and $a(u)$ that appear in the expressions of $\psi_t(u)$ and α, respectively. The related formulas are the same as the case where u runs through R.

Assuming differentiability of f we have the formula (2.4). For $a(u) = u^p$, the corresponding $f(u)$ is determined. Namely,

$$u^p = f'(f^{-1}(u)).$$

An additional requirement for f is that the domain of f should be the entire $[0.\infty)$. Hence, we have

$$f(u) = c_p u^{\frac{1}{1-p}}, \tag{3.2}$$

where $c_p = (1-p)^{1/(1-p)}$. Summing up, we define $\psi_t(u)$ for $u > 0$ by (2.2) such that

$$\psi_t(u) = f^{-1}(f(u) + t)$$

with the special choice of f given by (3.2).

We, therefore, have

$$f^{-1}(u) = (1-p)^{-1} u^{1-p}. \tag{3.3}$$

We are ready to define a transformation g_t^p acting on D_{00} by

$$(g_t \xi)(u) = \xi(c_p(\frac{u^{1-p}}{1-p} + t)^{1/(1-p)}) \sqrt{\frac{c_p}{1-p}(\frac{u^{1-p}}{1-p} + t)^{p/(1-p)} u^{-p}}. \tag{3.4}$$

Note that f is always positive and maps $(0, \infty)$ onto itself, in the ordinary order in the case $p < 1$, and in the reciprocal order in the case $p > 1$.

The exceptional case $p = 1$ is refered to the literature [5]. It has been well defined.

Then, we claim, still assuming $p \neq 1$, the following theorem.

Theorem 3.1.

i) g_t^p is a member of $O(D_{00})$ for every $t > 0$.

ii) The collection $\{g_t^p, t \geq 0\}$ forms a continuous semi-group with the product $g_t^p \cdot g_s^p = g_{t+s}^p$ for $t, s \geq 0$.

iii) The generator α^p of g_t^p is α^p given by (3.1).

Proof. Assertion i) comes from the structure of D_{00}. Assertions ii) and iii) can be proved by the actual computations. □

Definition. A continuous semi-group $g_t, t \geq 0$, each member of which comes from $\psi_t(u)$ is called a *half whisker*.

Theorem 3.2. *The collection of half whiskers $g_t^p, t \geq 0, p \in R$, generates a local Lie semi-group G_L:*

$$G_L = \text{generated by } \{g_{t_1}^{p_1} \cdots g_{t_n}^{p_n}\}$$

The definition of a local Lie group is found, e.g. in W. Miller, Jr. [10]. A semi-group is defined similarly.

4. Lie algebra

The collection $\{\alpha^p; p \in R\}$ generates a vector space $\mathbf{g}_L$. There is introduced the Lie product $[\cdot, \cdot]$. Note that the exceptional power $p = 1$ is now included. The whiskers introduced in Section 1.2 are considered as half whiskers by letting the parameter t run through $[0, \infty)$. In this sense, we identify in such a manner that

$$\alpha^0 = s, \quad \alpha^1 = \tau, \quad \alpha^2 = \kappa.$$

With these understanding, we have

Theorem 4.1. *The space* $\mathbf{g}_L$ *is a Lie algebra parameterized by* $p \in R$. *It is associated with the local Lie semi-group* G_L.

Proof. We have

$$[\alpha^p, \alpha^q] = (q-p)u^{p+q-1}\frac{d}{du} + \frac{1}{2}(q-p)u^{(p+q-2)}. \tag{4.1}$$

The result is $a(u) = (q-p)u^{p+q-1}$. This proves the theorem. □

In fact, we have an infinite dimensional Lie algebra, the base of which consists of one-parameter generators of half whiskers.

5. Concluding remarks

We shall propose a more general theory, where it is possible to propose many kinds of half whiskers. Namely, we may consider general infinitesimal generators, where the functions $a(u)$ in (2.3) or f in (2.2) are restricted so as to define subgroups of $O(D_{00})$.

For every p, the $(g_t^p)^*$ is a semigroup of μ-measure preserving transformations. We may, therefore, define a Gaussian process $X^p(t)$ in such a manner that

$$X^p(t) = \langle (g_t^p)^* x, \xi \rangle,$$

where $x \in E^*(\mu)$.

We have much freedom to choose ξ, in fact, we may choose the indicator function $\chi_{[0,1]}(u)$.

Acknowledgement

We are grateful to Professor I. Volovich who told us to remind Virasoro type Lie algebras in connection with the group $O(E)$. We are encouraged to think of the problem raised in [11] and the whiskers discussed in [4] and [5].

References

1. J. Aczél, Vorlesungen über Funktionalgleichungen und ihre Anwendungen, Birkhüser, 1960.
2. I.M. Gel'fand. Russian Math. Survey 29 (1974) 3-16.
3. I.M. Gel'fand, M.I. Graev and N.Ya. Vilenkin, Generalized functions. vol. 5, 1962 (Russian original), Academic Press, 1966.
4. T. Hida, Stationary stochastic processes, Princeton Univ. Press, 1970.

5. T. Hida, Brownian motion. Iwanami Pub. Co. 1975, in Japanese; english transl. Springer-Verlag, 1980.
6. T. Hida and Si Si, Lectures on white noise functionals. World Sci. Pub. Co. 2008.
7. A.A. Kirillov, Kähler structures on K-orbits of the group of diffepmorphisms of a circle. Functional Analysis and its Applications. vol.21, no.2. 42-45 (1986; english. 1987).
8. A.A. Kirillov and D.V. Yur'ev, Kähler geometry of the infinite dimensional homogeneous space $M = Diff_+(S^1)/Rot(S^1)$. Functional analysis and its Applications vol.21, no.4, (1987; english. 284-294).
9. P.A. Meyer et J.A. Yan, A propos des distributions. LNM 1247 (1987), 8-26.
10. W.Miller, Jr. Lie theory and special functions, Academic Press. 1968.
11. T. Shimizu and Si Si, Professor Takeyuki Hida's mathematical notes. informal publication 2004.

Chapter 12

Stopping Weyl Processes

R. L. Hudson

Mathematics Department, Loughborough University, Loughborough, Leicestershire
LE11 3TU, Great Britain
R.Hudson@lboro.ac.uk

It is shown that, in order that the multiplicativity rule $W_a^b W_b^c = W_a^c$, which essentially characterizes Weyl processes, continue to hold when the sure times $a < b < c$ are replaced by stop times, it is sufficient to use left stopping at the lower limit a and right stopping at the upper limit b in W_a^b.

Contents

1 Introduction 185
2 Characterization of unitary product integrals 187
3 Stop times 190
4 Factorizing product integrals at stop times 191
References 192

1. Introduction

The quantum notion of stop time was introduced in Ref. 3, where it was called Markov time, in the context of both Fock and non-Fock extremal universally invariant quantum stochastic calculus. It has been developed in the more general context of a filtered von Neumann algebra by a number of authors beginning with Barnett and Lyons[2] (see Ref. 8 for more recent references) and in the context of Fock quantum stochastic calculus by Parthasarathy and Sinha.[9] In all these works except Ref. 3 a distinction is made, when stopping a quantum process at a stop time, between *right* and *left* stopping. The stop time is in effect a projection-valued measure and a process is an operator-valued function of time. To stop the process one evaluates it at the stop time as a spectral integral against the projection valued measure as integrator, with the operator-valued function as integrand. If the latter does not commute with the former the integrator may be placed on either side of the integrand resulting in two different values for the integral. This problem is avoided in Ref. 3 by formulating the strong Markov property in such a way that the relevant integrand commutes with the integrator; see also Ref. 4 where this aspect is emphasised.

The Weyl operators $W(f), f \in L^2(\mathbb{R}_+)$, forming a representation of the canonical commutation relations,

$$W(f)W(g) = \exp\left(\frac{1}{2}i\mathrm{Im}\,\langle f,g\rangle\right) W(f+g), \quad f,g \in L^2(\mathbb{R}_+),$$

have from the beginning[7] played a motivating role in both Fock and non-Fock quantum stochastic calculus. Each Weyl operator $W(f)$ generates a *Weyl process* $(W(f)_t)_{t\in\mathbb{R}_+}$ with $W(f)_t \stackrel{\cdot}{=} W(f\chi_{[0,t]})$, where $\chi_{[0,t]}$ is the indicator function of $[0,t]$, which is the unique solution of the quantum stochastic differential equation

$$dW(f) = W(f)(fdA^\dagger - \bar{f}dA - \frac{\sigma^2}{2}|f|^2\,dT), \tag{1.1}$$

with initial condition $W(f)_0 = 1$, where the variance parameter $\sigma = 1$ in the Fock case and $\sigma > 1$ in the non-Fock case. It follows that if, for $a < b$ we define $W_a^b(f) = W(f\chi_{]a,b]})$, then $(W_a^b(f))_{0 \le a \le b}$ satisfies the evolution equation

$$W_a^b W_b^c = W_a^c \text{ whenever } a < b < c. \tag{1.2}$$

A more suggestive notation for $W_a^b(f)$ is as a product integral,

$$W_a^b(f) = \prod_a^b (1 + fdA^\dagger - \bar{f}dA - \frac{\sigma^2}{2}|f|^2\,dT)$$

meaning that $W_a^b(f)$ is the solution at $b > a$ of the differential equation (1.1) with initial condition $W(f)_a = 1$. Note that $W_a^b(f)$ is strongly continuous in $a, b \in \mathbb{R}_+$; this follows from the strong continuity of the Weyl operators $W(f)$ in $f \in L^2(\mathbb{R}_+)$. It is well defined when f is only locally square integrable; $f \in L^2_{\mathrm{loc}}(\mathbb{R}_+)$. More generally given such f and also a real-valued locally integrable function h on $\mathbb{R}_+$, then the family of unitary operators

$$W_a^b(f;h) = \prod_a^b \left(1 + fdA^\dagger - \bar{f}dA + (ih - \frac{\sigma^2}{2}|f|^2)dT\right)$$
$$= W_a^b(f)\gamma_a^b \tag{1.3}$$

where $\gamma_a^b = \prod_a^b (1 + ihdT) = \exp\left(i\int_a^b h(x)dx\right)$ also satisfies (1.2).

Conversely let there be given a strongly continuous family of unitary operators $(W_a^b)_{a<b\in\mathbb{R}_+}$ satisfying the evolution equation $W_a^b W_b^c = W_a^c$ whenever $a < b < c$ and such that each W_a^b belongs to the von Neumann algebra $\mathcal{N}_a^b$ generated by the Weyl operators $W(f)$.with $f \in L^2(\mathbb{R}_+)$ supported by the interval $]a,b]$. Then in the non-Fock case $\sigma > 1$ there exists a (locally) square integrable complex-valued function f on $\mathbb{R}_+$ and a locally integrable real-valued function h on $\mathbb{R}_+$ such that each $W_a^b = W_a^b(f;h)$. This is not true in the Fock case, when product integrals involving the conservation process can also arise. Although it is surely well known to be true in the non-Fock case, I do not know of a proof so one is given in Section 2. For a related result proved by similar methods see Ref. 6.

The main purpose of this paper is to obtain a generalisation of the evolution property (1.2) which will hold when the real numbers a, b and c are replaced by stop times. We shall see that (1.2) will hold for stop times a, b and c satisfying $a < b < c$, where $a < b$ means that every point in the spectrum of a does not exceed every point of the spectrum of b, if and only if W_a^b is defined for stop times $a < b$ in this sense by using left stopping at a and right stopping at b.

2. Characterization of unitary product integrals

From now on we take the quantum stochastic calculus to be the non-Fock extremal universally invariant version of Ref. 5 of variance $\sigma > 1$, so that the quantum Itô table is

$$
\begin{array}{c|ccc}
 & dA^\dagger & dA & dT \\
\hline
dA^\dagger & 0 & \beta^2 dT & 0 \\
dA & \alpha^2 dT & 0 & 0 \\
dT & 0 & 0 & 0
\end{array}
\qquad (2.1)
$$

where α and β are real numbers satisfying $\alpha^2 + \beta^2 = \sigma^2$, $\alpha^2 - \beta^2 = 1$.

Theorem 2.1. *Let there be given a family* $\left(W_a^b\right)_{a<b\in\mathbb{R}_+}$ *of unitary operators such that each* $W_a^b \in \mathcal{N}_a^b$, *the map* $[0,b[\ni a \mapsto W_a^b$ *is strongly continuous for each* $b > 0$, *the map* $]a,\infty[\ni b \mapsto W_a^b$ *is strongly continuous and the evolution equation (1.2) holds. Then there exists a locally square-integrable complex-valued function f and a locally integrable real valued function h such that each W_a^b is given by (1.3).*

Proof. The von Neumann algebras $\mathcal{N}_a^b$ are ampliations of von Neumann algebras in tensor factor Hilbert spaces in which factorisations the cyclic separating vacuum vector Ω is a product vector. Thus, taking vacuum expectations in the evolution equation $W_a^b W_b^c = W_a^c$, we deduce that the complex numbers $\Phi(a,b) = \langle \Omega, W_a^b \Omega \rangle$ satisfy the functional equation

$$\Phi(a,b)\Phi(b,c) = \Phi(a,c) \text{ whenever } a < b < c.$$

Moreover the complex-valued function Φ inherits continuity in its two arguments separately from the strong continuity of the family W_a^b. It follows Ref. 1 that Φ is of the form

$$\Phi(a,b) = \exp\left(\int_a^b \phi(x)\, dx\right) \qquad (2.2)$$

for some integrable complex-valued function ϕ. Since, in view of the unitarity of the W_a^b, each $|\Phi(a,b)| \leq 1$, we can put $\phi = -F + ih$ where F is nonnegative and h is real valued.

Now consider, for fixed a, the operator valued function M_a on $[a, \infty[$ where $M_a(b) = W_a^b - 1 - \int_a^b W_a^x \phi(x)\, dx$. The vacuum expectation of $M_a(b)$ is

$$\mathbb{E}[M_a(b)] = \left\langle \Omega, \left(W_a^b - 1 - \int_a^b W_a^x \phi(x)\, dx\right) \Omega \right\rangle$$

$$= \Phi(a,b) - 1 - \int_a^b \Phi(a,x)\phi(x)\, dx.$$

It follows from the fundamental theorem of calculus that $\mathbb{E}[M_a(b)] = 0$ for arbitrary $a < b$. For $a < b < c$ we have

$$M_a(c) = W_a^c - 1 - \int_a^c W_a^x \phi(x)\, dx$$

$$= W_a^b W_b^c - 1 - \int_a^b W_a^x \phi(x)\, dx - W_a^b \int_b^c W_b^x \phi(x)\, dx$$

$$= W_a^b \left(W_b^c - 1 - \int_b^c W_b^x \phi(x)\, dx\right) + W_a^b - 1 - \int_a^b W_a^x \phi(x)\, dx$$

$$= W_a^b M_b(c) + M_a(b).$$

Taking the vacuum conditional expectation at time b we find that

$$\mathbb{E}\left[M_a(c) \,|\, \mathcal{N}_a^b\right] = \mathbb{E}\left[W_a^b M_b(c) + M_a(b).\,|\, \mathcal{N}_a^b\right]$$

$$= W_a^b \mathbb{E}[M_b(c)] + M_a(b)$$

$$= M_a(b).$$

Hence the process $(M_a(b))_{b \geq a}$ is a martingale, and we can apply the martingale representation theorem of Ref. 5 and the fact that $M_a(b) = 0$ when $b = a$ to write each $M_a(b)$ in the form

$$M_a(b) = \int_a^b \left(E_a(s)\, dA^\dagger(s) + F_a(s) dA(s)\right)$$

for unique adapted, locally square-integrable processes E_a and F_a on $[a, \infty[$. Equivalently we can write

$$W_a^b = 1 - \int_a^b \left(E_a\, dA^\dagger + F_a dA + W_a^x \phi(x)\, dx\right). \tag{2.3}$$

Comparing two expressions

$$W_a^c = 1 - \int_a^c \left(E_a\, dA^\dagger + F_a dA + W_a^x \phi(x)\, dx\right),$$

$$W_a^c = W_a^b W_b^c = W_a^b - \int_b^c \left(W_a^b E_b\, dA^\dagger + W_a^b F_b dA + W_a^x \phi(x)\, dx\right)$$

we see that, for $x > b > a$

$$E_a(x) = W_a^b E_b(x), \quad F_a(x) = W_a^b F_b(x). \tag{2.4}$$

Equivalently, for $a < b < c$

$$(W_a^h)^{-1} E_a(c) = E_b(c), \quad (W_a^b)^{-1} F_a(c) = F_b(c).$$

Consider, for $n \in \mathbb{N}$,

$$\begin{aligned}(W_a^c)^{-1} E_a(c) &= \left(W_a^{c-\frac{1}{n}} W_{c-\frac{1}{n}}^c\right)^{-1} E_a(c) \\ &= \left(W_{c-\frac{1}{n}}^c\right)^{-1} \left(W_a^{c-\frac{1}{n}}\right)^{-1} E_a(c) \\ &= \left(W_{c-\frac{1}{n}}^c\right)^{-1} E_{c-\frac{1}{n}}(c).\end{aligned}$$

The latter is an element of $\mathcal{N}_{c-\frac{1}{n}}^c$. Hence the former also belongs to $\mathcal{N}_{c-\frac{1}{n}}^c$ and hence to $\bigcap_{n=1}^\infty \mathcal{N}_{c-\frac{1}{n}}^c = \mathbb{C}1$. Thus we can write $E_a(c) = W_a^c f_a(c)$ and similarly $F_a(c) = W_a^c g_a(c)$ for some complex-valued functions f_c and g_c. Making these substitutions into (2.4) we find that

$$W_a^x f_a(x) = W_a^b W_b^x f_b(x) = W_a^x f_b(x), \quad W_a^x g_a(x) = W_a^b W_b^x g_b(x) = W_a^x g_b(x),$$

hence we can write $f_a = f$ and $g_a = g$ independently of a. By local square integrability of the processes E_a and F_a, f and g are locally square-integrable. Thus (2.3) becomes

$$W_a^b = 1 - \int_a^b \left(W_a^x f(x) \, dA^\dagger(x) + W_a^x g(x) \, dA(x) + W_a^x \phi(x) \, dx\right)$$

and so the process W_a satisfies

$$dW_a = W_a(f \, dA^\dagger + g \, dA + \phi \, dT)$$

with initial condition $W_a^a = 1$. Finally, since each W_a^b is unitary, differentiating the process $W_a W_a^* = 1$ we find that

$$\begin{aligned}0 &= d(W_a W_a^*) \\ &= W_a(f \, dA^\dagger(x) + g \, dA + \phi \, dT) W_a^* \\ &\quad + W_a(\bar{g} \, dA^\dagger + \bar{f} \, dA + \bar{\phi} \, dT) W_a^* \\ &\quad + W_a(f \, dA^\dagger + g \, dA + \phi \, dT)(\bar{g} \, dA^\dagger(x) + \bar{f} \, dA + \bar{\phi} \, dT) W_a^* \\ &= W_a W_a^* \left((f + \bar{g}) dA^\dagger + (g + \bar{f}) dA + (\phi + \bar{\phi} + \beta^2 |f|^2 + \alpha^2 |g|^2) dT\right)\end{aligned}$$

and so, again using the fact that $W_a W_a^* = 1$, we must have

$$g = -\bar{f} \quad \text{and} \quad \phi = -\frac{\sigma^2}{2} |f|^2 + ih$$

for some real valued integrable h. Then W_a^b is given by (1.3). □

3. Stop times

A *stop time* is a right-continuous increasing family of projections $(E(\lambda))_{\lambda \in \mathbb{R}_+}$ labelled by the nonnegative real numbers with the property that each

$$E(\lambda) \in \mathcal{N}_\lambda = \left\{ (W(f)_t)_{t \le \lambda} : f \in L^2(\mathbb{R}_+) \right\}'', \qquad (3.1)$$

the von Neumann algebra generated by the Weyl processes $(W(f)_t)_{t \le \lambda}$ up to time λ. We shall always assume that the strong limit $\lim_{\lambda \to \infty} E(\lambda) = 1$. Thus a stop time is charaterised by the nonnegative self-adjoint operator $\tau = \int_{0-}^{\infty} \lambda dE(\lambda)$ which is affiliated to $\mathcal{N}$.

Stop times may be partially ordered either by prescribing that $\tau_1 \le \tau_2$ holds in the usual sense for self adjoint operators, that $\langle \psi, \tau_1 \psi \rangle \le \langle \psi, \tau_2 \psi \rangle$ for arbitrary ψ in the domain of τ_1, or, more usefully,[2] that $E_2(\lambda) \le E_1(\lambda)$ for all $\lambda \in \mathbb{R}_+$. In this chapter we shall use a partial ordering that is stronger than either of these, namely $\tau_1 \le \tau_2$ will mean that there is a real number $\mu \ge 0$ such that

$$E_1(\mu) = 1, \quad E_2(\mu) = 0; \qquad (3.2)$$

equivalently every element of the spectrum of τ_1 is majorised by every element of the spectrum of τ_2.

Following the intuition that a function $f(\tau)$ of the self-adjoint operator τ is defined as the spectral integral $\int_{0-}^{\infty} f(\lambda) dE(\lambda)$, given a process $X = (X(t))_{t \in \mathbb{R}_+}$ we may try to define the observable or random variable $X(\tau)$ got by stopping the process at τ as

$$X(\tau) = \int_{0-}^{\infty} X(\lambda) dE(\lambda)$$

whenever this exists in a suitable analytic sense, for example as the strong limit $\lim_P \sum X(\lambda_j)(E(\lambda_{j+1}) - E(\lambda_j))$ over partitions $P = (0, \lambda_1, \lambda_2, ..., \lambda_N = \lambda_P)$ of intervals $[0, \lambda_P]$ with $\lambda_P \to \infty$ on the domain for which such limits exist. A problem which arises with this definition, of the *right-stopped process*, is that it fails to be equal to the similarly defined *left-stopped process*

$$\int_{0-}^{\infty} dE(\lambda) X(\lambda) = \lim_P \sum (E(\lambda_j) - E(\lambda_{j-1})) X(\lambda_j)$$

or to the *double-stopped process*

$$\int_{0-}^{\infty} dE(\lambda) X(\lambda) dE(\lambda) = \lim_P \sum (E(\lambda_j) - E(\lambda_{j-1})) X(\lambda_j)(E(\lambda_j) - E(\lambda_{j-1})).$$

Compounding this problem of nonuniqueness, all three definitions have the defect that in general stopping the product of two commuting processes is not the same thing as the product of the stopped processe. These problems are avoided in the situation that each $X(t)$ is affiliated to the commutant $\mathcal{N}_t''$ of $\mathcal{N}_t$ so that each

$$X(\lambda_j)(E(\lambda_j) - E(\lambda_{j-1})) = (E(\lambda_j) - E(\lambda_{j-1})) X(\lambda_j)$$
$$= (E(\lambda_j) - E(\lambda_{j-1})) X(\lambda_j)(E(\lambda_j) - E(\lambda_{j-1})),$$

and in fact the stopped process can then be given a more direct definition in terms of factorising vectors $\psi = \psi_t \otimes \psi^t$, $\chi = \chi_t \otimes \chi^t$, $t \in \mathbb{R}_+$, namely the double exponential vectors $e(f, \bar{g}) = e(f) \otimes e(\bar{g})$, as

$$\langle \psi, X(\tau)\chi \rangle = \int_0^\infty \langle \psi^t, X(\tau)\chi^t \rangle \, d\langle \psi_t, E(t)\chi_t \rangle.$$

Though this procedure may seem artificial it allows a precise formulation of the strong Markov property,[3] that the processes $A^\dagger$ and A begin anew independently of the past at each stop time. Parthasarathy and Sinha[9] refined this idea in the case of Fock quantum stochastic calculus to show that the usual Fock space splitting at each sure time $t \in \mathbb{R}_+$, $\mathcal{H} = \mathcal{H}_t \otimes \mathcal{H}^t$ extends to stop times, $\mathcal{H} = \mathcal{H}_\tau \otimes \mathcal{H}^\tau$ in which each exponential vector $e(f)$ is a product vector. Their treatment can be extended straightforwardly to the non-Fock case.

4. Factorizing product integrals at stop times

We consider a family of product integrals

$$W_a^b(f; h) = \prod_a^b \left(1 + f dA^\dagger - \bar{f} dA + (ih - \frac{\sigma^2}{2}|f|^2) dT\right), \quad a < b,$$

where f is complex-valued square-integrable and h is real-valued and integrable, equivalently, by Theorem 1, a strongly continuous family of unitary operators $W_a^b \in \mathcal{N}_a^b$ satisfying the evolution equation (1.2). Our purpose is to study to what extent (1.2) when a, b and c are replaced by stop times.

For stop times $\tau_1 = \int_{0-}^\infty \lambda dE_1(\lambda)$, $\tau_2 = \int_{0-}^\infty \lambda dE_2(\lambda)$ with $\tau_1 < \tau_2$ in the sense (3.2) we define $W_{\tau_1}^{\tau_2}$ informally using left stopping at the lower limit τ_1 and right stopping at the upper limit τ_2. Then

$$W_{\tau_1}^{\tau_2} = \int_{0-}^\infty \int_{0-}^\infty dE_1(\lambda_1) W_{\lambda_1}^{\lambda_2} dE_2(\lambda_2)$$

$$= \int_{0-}^\mu \int_\mu^\infty dE_1(\lambda_1) W_{\lambda_1}^{\lambda_2} dE_2(\lambda_2)$$

$$= \int_{0-}^\mu \int_\mu^\infty dE_1(\lambda_1) W_{\lambda_1}^{\lambda_2} W_\mu^{\lambda_2} dE_2(\lambda_2)$$

$$= \int_{0-}^\mu dE_1(\lambda) W_\lambda^\mu \int_\mu^\infty W_\mu^\lambda dE_2(\lambda), \tag{4.1}$$

using the factorisation (1.2) with $a = \lambda_1$, $b = \mu$ and $c = \lambda_2$. The latter integrals are well defined in view of the unitarity and continuity of the family $(W_a^b)_{a \leq b}$ so we may take (4.1) as the rigorous definition of $W_{\tau_1}^{\tau_2}$. It is easy to see using (1.2) that this definition is independent of the choice of μ satisfying (3.2). Also it is clear that (1.2) holds for stop times a, b and c satisfying $a \leq b \leq c$, provided that b is sure.

Theorem 4.1. *(1.2) holds for arbitrary stop times a, b and c satisfying $a \leq b \leq c$ in the sense of (3.2).*

Proof. By replacing a by a sure time μ for which (3.2) holds with $\tau_1 = a$ and $\tau_2 = b$ and using the fact that (1.2) holds when b is sure we may assume without loss of generality that a is sure. Similarly we may assume that c is sure. Then, with

$$b = \int_{0-}^{\infty} \lambda dE(\lambda) = \int_{a}^{c} \lambda dE(\lambda)$$

we have

$$\begin{aligned}
W_a^b W_b^c &= \int_a^c W_a^{\lambda_1} dE(\lambda_1) \int_a^c dE(\lambda_2) W_{\lambda_2}^c \\
&= \int_a^c W_a^\lambda dE(\lambda) W_\lambda^c \\
&= \int_a^c W_a^\lambda W_\lambda^c dE(\lambda) \\
&= \int_a^c W_a^c dE(\lambda) \\
&= W_a^c \int_a^c dE(\lambda) \\
&= W_a^c
\end{aligned}$$

where we used (3.1) and the fact that $W_\lambda^c \in \mathcal{N}_\lambda^c$, together with the fact that (1.2) holds with b replaced by a sure time λ. $\square$

Acknowledgement

Conversations with Sylvia Pulmannová are acknowledged.

References

1. J Aczel, *Lectures on functional equations and their applications*, Academic Press (1966).
2. C Barnett and T Lyons, Stopping noncommutative processes, Math. Proc. Cambridge Philos. Soc **99**, 151-161 (1984)
3. R L Hudson, The strong Markov property for canonical Wiener processes, Jour. Funct. Anal. **34** 266-281 (1979).
4. R L Hudson, Stop times in Fock space quantum probability, Stochastics **79**, 383-391 (2007).
5. R L Hudson and J M Lindsay, A non-commutative martingale representation theorm for non-Fock quantum stochastic calculus, Jour. Funct. Anal **61**, 202-221 (1985).
6. R L Hudson and J M Lindsay, Uses of non-Fock quantum Brownian motion and a quantum martingale representation theorem, pp276-305 in *Quantum Probability II*, Proceedings, Heidelberg 1984, Springe Lecture Notes in mathematics **1136** (1985).
7. R L Hudson and R F Streater, Noncommutative martingales and stochastic integrals in Fock space, pp216-227, in *Stochastic processes in quantum theory and statistical physics*, Proceedings, Marseilles, 1982, Springer Lecture Notes in Physics, 173 (1983)

8. A Luczak and A A A Mohammed, Stochastic integration in finite von Neumann algebras, Studia Sci. Math. Hungar. **44**, 233-244 (2007)
9. K R Parthasarathy and K B Sinha, Stop times in Fock space stochastic calculus, PTRF **73**, 317-349 (1987).

Chapter 13

Karhunen-Loéve Expansion for Stochastic Convolution of Cylindrical Fractional Brownian Motions

Zongxia Liang

Department of Mathematical Sciences, Tsinghua University,
Beijing 100084, China
zliang@math.tsinghua.edu.cn

This chapter aims at firstly establishing a Karhunen-Loéve expansion for stochastic convolution of cylindrical fractional Brownian motion, then studying asymptotic behavior of conditional exponential moments of the stochastic convolution via this expansion.

Keywords: Cylindrical Brownian motion, Cylindrical fractional Brownian motion, Stochastic convolution, Conditional exponential moments

2000 AMS Subject Classification: Primary 60H05, 60G15; Secondary 60J65, 60F10

Contents

1 Introduction 195
2 Preliminaries 197
    2.1 Fractional Brownian motion 197
    2.2 Hypotheses on operators A and Φ 198
    2.3 Stochastic convolution W_A^H of cylindrical fractional Brownian motion 198
3 Karhunen-Loéve expansion of W_A^H 199
4 Evaluation of conditional exponential moments 203
References 206

1. Introduction

It is well-known that one of important questions related to the diffusion processes investigated by physicists in the context of statistics mechanics and quantum theory (see [8, 12]) is to compute the following Onsager-Machlup functional

$$J(\phi) = \log \lim_{\varepsilon \to 0} \frac{\mathbf{P}(\|x - \phi\| \leq \varepsilon)}{\mathbf{P}(\|W\| \leq \varepsilon)}$$

under a suitable norm $\|\cdot\|$ and some regularity conditions imposed on deterministic function ϕ, where $x(t)$ is a n-dimensional diffusion process driven by a standard Brownian motion W.

A rigorous mathematical treatment of this question was initiated by Stratonovich [15] and carried out by Ikeda and Watanabe [9], Takahashi and Watanabe [16], Fujita and Kotani [7], Zeitouni [18], Shepp and Zeitouni [14], Lyons and Zeitouni [10], Capitaine [4], Chaleyat-Maurel and Nualart [5] and other authors in various degrees of generality. They proved that the function $J(\phi)$ has the following expression

$$J(\phi) = -\frac{1}{2}\int_0^1 |\dot{\phi}(t) - f(\phi(t))|^2 dt - \frac{1}{2}\int_0^1 \nabla f(\phi(t))dt. \qquad (1.1)$$

We noticed that in order to get this expression (1.1) one has to evaluate the following conditional exponential expectation

$$\lim_{\varepsilon\to 0}\mathbf{E}\big[\exp\{\int_0^1 f(s)dW(s)\}\big|\|W\|\le\varepsilon\big]=1 \qquad (1.2)$$

for any function $f \in L^2([0,1])$ (see [10,14]).

Therefore the problem (1.2) has attracted a lot of interest recently. Bardina, Rovira and Tindel [2, 3], Moret and Nualart [11] extended the problem above to cylindrical Brownian motion $\mathbf{W}$(see [2, 3, 13]) and one-dimensional fractional Brownian motion B^H with Hurst parameter $H \in (0,1)$ (cf. [11]).

In this chapter we generalize the problem (1.2) to a cylindrical fractional Brownian motion $\mathbf{B^H}$(see section 2 below). More precisely, we investigate the following *limit*

$$\lim_{\varepsilon\to 0}\mathbf{E}\big[\exp\{\int_0^1 <f(s),d\mathbf{W}>\}\big|\|W_A^H\|_2\le\varepsilon\big] \qquad (1.3)$$

where $f \in L^2([0,1];\mathbb{U})$, $W_A^H(t) = \int_0^t \exp\{(t-s)A\}\Phi d\mathbf{B^H}(s)$ is a stochastic convolution of a cylindrical Brownian motion $\mathbf{B^H}$, A and Φ are operators on a real separable Hilbert space $\mathbb{U}$(see section 2 below). To the author's knowledge, there is no answer about this problem up to now. We will show that the *limit* (1.3) is not one in general, i.e., for any $L^2([0,1];\mathbb{U})$-valued function ϕ there is an orthogonal subset $\{\tilde{h}_{kj}\otimes e_j, k\ge 1, j\ge 1\}$ of $L^2([0,1];\mathbb{U})$ such that

$$\lim_{\varepsilon\to 0}\mathbf{E}\big[\exp\{\int_0^1 <\phi(s),d\mathbf{W}>\}\big|\|W_A^H\|_2\le\varepsilon\big]$$
$$= \exp\{\frac{1}{2}\sum_{k,j=1}^\infty <\phi,\tilde{h}_{kj}\otimes e_j>_{L^2([0,1];\mathbb{U})}^2\}$$

under some conditions (H1) and (H2) imposed on A and Φ (see section 2 below) and norm $\|\cdot\|_2$ in $L^2([0,1];\mathbb{U})$. We prove this result via a Karhunen-Loéve expansion of W_A^H. The problem (1.1) in this case remains unsolved.

This paper is organized as follows. In section 2 we collect basic facts about fractional Brownian motion that will be used in this paper. In section 3 we give the Karhunen-Loéve expansion of W_A^H. In section 4 we devote to proving main result of this paper.

2. Preliminaries

2.1. *Fractional Brownian motion*

In this subsection we collect some basic facts about fractional Brownian motion as well as stochastic integral w.r.t. it. We refer the readers to [1, 17].

Let $B^H = \{B_t^H, t \in [0,1]\}$ be an one-dimensional fractional Brownian motion with Hurst parameter $H \in (0,1)$. B^H has the following Wiener integral representation,

$$B_t^H = \int_0^t K(t,s) dW_s, \qquad (2.1)$$

where $W = \{W_s, s \in [0,1]\}$ is a standard Brownian motion, $K(t,s)$ is a kernel given by

$$K(t,s) = C_H (\frac{t}{s})^{H-\frac{1}{2}} (t-s)^{H-\frac{1}{2}} + s^{\frac{1}{2}-H} F(\frac{t}{s}),$$

C_H is a constant and $F(z) = C_H(\frac{1}{2}-H) \int_0^{z-1} \theta^{H-\frac{3}{2}}(1-(\theta+1)^{H-\frac{1}{2}}) d\theta$. In particular, $B^{\frac{1}{2}}$ is a standard Brownian motion. We denote by $\mathcal{E}$ the linear space of step functions on $[0,1]$. Let $\mathcal{H}$ be an Hilbert space of defined as the closure of $\mathcal{E}$ with respect to the scalar product

$$< I_{[0,t]}, I_{[0,s]} >_{\mathcal{H}} = R(t,s),$$

where $R(t,s) = \frac{1}{2}(t^{2H} + s^{2H} - |t-s|^{2H})$.

If $\phi(t) = \sum_{i=1}^n a_i I_{(t_i, t_{i+1}]}(t) \in \mathcal{E}$, $t_1, \cdots, t_n \in [0,1]$, $n \in \mathbf{N}$, $a_i \in \Re$, then we define its Wiener integral with respect to the fraction Brownian motion as follows,

$$\int_0^1 \varphi(s) dB_s^H = \sum_{i=1}^n a_i (B_{t_{i+1}}^H - B_{t_i}^H). \qquad (2.2)$$

The mapping $\phi(t) = \sum_{i=1}^n a_i I_{(t_i, t_{i+1}]} \to \int_0^1 \phi(s) dB_s^H$ is an isometry between $\mathcal{E}$ and the linear space span $\{B_t^H, t \in [0,1]\} \subset L^2(\Omega)$ and it can be extended to an isometry between $\mathcal{H}$ and the first Wiener chaos generated by $\{B_t^H, t \in [0,1]\}$. The image on an element $\phi \in \mathcal{H}$ by this isometry is called the Wiener integral of ϕ with respect to B^H.

Let $s < t$ and consider an operator K_t^* in $L^2([0,1])$,

$$(K_t^* \phi)(s) = K(t,s) \phi(s) + \int_s^t (\phi(r) - \phi(s)) \frac{\partial K}{\partial r}(r,s) dr.$$

Then $K_t^\star$ is an isometry between $\mathcal{H}$ and $L^2([0,1])$,

$$\int_0^t \varphi(s)dB_s^H = \int_0^t (K_t^\star\varphi)(s)dW_s$$

for every $t \in [0,1]$, and $\varphi I_{[0,t]} \in \mathcal{H}$ iff $K_t^\star\varphi \in L^2[0,1]$ (see [1, 17]). Moreover, if $H \geq \frac{1}{2}$, d $\phi, \varphi \in \mathcal{H}$ are such that $\int_0^t \int_0^t |\phi(s)\varphi(t)||t-s|^{2H-2}dsdt < +\infty$, then

$$<\phi,\varphi>_\mathcal{H} = H(2H-1)\int_0^t\int_0^t \phi(u)\varphi(v)|u-v|^{2H-2}dudv$$

$$= \mathbf{E}(\int_0^t \phi(s)dB_s^H \int_0^t \varphi(s)dB_s^H)$$

$$= \mathbf{E}\int_0^1 (K_t^\star\phi)(s)dW_s \int_0^1 (K_t^\star\varphi)(s)dW_s$$

$$= \int_0^1 (K_t^\star\phi)(s)(K_t^\star\varphi)(s)ds. \qquad (2.3)$$

2.2. Hypotheses on operators A and Φ

In this subsection we introduce hypotheses (H1) and (H2) on the operators A and Φ.

Let $\mathbb{U}$ be a real separable Hilbert space and $A : \mathcal{D}(A) \subset \mathbb{U} \to \mathbb{U}$ an unbounded operator on $\mathbb{U}$. The norm and scalar product in $\mathbb{U}$ will be denoted by $|\cdot|_\mathbb{U}$ and $<\cdot,\cdot>$ respectively. The L^2-norm in $L^2([0,1];\mathbb{U})$ will be denoted by $\|\cdot\|_2$. Throughout this paper we assume that

(H1) The operator A generates a self adjoint C_0–semigroup $\{\exp(tA); t \geq 0\}$, of negative type. Moreover, there exists a complete orthogonal system $\{e_j; j \geq 1\}$ which diagonalizes A. We will denote by $\{-\alpha_j, j \geq 1\}$ the corresponding set of eigenvalues and we also suppose that $0 < \alpha_1 < \alpha_2 < \cdots < \alpha_n \to +\infty$ as $n \to +\infty$ for convenience.

(H2) The operator Φ is a bounded linear operator on $\mathbb{U}$. Φ is of non-negative type, and is diagonal when expressed in the orthogonal basis $\{e_j; j \geq 1\}$. Denote by $\{\beta_j; j \geq 1\}$ the corresponding set of eigenvalues. $\{\beta_j; j \geq 1\}$ satisfies

$$\sum_{j=1}^\infty \frac{\beta_j^2}{\alpha_j^{2H-1}(1+2\alpha_j)} < +\infty, \qquad (2.4)$$

and the Hurst parameter H is in $[\frac{1}{2}, 1)$.

2.3. Stochastic convolution W_A^H of cylindrical fractional Brownian motion

In this subsection we introduce cylindrical fractional Brownian motion and its stochastic convolution.

Let $(\Omega, \mathfrak{F}, \mathfrak{F}_t, \mathbf{P})$ be a probability space. $\{W^i(t); t \in [0,1], i \geq 1\}$ is a sequence of mutually independent Brownian motions on this probability space and is adapted to $\mathfrak{F}_t$. The cylindrical Brownian motion $\mathbf{W}$ and cylindrical fractional Brownian motion $\mathbf{B}^H$ (see [2, 13, 17]) on $\mathbb{U}$ are defined by the following formal series'

$$\mathbf{W}(t) = \sum_{i=1}^{\infty} W^i(t) e_i \quad \text{and} \tag{2.5}$$

$$\mathbf{B}^H(t) = \sum_{i=1}^{\infty} B_i^H(t) e_i \tag{2.6}$$

respectively, where $\{B_i^H, i \geq 1\}$ is a family of mutually independent fractional Brownian motions with the Hurst parameter $H \in (0, 1)$, that is, for $i \geq 1$

$$B_i^H(t) = \int_0^t K(t, s) dW^i(s).$$

We define the stochastic convolution W_A^H of cylindrical fractional Brownian motion $\mathbf{B}^H$ by

$$W_A^H(t) = \int_0^t \exp\{(t-s)A\} \Phi d\mathbf{B}^H(s)$$

$$\equiv \sum_{j=1}^{\infty} \beta_j \Big(\int_0^t \exp\{-(t-s)\alpha_j\} dB_j^H(s) \Big) e_j. \tag{2.7}$$

We will prove the series (2.7) is convergent in $L^2(\Omega \times [0,1] \times \mathbb{U})$ under the assumption (H1) and (H2) in next section.

3. Karhunen-Loéve expansion of W_A^H

Let $X(t) = \int_0^t \exp\{-\alpha(t-s)\} dW_s^H$ with $\alpha > 0$, and W^H is an one-dimensional fractional Brownian motion with Hurst parameter $H \in [\frac{1}{2}, 1)$. Then by (2.3)

$\mathbf{E}X(t) = 0$,

$\{X(t), t \in [0,1]\}$ is a Gaussian process,

$$\mathbf{E}X^2(t) = \int_0^t \int_0^t \exp\{-\alpha(2t-u-v)\}|u-v|^{2H-2} du dv \leq \frac{\Gamma(2H-1)}{\alpha^{2H}} < +\infty, \tag{3.1}$$

where Γ denotes the Euler function. Moreover

$$\mathbf{E}|X(t) - X(s)|^2 \leq 2\mathbf{E}[\int_s^t \exp\{-\alpha(t-u)\}dW_u^H]^2$$

$$+ 2\mathbf{E}[\int_0^s (\exp\{-\alpha(t-u)\} - \exp\{-\alpha(s-u)\})dW_u^H]^2$$

$$= 2\int_0^1 \int_0^1 I_{[s,t]\times[s,t]}(u,v) \exp\{-\alpha(2t-u-v)\}|u-v|^{2H-2}dudv$$

$$+ 2\int_0^1 \int_0^1 I_{[0,s]\times[0,s]}(u,v)(\exp\{-\alpha(t-u)\} - \exp\{-\alpha(s-u)\})$$

$$\times (\exp\{-\alpha(t-v)\} - \exp\{-\alpha(s-v)\})|u-v|^{2H-2}dudv$$

$$\to 0 \quad \text{as} \quad t-s \to 0. \qquad (3.2)$$

Its covariance function $B(t,s) = \mathbf{E}[X(t)X(s)]$ is thus a continuous, symmetric, non-negative definite kernel on $[0,1]^2$. The following integral operator

$$\mathbb{T}(f)(t) = \int_0^1 B(t,s)f(s)ds$$

is compact and self adjoint, and admits a complete orthogonal base of $L^2([0,1])$ consisting of eigenfunctions $\phi_n(t)$ of $\mathbb{T}$ with corresponding eigenvalues λ_n. By Mercer's theorem (see [6], p.117)

$$B(t,s) = \sum_{j=1}^{\infty} \lambda_j \phi_j(t) \phi_j(s).$$

Let $\xi_n(\omega) = \int_0^1 X(t)\phi_n(t)dt$. Then

$$\mathbf{E}[\xi_n \xi_m] = \int_0^1 \int_0^1 \mathbf{E}[X(t)X(s)]\phi_n(t)\phi_m(s)dtds$$

$$= \int_0^1 \int_0^1 B(t,s)\phi_n(t)\phi_m(s)dtds$$

$$= \int_0^1 \phi_n(t)\mathbb{T}\phi_m(t)dt$$

$$= \lambda_m \int_0^1 \phi_n(t)\phi_m(t)dt$$

$$= \delta_{\{n=m\}}\lambda_m. \qquad (3.3)$$

Therefore $\{\xi_n, n \geq 1\}$ forms a sequence of zero-mean, independent Gaussian random variables with variance λ_n. Further

$$\mathbf{E}|X(t) - \sum_{k=1}^n \xi_k \phi_k(t)|^2 = B(t,t) - \sum_{j=1}^n \lambda_j \phi_j^2(t) \to 0 (\text{as} \quad n \to +\infty).$$

That is,
$$X(t) = \sum_{k=1}^{\infty} \xi_k \phi_k(t) = \sum_{k=1}^{\infty} \sqrt{\lambda_k} z_k \phi_k(t), \qquad (3.4)$$

where $z_k = \frac{1}{\sqrt{\lambda_k}} \xi_k$ are standard normal random variables. Without generality we assume that $\lambda_1 \geq \lambda_2 \geq \cdots \geq \lambda_n \to 0$ as $n \to +\infty$. Hence if we replace X by $X_j(t) = \int_0^t \exp\{-\alpha_j(t-s)\} dB_j^H(s)$, then by (3.4) we can define the stochastic convolution W_A^H of cylindrical fractional Brownian motion B^H by

$$W_A^H(t) = \sum_{j,k=1}^{\infty} \beta_j \sqrt{\lambda_{kj}} z_{kj} (\phi_{kj}(t) \otimes e_j) \qquad (3.5)$$

if the series (3.5) is convergent, where

$$\lambda_{kj} = \mathbf{E}\xi_{kj}$$
$$\xi_{kj} = \int_0^1 X_j(t) \phi_{kj}(t) dt$$
$$z_{kj} = \frac{1}{\sqrt{\lambda_{kj}}} \xi_{kj}$$

ϕ_{kj} is an eigenfunction corresponding to λ_{kj}.

The main result of this section is the following.

Theorem 3.1. *Assume that the (H1) and (H2) hold. Then the stochastic convolution W_A^H of cylindrical fractional Brownian motion B^H has Karhunen-Loéve expansion (3.5). That is, the series (3.5) is convergent in $L^2(\Omega \times [0,1] \times \mathbb{U})$.*

Proof. Since $\{\phi_{kj} \otimes e_j, k \geq 1, j \geq 1\}$ forms a complete orthogonal base of $L^2([0,1]; \mathbb{U})$ such that for any $k, j \geq 1$

$$Cov(<W_A^H, \phi_{kj} \otimes e_j>_{L^2([0,1];\mathbb{U})}) = \beta_j^2 \lambda_{kj},$$

we have

$$\|W_A^H\|_2^2 = \sum_{k,j=1}^{\infty} \beta_j^2 \lambda_{kj} z_{kj}^2,$$

$$\|W_A^H\|_{L^2(\Omega \times [0,1] \times \mathbb{U})}^2 = \sum_{k,j=1}^{\infty} \beta_j^2 \lambda_{kj}. \qquad (3.6)$$

Therefore, we only need to prove that

$$\sum_{k,j=1}^{\infty} \beta_j^2 \lambda_{kj} < +\infty. \qquad (3.7)$$

For every $j \geq 1$, noting that $\{\phi_{kj}, k \geq 1\}$ is a complete orthogonal base of $L^2([0,1])$, by (2.4) and the Parseval equality we have

$$\sum_{k=1}^{\infty} \lambda_{kj} = \mathbf{E}[\sum_{k=1}^{\infty} <X_j, \phi_{kj}>^2]$$
$$= \mathbf{E}\|X_j\|_{L^2([0,1])}^2$$
$$= \int_0^1 \mathbf{E}X_j^2(t)dt$$
$$= \int_0^1 [\int_0^t \int_0^t \exp\{-\alpha_j(2t-u-v)\}|u-v|^{2H-2}dudv]dt. \qquad (3.8)$$

We now calculate the last integral. Using the Fubini Theorem

$$\int_0^t \int_0^t \exp\{-\alpha_j(2t-u-v)\}|u-v|^{2H-2}dudv$$
$$= \int_0^t \int_0^t I_{\{u \geq v\}} \exp\{-\alpha_j(2t-u-v)\}|u-v|^{2H-2}dudv$$
$$+ \int_0^t \int_0^t I_{\{u \leq v\}} \exp\{-\alpha_j(2t-u-v)\}|u-v|^{2H-2}dudv$$
$$\equiv J_1(t) + J_2(t), \qquad (3.9)$$

$$J_1(t) = \int_0^t dt_1 \int_0^{t_1} e^{-\alpha_j(2t-t_1-s_1)}(t_1-s_1)^{2H-2}ds_1$$
$$= e^{-2\alpha_j t} \int_0^t e^{\alpha_j t_1}[\int_0^{t_1} e^{\alpha_j s_1}(t_1-s_1)^{2H-2}ds_1]dt_1$$
$$= e^{-2\alpha_j t} \int_0^t e^{2\alpha_j t_1} \alpha_j^{1-2H}[\int_0^{\alpha_j t_1} e^{-x}x^{2H-2}dx]dt_1$$
$$\leq e^{-2\alpha_j t}\Gamma(2H-1)\int_0^t e^{2\alpha_j t_1}dt_1 \cdot \alpha_j^{1-2H}$$
$$= \frac{\Gamma(2H-1)}{2\alpha_j^{2H}}(1-e^{-2\alpha_j t}). \qquad (3.10)$$

Similarly

$$J_2(t) \leq \frac{\Gamma(2H-1)}{2\alpha_j^{2H}}(1-e^{-2\alpha_j t}). \qquad (3.11)$$

By the inequalities (3.8)-(3.11)

$$\sum_{k=1}^{\infty} \lambda_{kj} = \int_0^1 [J_1(t) + J_2(t)]dt$$

$$\leq \frac{\Gamma(2H-1)}{\alpha_j^{2H}} \int_0^1 (1 - e^{-2\alpha_j t})dt$$

$$= \frac{\Gamma(2H-1)}{\alpha_j^{2H}}(2\alpha_j - 1 + e^{-2\alpha_j})/2\alpha_j$$

$$\leq 2\Gamma(2H-1)\frac{1}{\alpha_j^{2H-1}(1+2\alpha_j)}, \qquad (3.12)$$

where we have used an elementary inequality: $x - 1 + e^{-x} \leq \frac{x^2}{1+x}$ for $x \geq 0$. Hence

$$\sum_{k,j=1}^{\infty} \beta_j^2 \lambda_{kj} \leq 2\Gamma(2H-1) \sum_{j=1}^{\infty} \frac{\beta_j^2}{\alpha_j^{2H-1}(1+2\alpha_j)} < +\infty \qquad (3.13)$$

by (2.4). We thus complete the proof.

4. Evaluation of conditional exponential moments

In this section we will use the Karhunen-Loéve expansion (3.5) of W_A^H to compute the *limit* $\lim_{\varepsilon \to 0} \mathbf{E}\{\exp\{[\int_0^1 < l(s), d\mathbf{W}(s) >]|\|W_A^H\|_2 \leq \varepsilon\}$ for any function $l \in L^2([0,1]; \mathbb{U})$. Where $\mathbf{W}$ is cylindrical Brownian motion. Before doing this we first present a lemma that we will use in this section.

We denote here by ℓ^2 a set of sequences of real numbers $\{a_i, i \geq 1\}$ such that $\sum_{i \geq 1} a_i^2 < +\infty$.

Lemma 4.1. *(see [2] Lemma 2.4) Let $\{z_n, n \geq 1\}$ be a sequence of independent $N(0,1)$ random variables defined on $(\Omega, \mathfrak{F}, \mathbf{P})$, and $\{\eta_i, i \geq 1\}$ and $\{v_i, i \geq 1\}$ two ℓ^2 sequences of real numbers such that $\eta_i > 0$ for any $i \geq 1$. Then*

$$\lim_{\varepsilon \to 0} \mathbf{E}\Big[\exp\Big(\sum_{i=1}^{\infty} z_i v_i\Big) \Big| \sum_{i=1}^{\infty} \eta_i^2 z_i^2 \leq \varepsilon^2\Big] = 1. \qquad (4.1)$$

Now we state the main result of this paper.

Theorem 4.1. *Let $l \in L^2([0,1]; \mathbb{U})$, the (H1) and (H2) hold. Then there exists an orthogonal subset $\{\widetilde{h}_{kj}\}$ of $L^2([0,1])$ such that*

$$\lim_{\varepsilon \to 0} \mathbf{E}\{\exp\{[\int_0^1 < l(s), d\mathbf{W}(s) >]\}|\|W_A^H\|_2 \leq \varepsilon\}$$

$$= \exp\{\frac{1}{2} \sum_{k,j=1}^{\infty} < l, \widetilde{h}_{kj} \otimes e_j >_{L^2([0,1]; \mathbb{U})}^2\}. \qquad (4.2)$$

Proof. For $\lambda_{kj} \neq 0$ we let

$$h_{kj}(s) \equiv \frac{1}{\sqrt{\lambda_{kj}}} \int_s^1 K_t^*(e^{-\alpha_j(t-\cdot)})(s)\phi_{kj}(t)dt, \quad s \in [0,1].$$

Then $\{h_{kj}, k \geq 1\}$ is an orthogonal subset of $L^2([0,1])$.

Indeed, for $n, m \geq 1$, by (2.4)

$$< h_{nj}, h_{mj} >_{L^2([0,1])}$$

$$= \frac{1}{\sqrt{\lambda_{nj}\lambda_{mj}}} \int_0^1 [\int_s^1 K_t^*(e^{-\alpha_j(t-\cdot)})(s)\phi_{nj}(t)dt] \cdot [\int_s^1 K_t^*(e^{-\alpha_j(t-\cdot)})(s)\phi_{mj}(t)dt]ds$$

$$= \frac{1}{\sqrt{\lambda_{nj}\lambda_{mj}}} \int_0^1 \int_0^1 [\int_0^{t \wedge u} K_t^*(e^{-\alpha_j(t-\cdot)})(s)K_u^*(e^{-\alpha_j(u-\cdot)})(s)ds]\phi_{nj}(t)\phi_{mj}(u)dtdu$$

$$= \frac{1}{\sqrt{\lambda_{nj}\lambda_{mj}}} \int_0^1 \int_0^1 \mathbf{E}[X_j(t)X_j(u)]\phi_{nj}(t)\phi_{mj}(u)dtdu$$

$$= \sqrt{\frac{\lambda_{mj}}{\lambda_{nj}}} < \phi_{nj}, \phi_{mj} >_{L^2([0,1])} = \sqrt{\frac{\lambda_{mj}}{\lambda_{nj}}} \delta_{\{n=m\}}, \qquad (4.3)$$

that is, $< h_{nj}, h_{mj} >_{L^2([0,1])} = 0$ for $n \neq m$ and $\|h_{nj}\|_{L^2([0,1])} = 1$.

If we denote by $I(f)$ the Ito integral of f, then we have

$$z_{kj} = I(h_{kj}).$$

In fact,

$$z_{kj} = \frac{1}{\lambda_{kj}}\xi_{kj}$$

$$= \frac{1}{\lambda_{kj}} \int_0^1 [\int_0^t K_t^*(e^{-\alpha_j(t-\cdot)})(s)dW^j(s)]\phi_{kj}(t)dt$$

$$= \int_0^1 [\frac{1}{\lambda_{kj}} \int_s^1 K_t^*(e^{-\alpha_j(t-\cdot)})(s)\phi_{kj}(t)dt]dW^j(s)$$

$$= \int_0^1 h_{kj}(s)dW^j(s)$$

$$= I(h_{kj}). \qquad (4.4)$$

Since $L^2([0,1])$ is a real separable Hilbert space, we can find an orthogonal subset $\{\widetilde{h}_{kj}, k \geq 1\}$ of $L^2([0,1])$ such that $\{h_{kj}, \widetilde{h}_{nj}, k \geq 1, n \geq 1\}$ becomes a complete orthogonal base of $L^2([0,1])$. Hence $\{h_{kj} \otimes e_j, \widetilde{h}_{nj} \otimes e_j, k \geq 1, n \geq 1, j \geq 1\}$ is a complete orthogonal base of $L^2([0,1]; \mathbb{U})$. And for any $l \in L^2([0,1]; H)$ we have

$$l(s) = \sum_{k,j=1} < l, h_{kj} \otimes e_j > h_{kj}(s) \otimes e_j + \sum_{n,j=1} < l, \widetilde{h}_{nj} \otimes e_j > \widetilde{h}_{nj}(s) \otimes e_j.$$

So

$$\int_0^1 < l, d\mathbf{W}(s) > = \sum_{k,j=1} < l, h_{kj} \otimes e_j > z_{kj} + \sum_{n,j=1} < l, \widetilde{h}_{nj} \otimes e_j > \widetilde{z}_{nj}, \qquad (4.5)$$

where $\widetilde{z}_{kj} = I(\widetilde{h}_{kj})$. Moreover,

$$\mathbf{E} z_{mk} \widetilde{z}_{nj}$$
$$= \mathbf{E}[\int_0^1 h_{mk}(s) dW^k(s) \int_0^1 \widetilde{h}_{nj}(s) dW^j(s)]$$
$$= \int_0^1 h_{mk}(s) \widetilde{h}_{nj}(s) ds$$
$$= 0 \tag{4.6}$$

for $n, m, k, j \geq 1$. Therefore $\{z_{mk}, \widetilde{z}_{nj}\}$ is a family of independent standard normal random variables. Consequently,

$$\mathbf{E}\{\exp\{[\int_0^1 <l(s), d\mathbf{W}(s)>]\}|\|W_A^H\|_2 \leq \varepsilon\}$$
$$= \mathbf{E}\Big[\exp\{\sum_{k,j=1} <l, h_{kj} \otimes e_j > z_{kj}\}$$
$$\times \exp\{\sum_{n,j=1} <l, \widetilde{h}_{nj} \otimes e_j > \widetilde{z}_{nj}\}| \sum_{k,j=1} \beta_j^2 \lambda_{kj} z_{kj}^2 \leq \varepsilon^2\Big]$$
$$= \mathbf{E}[\exp\{\sum_{n,j=1} <l, \widetilde{h}_{nj} \otimes e_j > \widetilde{z}_{nj}\}]$$
$$\times \mathbf{E}\Big[\exp\{\sum_{k,j=1} <l, h_{kj} \otimes e_j > z_{kj}\}| \sum_{k,j=1} \beta_j^2 \lambda_{kj} z_{kj}^2 \leq \varepsilon^2\Big]$$
$$\tag{4.7}$$

By using the Bessel inequality and (3.13)

$$\sum_{k,j=1} <l, h_{kj} \otimes e_j >^2 \leq \|l\|_{L^2([0,1];U)}^2 \text{ and}$$

$$\sum_{k,j=1}^\infty \beta_j^2 \lambda_{kj} \leq 2\Gamma(2H-1) \sum_{j=1}^\infty \frac{\beta_j^2}{\alpha_j^{2H-1}(1+2\alpha_j)} < +\infty. \tag{4.8}$$

By Lemma 4.1 it follows that

$$\lim_{\varepsilon \to 0} \mathbf{E}\Big[\exp\{\sum_{k,j=1} <l, h_{kj} \otimes e_j > z_{kj}\}| \sum_{k,j=1} \beta_j^2 \lambda_{kj} z_{kj}^2 \leq \varepsilon^2\Big] = 1. \tag{4.9}$$

A combination of (4.7),(4.8) and (4.9) yields

$$\lim_{\varepsilon \to 0} \mathbf{E}\{\exp\{[\int_0^1 <l(s), d\mathbf{W}(s)>]\}|\|W_A^H\|_2 \leq \varepsilon\}$$
$$= \mathbf{E}[\exp\{\sum_{n,j=1} <l, \widetilde{h}_{nj} \otimes e_j > \widetilde{z}_{nj}\}]$$
$$= \exp\{\frac{1}{2} \sum_{n,j=1}^\infty <l, \widetilde{h}_{nj} \otimes e_j >^2_{L^2([0,1];U)}\}. \tag{4.10}$$

The proof is thus complete.

Acknowledgement

This work is supported by Project 10771114 of NSFC, Project 20060003001 of SRFDP, and SRF for ROCS, SEM, and the Korea Foundation for Advanced Studies. The author would like to thank the institutions for the generous financial support. He is also very grateful to College of Social Sciences and College of Engineering at Seoul National University for providing excellent working conditions for him.

References

1. Alos, E., Mazet,O., Nualart, D.: Stochastic calculus with respect to Gaussian processes. Ann.Probab. **29**, 766-801(1999).
2. Bardina,X., Rovira,C., Tindel,S.: Onsager-Machlup functional for stochastic evolution equations. Ann.I. Poincare-PR. **39**, 69-93(2003).
3. Bardina,X., Rovira,C., Tindel,S.: Onsager-Machlup functional for stochastic evolution equations in a class of norms. Stochastic analysis and applications. **21**, 1231-1253(2002).
4. Capitaine, M.: Onsager-Machlup functional for some smooth norms on Wiener space. Probab. Theory Related Fields. **102**, 189-201(1995).
5. Chaleyat-Maurel, M., Nualart,D.: The Onsager-Machlup functional for a class of anticipating processes. Probab. Theory Related Fields. **94**, 247-270(1992).
6. Courant,R., Hilbert, D.: Methods of mathematical physics. John Wiley Inc., 1989.
7. Fujita, T. and Kotani. S.: The Onsager-Machlup function for diffusion processes. J.Math. Kyoto Univ. **22**, 115-130(1982).
8. Graham, R. : Path integral formulation of general diffusion process. Z. Phys. B. **26**, 281-290(1979).
9. Ikeda, N., Watanabe, S.: Stochastic differential equations and diffusion processes. Wiley, New York, 1980.
10. Lyons, T. and Zeitouni O.: Conditional exponential moments for iterated wiener integrals. Ann.Probab. **27** , 1738-1749(1999).
11. Moret, S., Nualart, D: Onsager- Machlup functional for the fractional Brownian motion.Probab. Theory Related Fields. **124**, 227-260(2002).
12. Onsager, L., Machlup, S. : Fluctuations and irreversible processes. I.II. Phys.Rev. **91**, 1505-1512, 1512-1515(1953).
13. Prato,G.D., Zabczyk, J.: Stochastic equations in infinite dimensions. Cambridge University Press, 1992.
14. Shepp, L.A., Zeitouni, O.: A note on conditional exponential moments and Onsager-Machlup functionals. Ann.Probab. **20**, 652-654(1992).
15. Stratanovich, R.L.: On the probability function of diffusion processes. Selected Transl. Math. Statist. Probab. **10**, 273-286.
16. Takahashi, Y., Watanabe, S.: The probability functionals (Onsager- Machlup functions) of diffusion processes. Stochastic Integrals. Lecture Notes in Math., **851**, 433-463(1981). Springer, Berlin.
17. Tindel,S., Tudor, C.A., Viens, F.: Stochastic evolution equations with fractional Brownian motion. Probab. Theory Related Fields. **127**, 186-204(2003).
18. Zeitouni, O. : On the Onsager-Machlup functional of diffusion processes around non C^2 curves. Ann.Probab. **17**, 1037-1054(1989).

Chapter 14

Stein's Method Meets Malliavin Calculus: A Short Survey With New Estimates

Ivan Nourdin[*] and Giovanni Peccati[†]

Université Paris VI and Université Paris Ouest

This is an overview of some recent techniques involving the Malliavin calculus of variations and the so-called 'Stein's method' for the Gaussian approximations of probability distributions. Special attention is devoted to establishing explicit connections with the classic method of moments: in particular, interpolation techniques are used in order to deduce some new estimates for the moments of random variables belonging to a fixed Wiener chaos. As an illustration, a class of central limit theorems associated with the quadratic variation of a fractional Brownian motion is studied in detail.

Keywords: Central limit theorems, fractional Brownian motion, isonormal Gaussian processes, Malliavin calculus, multiple integrals, Stein's method

2000 AMS Subject Classification: 60F05, 60G15, 60H05, 60H07

Contents

1 Introduction 208
  1.1 Stein's heuristic and method 208
  1.2 The role of Malliavin calculus 209
  1.3 Beyond the method of moments 210
  1.4 An overview of the existing literature 210
2 Preliminaries 211
  2.1 Isonormal Gaussian processes 212
  2.2 Chaos, hypercontractivity and products 214
  2.3 The language of Malliavin calculus 215
3 One-dimensional approximations 217
  3.1 Stein's lemma for normal approximations 217
  3.2 General bounds on the Kolmogorov distance 219
  3.3 Wiener chaos and the fourth moment condition 220
  3.4 Quadratic variation of the fractional Brownian motion, part one 223
  3.5 The method of (fourth) moments: explicit estimates via interpolation .. 226
4 Multidimensional case 228

---

[*]Laboratoire de Probabilités et Modèles Aléatoires, Université Pierre et Marie Curie, Boîte courrier 188, 4 Place Jussieu, 75252 Paris Cedex 5, France. Email: ivan.nourdin@upmc.fr

[†]Equipe Modal'X, Université Paris Ouest – Nanterre la Défense, 200 Avenue de la Rpublique, 92000 Nanterre, and LSTA, Université Paris VI, France. Email: giovanni.peccati@gmail.com

4.1	Main bounds	229
4.2	Quadratic variation of fractional Brownian motion, continued	232
References		234

1. Introduction

This survey deals with the powerful interaction of two probabilistic techniques, namely the *Stein's method* for the normal approximation of probability distributions, and the *Malliavin calculus* of variations. We will first provide an intuitive discussion of the theory, as well as an overview of the literature developed so far.

1.1. *Stein's heuristic and method*

We start with an introduction to Stein's method based on moments computations. Let $N \sim \mathcal{N}(0,1)$ be a standard Gaussian random variable. It is well-known that the (integer) moments of N, noted $\mu_p := E(N^p)$ for $p \geq 1$, are given by: $\mu_p = 0$ if p is odd, and $\mu_p = (p-1)!! := p!/(2^{p/2}(p/2)!)$ if p is even. A little inspection reveals that the sequence $\{\mu_p : p \geq 1\}$ is indeed completely determined by the recurrence relation:

$$\mu_1 = 0, \quad \mu_2 = 1, \quad \text{and} \quad \mu_p = (p-1) \times \mu_{p-2}, \quad \text{for every } p \geq 3. \tag{1.1}$$

Now (for $p \geq 0$) introduce the notation $f_p(x) = x^p$, so that it is immediate that the relation (1.1) can be restated as

$$E[N \times f_{p-1}(N)] = E[f'_{p-1}(N)], \quad \text{for every } p \geq 1. \tag{1.2}$$

By using a standard argument based on polynomial approximations, one can easily prove that relation (1.2) continues to hold if one replaces f_p with a sufficiently smooth function f (e.g. any C^1 function with a sub-polynomial derivative will do). Now observe that a random variable Z verifying $E[Zf_{p-1}(Z)] = E[f'_{p-1}(Z)]$ for every $p \geq 1$ is necessarily such that $E(Z^p) = \mu_p$ for every $p \geq 1$. Also, recall that the law of a $\mathcal{N}(0,1)$ random variable is uniquely determined by its moments. By combining these facts with the previous discussion, one obtains the following characterization of the (standard) normal distribution, which is universally known as 'Stein's Lemma': *a random variable Z has a $\mathcal{N}(0,1)$ distribution if and only if*

$$E[Zf(Z) - f'(Z)] = 0, \tag{1.3}$$

for every smooth function f. Of course, one needs to better specify the notion of 'smooth function' – a rigorous statement and a rigorous proof of Stein's Lemma are provided at Point 3 of Lemma 3.1 below.

A far-reaching idea developed by Stein (starting from the seminal paper [36]) is the following: in view of Stein's Lemma and given a generic random variable Z, one can measure the distance between the laws of Z and $N \sim \mathcal{N}(0,1)$, by assessing the distance from zero of the quantity $E[Zf(Z) - f'(Z)]$, for every f belonging to a 'sufficiently large' class of smooth functions. Rather surprisingly, this somewhat heuristic approach to probabilistic approximations can be made rigorous by using ordinary differential equations. Indeed, one of the main findings of [36] and [37] is that bounds of the following type hold in great generality:

$$d(Z, N) \leq C \times \sup_{f \in \mathcal{F}} |E[Zf(Z) - f'(Z)]|, \qquad (1.4)$$

where: (i) Z is a generic random variable, (ii) $N \sim \mathcal{N}(0,1)$, (iii) $d(Z,N)$ indicates an appropriate distance between the laws of Z and N (for instance, the Kolmogorov, or the total variation distance), (iv) $\mathcal{F}$ is some appropriate class of smooth functions, and (v) C is a universal constant. The case where d is equal to the Kolmogorov distance, noted d_{Kol}, is worked out in detail in the forthcoming Section 3.1: we anticipate that, in this case, one can take $C = 1$, and $\mathcal{F}$ equal to the collection of all bounded Lipschitz functions with Lipschitz constant less or equal to 1.

Of course, the crucial issue in order to put Stein-type bounds into effective use, is how to assess quantities having the form of the right-hand side of (1.4). In the last thirty years, an impressive panoply of approaches has been developed in this direction: the reader is referred to the two surveys by Chen and Shao [7] and Reinert [33] for a detailed discussion of these contributions. In this chapter, we shall illustrate how one can effectively estimate a quantity such as the right-hand side of (1.4), whenever the random variable Z can be represented as a regular functional of a generic and possibly infinite-dimensional Gaussian field. Here, the correct notion of regularity is related to Malliavin-type operators.

1.2. The role of Malliavin calculus

All the definitions concerning Gaussian analysis and Malliavin calculus used in the Introduction will be detailed in the subsequent Section 2. Let $X = \{X(h) : h \in \mathfrak{H}\}$ be an isonormal Gaussian process over some real separable Hilbert space $\mathfrak{H}$. Suppose Z is a centered functional of X, such that $E(Z) = 0$ and Z is differentiable in the sense of Malliavin calculus. According to the Stein-type bound (1.4), in order to evaluate the distance between the law of Z and the law of a Gaussian random variable $N \sim \mathcal{N}(0,1)$, one must be able to assess the distance between the two quantities $E[Zf(Z)]$ and $E[f'(Z)]$. The main idea developed in [18], and later in the references [19,20,22,23], is that the needed estimate can be realized by using the following consequence of the *integration by parts formula* of Malliavin calculus: for every f sufficiently smooth (see Section 2.3 for a more precise statement),

$$E[Zf(Z)] = E[f'(Z)\langle DZ, -DL^{-1}Z\rangle_{\mathfrak{H}}], \qquad (1.5)$$

where D is the Malliavin derivative operator, L^{-1} is the pseudo-inverse of the Ornstein-Uhlenbeck generator, and $\langle \cdot, \cdot \rangle_{\mathfrak{H}}$ is the inner product of $\mathfrak{H}$. It follows from (1.5) that, if the derivative f' is bounded, then the distance between $E[Zf(Z)]$ and $E[f'(Z)]$ is controlled by the $L^1(\Omega)$-norm of the random variable

$1 - \langle DZ, -DL^{-1}Z\rangle_{\mathfrak{H}}$. For instance, in the case of the Kolmogorov distance, one obtains that, for every centered and Malliavin differentiable random variable Z,

$$d_{Kol}(Z, N) \leq E|1 - \langle DZ, -DL^{-1}Z\rangle_{\mathfrak{H}}|. \tag{1.6}$$

We will see in Section 3.3 that, in the particular case where $Z = I_q(f)$ is a multiple Wiener-Itô integral of order $q \geq 2$ (that is, Z is an element of the qth Wiener chaos of X) with unit variance, relation (1.6) yields the neat estimate

$$d_{Kol}(Z, N) \leq \sqrt{\frac{q-1}{3q}} \times |E(Z^4) - 3|. \tag{1.7}$$

Note that $E(Z^4) - 3$ is just the fourth cumulant of Z, and that the fourth cumulant of N equals zero. We will also show that the combination of (1.6) and (1.7) allows to recover (and refine) several characterizations of CLTs on a fixed Wiener chaos – as recently proved in [26] and [27].

1.3. Beyond the method of moments

The estimate (1.7), specially when combined with the findings of [23] and [31] (see Section 4), can be seen as a drastic simplification of the so-called 'method of moments and cumulants' (see Major [13] for a classic discussion of this method in the framework of Gaussian analysis). Indeed, such a relation implies that, if $\{Z_n : n \geq 1\}$ is a sequence of random variables with unit variance belonging to a fixed Wiener chaos, then, in order to prove that Z_n converges in law to $N \sim \mathcal{N}(0, 1)$, it is sufficient to show that $E(Z_n^4)$ converges to $E(N^4) = 3$. Again by virtue of (1.7), one also has that the rate of convergence of $E(Z_n^4)$ to 3 determines the 'global' rate convergence in the Kolmogorov distance. In order to further characterize the connections between our techniques and moments computations, in Proposition 3.1 we will deduce some new estimates, implying that (for Z with unit variance and belonging to a fixed Wiener chaos), for every integer $k \geq 3$ the quantity $|E[Z^k] - E[N^k]|$ is controlled (up to an explicit universal multiplicative constant) by the square root of $|E[Z^4] - E[N^4]|$. This result is obtained by means of an interpolation technique, recently used in [22] and originally introduced by Talagrand – see, e.g., [38].

1.4. An overview of the existing literature

The present survey is mostly based on the three references [18], [22] and [23], dealing with upper bounds in the one-dimensional and multi-dimensional approximations of regular functionals of general Gaussian fields (strictly speaking, the papers [18] and [22] also contain results on non-normal approximations, related, e.g., to the Gamma law). However, since the appearance of [18], several related works have been written, which we shall now shortly describe.

- Our paper [19] is again based on Stein's method and Malliavin calculus, and deals with the problem of determining optimal rates of convergence. Some results bear connections with one-term Edgeworth expansions.
- The paper [3], by Breton and Nourdin, completes the analysis initiated in Section 4 of [18], concerning the obtention of Berry-Esséen bounds associated with the so-called Breuer-Major limit theorems (see [5]). The case of non-Gaussian limit laws (of the Rosenblatt type) is also analyzed.
- In [20], by Nourdin, Peccati and Reinert, one can find an application of Stein's method and Malliavin calculus to the derivation of second order Poincaré inequalities on Wiener space. This also refines the CLTs on Wiener chaos proved in [26] and [27].
- One should also mention our paper [16], where we deduce a characterization of non-central limit theorems (associated with Gamma laws) on Wiener chaos. The main findings of [16] are refined in [18] and [22], again by means of Stein's method.
- The work [24], by Nourdin and Viens, contains an application of (1.5) to the estimate of densities and tail probabilities associated with functionals of Gaussian processes, like for instance quadratic functionals or suprema of continuous-time Gaussian processes on a finite interval.
- The findings of [24] have been further refined by Viens in [40], where one can also find some applications to polymer fluctuation exponents.
- The paper [4], by Breton, Nourdin and Peccati, contains some statistical applications of the results of [24], to the construction of confidence intervals for the Hurst parameter of a fractional Brownian motion.
- Reference [2], by Bercu, Nourdin and Taqqu, contains some applications of the results of [18] to almost sure CLTs.
- In [21], by Nourdin, Peccati and Reinert, one can find an extension of the ideas introduced in [18] to the framework of functionals of Rademacher sequences. To this end, one must use a discrete version of Malliavin calculus (see Privault [32]).
- Reference [29], by Peccati, Solé, Taqqu and Utzet, concerns a combination of Stein's method with a version of Malliavin calculus on the Poisson space (as developed by Nualart and Vives in [28]).
- Reference [22], by Nourdin, Peccati and Reinert, contains an application of Stein's method, Malliavin calculus and the 'Lindeberg invariance principle', to the study of universality results for sequences of homogenous sums associated with general collections of independent random variables.

2. Preliminaries

We shall now present the basic elements of Gaussian analysis and Malliavin calculus that are used in this chapter. The reader is referred to the monograph by Nualart

[25] for any unexplained definition or result.

2.1. Isonormal Gaussian processes

Let $\mathfrak{H}$ be a real separable Hilbert space. For any $q \geq 1$, we denote by $\mathfrak{H}^{\otimes q}$ the qth tensor product of $\mathfrak{H}$, and by $\mathfrak{H}^{\odot q}$ the associated qth *symmetric* tensor product; plainly, $\mathfrak{H}^{\otimes 1} = \mathfrak{H}^{\odot 1} = \mathfrak{H}$.

We write $X = \{X(h), h \in \mathfrak{H}\}$ to indicate an *isonormal Gaussian process* over $\mathfrak{H}$. This means that X is a centered Gaussian family, defined on some probability space $(\Omega, \mathcal{F}, P)$, and such that $E[X(g)X(h)] = \langle g, h \rangle_{\mathfrak{H}}$ for every $g, h \in \mathfrak{H}$. Without loss of generality, we also assume that $\mathcal{F}$ is generated by X.

The concept of an isonormal Gaussian process dates back to Dudley's paper [10]. As shown in the forthcoming five examples, this general notion may be used to encode the structure of many remarkable Gaussian families.

Example 2.1 (Euclidean spaces). Fix an integer $d \geq 1$, set $\mathfrak{H} = \mathbb{R}^d$ and let $(e_1, ..., e_d)$ be an orthonormal basis of $\mathbb{R}^d$ (with respect to the usual Euclidean inner product). Let $(Z_1, ..., Z_d)$ be a Gaussian vector whose components are i.i.d. $N(0, 1)$. For every $h = \sum_{j=1}^{d} c_j e_j$ (where the c_j are real and uniquely defined), set $X(h) = \sum_{j=1}^{d} c_j Z_j$ and define $X = \{X(h) : h \in \mathbb{R}^d\}$. Then, X is an isonormal Gaussian process over $\mathbb{R}^d$ endowed with its canonical inner product.

Example 2.2 (Gaussian measures). Let $(A, \mathcal{A}, \nu)$ be a measure space, where ν is positive, σ-finite and non-atomic. Recall that a (real) *Gaussian random measure* over $(A, \mathcal{A})$, with control ν, is a centered Gaussian family of the type

$$G = \{G(B) : B \in \mathcal{A}, \nu(B) < \infty\},$$

satisfying the relation: for every $B, C \in \mathcal{A}$ of finite ν-measure, $E[G(B)G(C)] = \nu(B \cap C)$. Now consider the Hilbert space $\mathfrak{H} = L^2(A, \mathcal{A}, \nu)$, with inner product $\langle h, h' \rangle_{\mathfrak{H}} = \int_A h(a) h'(a) \nu(da)$. For every $h \in \mathfrak{H}$, define $X(h) = \int_A h(a) G(da)$ to be the Wiener-Itô integral of h with respect to G. Then, $X = \{X(h) : h \in L^2(Z, \mathcal{Z}, \nu)\}$ defines a centered Gaussian family with covariance given by $E[X(h)X(h')] = \langle h, h' \rangle_{\mathfrak{H}}$, thus yielding that X is an isonormal Gaussian process over $L^2(A, \mathcal{A}, \nu)$. For instance, by setting $A = [0, +\infty)$ and ν equal to the Lebesgue measure, one obtains that the process $W_t = G([0, t))$, $t \geq 0$, is a standard Brownian motion started from zero (of course, in order to meet the usual definition of a Brownian motion, one has also to select a continuous version of W), and X coincides with the $L^2(\Omega)$-closed linear Gaussian space generated by W.

Example 2.3 (Isonormal spaces derived from covariances).
Let $Y = \{Y_t : t \geq 0\}$ be a real-valued centered Gaussian process indexed by the positive axis, and set $R(s,t) = E[Y_s Y_t]$ to be the covariance function of Y. One can embed Y into some isonormal Gaussian process as follows: (i) define $\mathscr{E}$ as the collection of all finite linear combinations of indicator functions of the type $\mathbf{1}_{[0,t]}$,

$t \geq 0$; (ii) define $\mathfrak{H} = \mathfrak{H}_R$ to be the Hilbert space given by the closure of $\mathscr{E}$ with respect to the inner product

$$\langle f, h \rangle_R := \sum_{i,j} a_i c_j R(s_i, t_j),$$

where $f = \sum_i a_i \mathbf{1}_{[0,s_i]}$ and $h = \sum_j c_j \mathbf{1}_{[0,t_j]}$ are two generic elements of $\mathscr{E}$; (iii) for $h = \sum_j c_j \mathbf{1}_{[0,t_j]} \in \mathscr{E}$, set $X(h) = \sum_j c_j Y_{t_j}$; (iv) for $h \in \mathfrak{H}_R$, set $X(h)$ to be the $L^2(P)$ limit of any sequence of the type $X(h_n)$, where $\{h_n\} \subset \mathscr{E}$ converges to h in $\mathfrak{H}_R$. Note that such a sequence $\{h_n\}$ necessarily exists and may not be unique (however, the definition of $X(h)$ does not depend on the choice of the sequence $\{h_n\}$). Then, by construction, the Gaussian space $\{X(h) : h \in \mathfrak{H}_R\}$ is an isonormal Gaussian process over $\mathfrak{H}_R$. See Chapter 1 of Janson [12] or Nualart [25], as well as the forthcoming Section 3.4, for more details on this construction.

Example 2.4 (Even functions and symmetric measures). Other classic examples of isonormal Gaussian processes (see, e.g., [6,11,13]) are given by objects of the type

$$X_\beta = \{X_\beta(\psi) : \psi \in \mathfrak{H}_{E,\beta}\},$$

where β is a real non-atomic symmetric measure on $(-\pi, \pi]$ (that is, $\beta(dx) = \beta(-dx)$), and

$$\mathfrak{H}_{E,\beta} = L^2_E((-\pi, \pi], d\beta) \tag{2.1}$$

stands for the collection of all *real* linear combinations of complex-valued even functions that are square-integrable with respect to β (recall that a function ψ is even if $\overline{\psi(x)} = \psi(-x)$). The class $\mathfrak{H}_{E,\beta}$ is a real Hilbert space, endowed with the inner product

$$\langle \psi_1, \psi_2 \rangle_\beta = \int_{-\pi}^{\pi} \psi_1(x) \psi_2(-x) \beta(dx) \in \mathbb{R}. \tag{2.2}$$

This type of construction is used in the spectral theory of time series.

Example 2.5 (Gaussian Free Fields). Let $d \geq 2$ and let D be a domain in $\mathbb{R}^d$. Denote by $H_s(D)$ the space of real-valued continuous and continuously differentiable functions on $\mathbb{R}^d$ that are supported on a compact subset of D (note that this implies that the first derivatives of the elements of $H_s(D)$ are square-integrable with respect to the Lebesgue measure). Write $H(D)$ in order to indicate real Hilbert space obtained as the closure of $H_s(D)$ with respect to the inner product $\langle f, g \rangle = \int_{\mathbb{R}^d} \nabla f(x) \cdot \nabla g(x) dx$, where ∇ is the gradient. An isonormal Gaussian process of the type $X = \{X(h) : h \in H(D)\}$ is called a *Gaussian Free Field* (GFF). The reader is referred to the survey by Sheffield [35] for a discussion of the emergence of GFFs in several areas of modern probability. See, e.g., Rider and Virág [34] for a connection with the 'circular law' for Gaussian non-Hermitian random matrices.

Remark 2.1. An isonormal Gaussian process is simply an isomorphism between a centered $L^2(\Omega)$-closed linear Gaussian space and a real separable Hilbert space $\mathfrak{H}$. Now, fix a generic centered $L^2(\Omega)$-closed linear Gaussian space, say $\mathcal{G}$. Since $\mathcal{G}$ is itself a real separable Hilbert space (with respect to the usual $L^2(\Omega)$ inner product) it follows that $\mathcal{G}$ can always be (trivially) represented as an isonormal Gaussian process, by setting $\mathfrak{H} = \mathcal{G}$. Plainly, the subtlety in the use of isonormal Gaussian processes is that one has to select an isomorphism that is well-adapted to the specific problem one wants to tackle.

2.2. Chaos, hypercontractivity and products

We now fix a generic isonormal Gaussian process $X = \{X(h), h \in \mathfrak{H}\}$, defined on some space $(\Omega, \mathcal{F}, P)$ such that $\sigma(X) = \mathcal{F}$.

Wiener chaos. For every $q \geq 1$, we write $\mathcal{H}_q$ in order to indicate the qth *Wiener chaos* of X. We recall that $\mathcal{H}_q$ is the closed linear subspace of $L^2(\Omega, \mathcal{F}, P)$ generated by the random variables of the type $H_q(X(h))$, where $h \in \mathfrak{H}$ is such that $\|h\|_{\mathfrak{H}} = 1$, and H_q stands for the qth *Hermite polynomial*, defined as

$$H_q(x) = (-1)^q e^{\frac{x^2}{2}} \frac{d^q}{dx^q} e^{-\frac{x^2}{2}}, \quad x \in \mathbb{R}, \quad q \geq 1. \tag{2.3}$$

We also use the convention $\mathcal{H}_0 = \mathbb{R}$. For any $q \geq 1$, the mapping

$$I_q(h^{\otimes q}) = q! H_q(X(h)) \tag{2.4}$$

can be extended to a linear isometry between the symmetric tensor product $\mathfrak{H}^{\odot q}$ equipped with the modified norm $\sqrt{q!} \, \|\cdot\|_{\mathfrak{H}^{\otimes q}}$ and the qth Wiener chaos $\mathcal{H}_q$. For $q = 0$, we write $I_0(c) = c$, $c \in \mathbb{R}$.

Remark 2.2. When $\mathfrak{H} = L^2(A, \mathcal{A}, \nu)$, the symmetric tensor product $\mathfrak{H}^{\odot q}$ can be identified with the Hilbert space $L_s^2(A^q, \mathcal{A}^q, \nu^q)$, which is defined as the collection of all symmetric functions on A^q that are square-integrable with respect to ν^q. In this case, it is well-known that the random variable $I_q(h)$, $h \in \mathfrak{H}^{\odot q}$, coincides with the (multiple) *Wiener-Itô integral*, of order q, of h with respect to the Gaussian measure $B \mapsto X(\mathbf{1}_B)$, where $B \in \mathcal{A}$ has finite ν-measure. See chapter 1 of [25] for more details on this point.

Hypercontractivity. Random variables living in a fixed Wiener chaos are hypercontractive. More precisely, assume that Z belongs to the qth Wiener chaos $\mathcal{H}_q$ ($q \geq 1$). Then, Z has a finite variance by construction and, for all $p \in [2, +\infty)$, one has the following estimate (see Theorem 5.10 in [12] for a proof):

$$E(|Z|^p) \leq (p-1)^{pq/2} E(Z^2)^{p/2}. \tag{2.5}$$

In particular, if $E(Z^2) = 1$, one has that $E(|Z|^p) \leq (p-1)^{pq/2}$. For future use, we also observe that, for every $q \geq 1$, the mapping $p \mapsto (p-1)^{pq/2}$ is strictly increasing on $[2, +\infty)$.

Chaotic decompositions. It is well-known (*Wiener chaos decomposition*) that the space $L^2(\Omega, \mathcal{F}, P)$ can be decomposed into the infinite orthogonal sum of the spaces $\mathcal{H}_q$. It follows that any square-integrable random variable $Z \in L^2(\Omega, \mathcal{F}, P)$ admits the following chaotic expansion

$$Z = \sum_{q=0}^{\infty} I_q(f_q), \qquad (2.6)$$

where $f_0 = E(Z)$, and the kernels $f_q \in \mathfrak{H}^{\odot q}$, $q \geq 1$, are uniquely determined. For every $q \geq 0$, we also denote by J_q the orthogonal projection operator on $\mathcal{H}_q$. In particular, if $Z \in L^2(\Omega, \mathcal{F}, P)$ is as in (2.6), then $J_q(Z) = I_q(f_q)$ for every $q \geq 0$.

Contractions. Let $\{e_k, k \geq 1\}$ be a complete orthonormal system in $\mathfrak{H}$. Given $f \in \mathfrak{H}^{\odot p}$ and $g \in \mathfrak{H}^{\odot q}$, for every $r = 0, \ldots, p \wedge q$, the *contraction* of f and g of order r is the element of $\mathfrak{H}^{\otimes(p+q-2r)}$ defined by

$$f \otimes_r g = \sum_{i_1, \ldots, i_r = 1}^{\infty} \langle f, e_{i_1} \otimes \ldots \otimes e_{i_r}\rangle_{\mathfrak{H}^{\otimes r}} \otimes \langle g, e_{i_1} \otimes \ldots \otimes e_{i_r}\rangle_{\mathfrak{H}^{\otimes r}}. \qquad (2.7)$$

Notice that $f \otimes_r g$ is not necessarily symmetric: we denote its symmetrization by $f \widetilde{\otimes}_r g \in \mathfrak{H}^{\odot(p+q-2r)}$. Moreover, $f \otimes_0 g = f \otimes g$ equals the tensor product of f and g while, for $p = q$, one has that $f \otimes_q g = \langle f, g \rangle_{\mathfrak{H}^{\otimes q}}$. In the particular case where $\mathfrak{H} = L^2(A, \mathcal{A}, \nu)$, one has that $\mathfrak{H}^{\odot q} = L_s^2(A^q, \mathcal{A}^q, \nu^q)$ (see Remark 2.2) and the contraction in (2.7) can be written in integral form as

$$(f \otimes_r g)(t_1, \ldots, t_{p+q-2r}) = \int_{A^r} f(t_1, \ldots, t_{p-r}, s_1, \ldots, s_r)$$
$$\times g(t_{p-r+1}, \ldots, t_{p+q-2r}, s_1, \ldots, s_r) d\nu(s_1) \ldots d\nu(s_r).$$

Multiplication. The following multiplication formula is well-known: if $f \in \mathfrak{H}^{\odot p}$ and $g \in \mathfrak{H}^{\odot q}$, then

$$I_p(f) I_q(g) = \sum_{r=0}^{p \wedge q} r! \binom{p}{r} \binom{q}{r} I_{p+q-2r}(f \widetilde{\otimes}_r g). \qquad (2.8)$$

Note that (2.8) gives an immediate proof of the fact that multiple Wiener-Itô integrals have finite moments of every order.

2.3. The language of Malliavin calculus

We now introduce some basic elements of the Malliavin calculus with respect to the isonormal Gaussian process X.

Malliavin derivatives. Let $\mathcal{S}$ be the set of all *cylindrical random variables* of the type

$$Z = g(X(\phi_1), \ldots, X(\phi_n)), \qquad (2.9)$$

where $n \geq 1$, $g : \mathbb{R}^n \to \mathbb{R}$ is an infinitely differentiable function with compact support and $\phi_i \in \mathfrak{H}$. The Malliavin derivative of Z with respect to X is the element of $L^2(\Omega, \mathfrak{H})$ defined as

$$DZ = \sum_{i=1}^{n} \frac{\partial g}{\partial x_i} (X(\phi_1), \ldots, X(\phi_n)) \phi_i.$$

By iteration, one can define the mth derivative $D^m Z$, which is an element of $L^2(\Omega, \mathfrak{H}^{\odot m})$, for every $m \geq 2$. For $m \geq 1$ and $p \geq 1$, $\mathbb{D}^{m,p}$ denotes the closure of $\mathcal{S}$ with respect to the norm $\|\cdot\|_{m,p}$, defined by the relation

$$\|Z\|_{m,p}^p = E[|Z|^p] + \sum_{i=1}^{m} E\left(\|D^i Z\|_{\mathfrak{H}^{\otimes i}}^p\right).$$

The chain rule. The Malliavin derivative D verifies the following chain rule. If $\varphi : \mathbb{R}^d \to \mathbb{R}$ is continuously differentiable with bounded partial derivatives and if $Z = (Z_1, \ldots, Z_d)$ is a vector of elements of $\mathbb{D}^{1,2}$, then $\varphi(Z) \in \mathbb{D}^{1,2}$ and

$$D\varphi(Z) = \sum_{i=1}^{d} \frac{\partial \varphi}{\partial x_i}(Z) DZ_i. \qquad (2.10)$$

A careful application, e.g., of the multiplication formula (2.8) shows that (2.10) continues to hold whenever the function φ is a polynomial in d variables. Note also that a random variable Z as in (2.6) is in $\mathbb{D}^{1,2}$ if and only if $\sum_{q=1}^{\infty} q \|J_q(Z)\|_{L^2(\Omega)}^2 < \infty$ and, in this case, $E\left(\|DZ\|_{\mathfrak{H}}^2\right) = \sum_{q=1}^{\infty} q \|J_q(Z)\|_{L^2(\Omega)}^2$. If $\mathfrak{H} = L^2(A, \mathcal{A}, \nu)$ (with ν non-atomic), then the derivative of a random variable Z as in (2.6) can be identified with the element of $L^2(A \times \Omega)$ given by

$$D_x Z = \sum_{q=1}^{\infty} q I_{q-1} (f_q(\cdot, x)), \quad x \in A. \qquad (2.11)$$

The divergence operator. We denote by δ the adjoint of the operator D, also called the *divergence operator*. A random element $u \in L^2(\Omega, \mathfrak{H})$ belongs to the domain of δ, noted Domδ, if and only if it verifies $|E\langle DZ, u\rangle_{\mathfrak{H}}| \leq c_u \|Z\|_{L^2(\Omega)}$ for any $Z \in \mathbb{D}^{1,2}$, where c_u is a constant depending only on u. If $u \in$ Domδ, then the random variable $\delta(u)$ is defined by the duality relationship

$$E(Z\delta(u)) = E(\langle DZ, u\rangle_{\mathfrak{H}}), \qquad (2.12)$$

which holds for every $Z \in \mathbb{D}^{1,2}$.

Ornstein-Uhlenbeck operators. The operator L, known as the *generator of the Ornstein-Uhlenbeck semigroup*, is defined as $L = \sum_{q=0}^{\infty} -q J_q$. The domain of L is

$$\mathrm{Dom} L = \{Z \in L^2(\Omega) : \sum_{q=1}^{\infty} q^2 \|J_q(Z)\|_{L^2(\Omega)}^2 < \infty\} = \mathbb{D}^{2,2}.$$

There is an important relation between the operators D, δ and L (see, e.g., Proposition 1.4.3 in [25]): a random variable Z belongs to $\mathbb{D}^{2,2}$ if and only if $Z \in \mathrm{Dom}\,(\delta D)$ (i.e. $Z \in \mathbb{D}^{1,2}$ and $DZ \in \mathrm{Dom}\,\delta$) and, in this case,

$$\delta DZ = -LZ. \qquad (2.13)$$

For any $Z \in L^2(\Omega)$, we define $L^{-1}Z = \sum_{q=1}^{\infty} -\frac{1}{q} J_q(Z)$. The operator L^{-1} is called the *pseudo-inverse* of L. For any $Z \in L^2(\Omega)$, we have that $L^{-1}Z \in \mathrm{Dom}\,L$, and

$$LL^{-1}Z = Z - E(Z). \qquad (2.14)$$

An important string of identities. Finally, let us mention a chain of identities playing a crucial role in the sequel. Let $f : \mathbb{R} \to \mathbb{R}$ be a C^1 function with bounded derivative, and let $F, Z \in \mathbb{D}^{1,2}$. Assume moreover that $E(Z) = 0$. By using successively (2.14), (2.13) and (2.10), one deduces that

$$\begin{aligned} E(Zf(F)) &= E(LL^{-1}Z \times f(F)) = E(\delta D(-L^{-1}Z) \times f(F)) \\ &= E(\langle Df(F), -DL^{-1}Z\rangle_{\mathfrak{H}}) \\ &= E(f'(F)\langle DF, -DL^{-1}Z\rangle_{\mathfrak{H}}). \end{aligned} \qquad (2.15)$$

We will shortly see that the fact $E(Zf(F)) = E(f'(F)\langle DF, -DL^{-1}Z\rangle_{\mathfrak{H}})$ constitutes a fundamental element in the connection between Malliavin calculus and Stein's method.

3. One-dimensional approximations

3.1. Stein's lemma for normal approximations

Originally introduced in the path-breaking paper [36], and then further developed in the monograph [37], *Stein's method* can be roughly described as a collection of probabilistic techniques, allowing to characterize the approximation of probability distributions by means of differential operators. As already pointed out in the Introduction, the two surveys [7] and [33] provide a valuable introduction to this very active area of modern probability. In this section, we are mainly interested in the use of Stein's method for the *normal approximation* of the laws of real-valued random variables, where the approximation is performed with respect to the *Kolmogorov distance*. We recall that the Kolmogorov distance between the laws of two real-valued random variables Y and Z is defined by

$$d_{Kol}(Y, Z) = \sup_{z \in \mathbb{R}} |P(Y \leq z) - P(Z \leq z)|.$$

The reader is referred to [18] for several extensions of the results discussed in this survey to other distances between probability measures, such as, e.g., the *total variation distance*, or the *Wasserstein distance*. The following statement, containing all the elements of Stein's method that are needed for our discussion, can be traced back to Stein's original contribution [36].

Lemma 3.1. *Let $N \sim \mathcal{N}(0,1)$ be a standard Gaussian random variable.*

1. Fix $z \in \mathbb{R}$, and define $f_z : \mathbb{R} \to \mathbb{R}$ as

$$f_z(x) = e^{\frac{x^2}{2}} \int_{-\infty}^{x} \left(\mathbf{1}_{(-\infty,z]}(a) - P(N \le z) \right) e^{-\frac{a^2}{2}} da, \quad x \in \mathbb{R}. \qquad (3.1)$$

Then, f_z is continuous on $\mathbb{R}$, bounded by $\sqrt{2\pi}/4$, differentiable on $\mathbb{R} \setminus \{z\}$, and verifies moreover

$$f_z'(x) - x f_z(x) = \mathbf{1}_{(-\infty,z]}(x) - P(N \le z) \quad \text{for all } x \in \mathbb{R} \setminus \{z\}. \qquad (3.2)$$

One has also that f_z is Lipschitz, with Lipschitz constant less or equal to 1.

2. Let Z be a generic random variable. Then,

$$d_{Kol}(Z, N) \le \sup_{f} |E[Zf(Z) - f'(Z)]|, \qquad (3.3)$$

where the supremum is taken over the class of all Lipschitz functions that are bounded by $\sqrt{2\pi}/4$ and whose Lipschitz constant is less or equal to 1.

3. Let Z be a generic random variable. Then, $Z \sim \mathcal{N}(0,1)$ if and only if $E[Zf(Z) - f'(Z)] = 0$ for every continuous and piecewise differentiable function f verifying the relation $E|f'(N)| < \infty$.

Proof. (Point 1) We shall only prove that f_z is Lipschitz and we will evaluate its constant (the proof of the remaining properties is left to the reader). We have, for $x \ge 0$, $x \ne z$:

$$|f_z'(x)| = \left| \mathbf{1}_{(-\infty,z]}(x) - P(N \le z) + xe^{\frac{x^2}{2}} \int_{-\infty}^{x} \left(\mathbf{1}_{(-\infty,z]}(a) - P(N \le z) \right) e^{-\frac{a^2}{2}} da \right|$$

$$\underset{(*)}{=} \left| \mathbf{1}_{(-\infty,z]}(x) - P(N \le z) - xe^{\frac{x^2}{2}} \int_{x}^{+\infty} \left(\mathbf{1}_{(-\infty,z]}(a) - P(N \le z) \right) e^{-\frac{a^2}{2}} da \right|$$

$$\le \|\mathbf{1}_{(-\infty,z]}(\cdot) - P(N \le z)\|_{\infty} \left(1 + xe^{\frac{x^2}{2}} \int_{x}^{+\infty} e^{-\frac{a^2}{2}} da \right)$$

$$\le 1 + e^{\frac{x^2}{2}} \int_{x}^{+\infty} a e^{-\frac{a^2}{2}} da = 2.$$

Observe that identity $(*)$ holds since

$$0 = E\big(\mathbf{1}_{(-\infty,z]}(N) - P(N \le z)\big) = \frac{1}{\sqrt{2\pi}} \int_{-\infty}^{+\infty} \left(\mathbf{1}_{(-\infty,z]}(a) - P(N \le z) \right) e^{-\frac{a^2}{2}} da.$$

For $x \le 0$, $x \ne z$, we can write

$$|f_z'(x)| = \left| \mathbf{1}_{(-\infty,z]}(x) - P(N \le z) + xe^{\frac{x^2}{2}} \int_{-\infty}^{x} \left(\mathbf{1}_{(-\infty,z]}(a) - P(N \le z) \right) e^{-\frac{a^2}{2}} da \right|$$

$$\le \|\mathbf{1}_{(-\infty,z]}(\cdot) - P(N \le z)\|_{\infty} \left(1 + |x| e^{\frac{x^2}{2}} \int_{-\infty}^{x} e^{-\frac{a^2}{2}} da \right)$$

$$\le 1 + e^{\frac{x^2}{2}} \int_{-\infty}^{x} |a| e^{-\frac{a^2}{2}} da = 2.$$

Hence, we have shown that f_z is Lipschitz with Lipschitz constant bounded by 2. For the announced refinement (that is, the constant is bounded by 1), we refer the reader to Lemma 2.2 in Chen and Shao [7].

(*Point 2*) Take expectations on both sides of (3.2) with respect to the law of Z. Then, take the supremum over all $z \in \mathbb{R}$, and exploit the properties of f_z proved at Point 1. ●

(*Point 3*) If $Z \sim \mathcal{N}(0,1)$, a simple application of the Fubini theorem (or, equivalently, an integration by parts) yields that $E[Zf(Z)] = E[f'(Z)]$ for every smooth f. Now suppose that $E[Zf(Z) - f'(Z)] = 0$ for every function f as in the statement, so that this equality holds in particular for $f = f_z$ and for every $z \in \mathbb{R}$. By integrating both sides of (3.2) with respect to the law of Z, this yields that $P(Z \leq z) = P(N \leq z)$ for every $z \in \mathbb{R}$, and therefore that Z and N have the same law.

Remark 3.3. Formulae (3.2) and (3.3) are known, respectively, as *Stein's equation* and *Stein's bound*. As already evoked in the Introduction, Point 3 in the statement of Lemma 3.1 is customarily referred to as *Stein's lemma*.

3.2. General bounds on the Kolmogorov distance

We now face the problem of establishing a bound on the normal approximation of a centered and Malliavin-differentiable random variable. The next statement contains one of the main findings of [18].

Theorem 3.1. *(See [18]) Let* $Z \in \mathbb{D}^{1,2}$ *be such that* $E(Z) = 0$ *and* $\mathrm{Var}(Z) = 1$. *Then, for* $N \sim \mathcal{N}(0,1)$,

$$d_{Kol}(Z, N) \leq \sqrt{\mathrm{Var}(\langle DZ, -DL^{-1}Z \rangle_{\mathfrak{H}})}. \tag{3.4}$$

Proof. In view of (3.3), it is enough to prove that, for every Lipschitz function f with Lipschitz constant less or equal to 1, one has that the quantity $|E[Zf(Z) - f'(Z)]|$ is less or equal to the right-hand side of (3.4). Start by considering a function $f : \mathbb{R} \to \mathbb{R}$ which is C^1 and such that $\|f'\|_\infty \leq 1$. Relation (2.15) yields

$$E(Zf(Z)) = E(f'(Z)\langle DZ, -DL^{-1}Z \rangle_{\mathfrak{H}}),$$

so that

$$\begin{aligned}|E(f'(Z)) - E(Zf(Z))| &= |E(f'(Z)(1 - \langle DZ, -DL^{-1}Z \rangle_{\mathfrak{H}}))| \\ &\leq E|1 - \langle DZ, -DL^{-1}Z \rangle_{\mathfrak{H}}|.\end{aligned}$$

By a standard approximation argument (e.g. by using a convolution with an approximation of the identity), one sees that the inequality $|E(f'(Z)) - E(Zf(Z))| \leq E|1 - \langle DZ, -DL^{-1}Z \rangle_{\mathfrak{H}}|$ continues to hold when f is Lipschitz with constant less or equal to 1. Hence, by combining the previous estimates with (3.3), we infer that

$$d_{Kol}(Z, N) \leq E|1 - \langle DZ, -DL^{-1}Z \rangle_{\mathfrak{H}}| \leq \sqrt{E(1 - \langle DZ, -DL^{-1}Z \rangle_{\mathfrak{H}})^2}.$$

Finally, the desired conclusion follows by observing that, if one chooses $f(z) = z$ in (2.15), then one obtains

$$E(\langle DZ, -DL^{-1}Z\rangle_{\mathfrak{H}}) = E(Z^2) = 1, \tag{3.5}$$

so that $E\left[\left(1 - \langle DZ, -DL^{-1}Z\rangle_{\mathfrak{H}}\right)^2\right] = \operatorname{Var}(\langle DZ, -DL^{-1}Z\rangle_{\mathfrak{H}}).$

Remark 3.4. By using the standard properties of conditional expectations, one sees that (3.4) also implies the 'finer' bound

$$d_{Kol}(Z, N) \le \sqrt{\operatorname{Var}(g(Z))}, \tag{3.6}$$

where $g(Z) = E[\langle DZ, -DL^{-1}Z\rangle_{\mathfrak{H}}|Z]$. In general, it is quite difficult to obtain an explicit expression of the function g. However, if some crude estimates on g are available, then one can obtain explicit upper and lower bounds for the densities and the tail probabilities of the random variable Z. The reader is referred to Nourdin and Viens [24] and Viens [40] for several results in this direction, and to Breton et al. [4] for some statistical applications of these ideas.

3.3. Wiener chaos and the fourth moment condition

In this section, we will apply Theorem 3.1 to *chaotic* random variables, that is, random variables having the special form of multiple Wiener-Itô integrals of some fixed order $q \ge 2$. As announced in the Introduction, this allows to recover and refine some recent characterizations of CLTs on Wiener chaos (see [26,27]). We begin with a technical lemma.

Lemma 3.2. *Fix an integer $q \ge 1$, and let $Z = I_q(f)$ (with $f \in \mathfrak{H}^{\odot q}$) be such that $\operatorname{Var}(Z) = E(Z^2) = 1$. The following three identities are in order:*

$$\frac{1}{q}\|DZ\|_{\mathfrak{H}}^2 - 1 = q\sum_{r=1}^{q-1}(r-1)!\binom{q-1}{r-1}^2 I_{2q-2r}(f\widetilde{\otimes}_r f), \tag{3.7}$$

$$\operatorname{Var}\left(\frac{1}{q}\|DZ\|_{\mathfrak{H}}^2\right) = \sum_{r=1}^{q-1}\frac{r^2}{q^2}r!^2\binom{q}{r}^4(2q-2r)!\|f\widetilde{\otimes}_r f\|_{\mathfrak{H}^{\otimes 2q-2r}}^2, \tag{3.8}$$

and

$$E(Z^4) - 3 = \frac{3}{q}\sum_{r=1}^{q-1}rr!^2\binom{q}{r}^4(2q-2r)!\|f\widetilde{\otimes}_r f\|_{\mathfrak{H}^{\otimes 2q-2r}}^2. \tag{3.9}$$

In particular,

$$\operatorname{Var}\left(\frac{1}{q}\|DZ\|_{\mathfrak{H}}^2\right) \le \frac{q-1}{3q}(E(Z^4) - 3). \tag{3.10}$$

Proof. Without loss of generality, we can assume that $\mathfrak{H}$ is equal to $L^2(A, \mathcal{A}, \nu)$, where $(A, \mathcal{A})$ is a measurable space and ν a σ-finite measure without atoms. For any $a \in A$, we have $D_a Z = q I_{q-1}(f(\cdot, a))$ so that

$$\frac{1}{q}\|DZ\|_{\mathfrak{H}}^2 = q \int_A I_{q-1}(f(\cdot, a))^2 \nu(da)$$

$$= q \int_A \sum_{r=0}^{q-1} r! \binom{q-1}{r}^2 I_{2q-2-2r}(f(\cdot, a) \otimes_r f(\cdot, a)) \nu(da) \quad \text{by (2.8)}$$

$$= q \sum_{r=0}^{q-1} r! \binom{q-1}{r}^2 I_{2q-2-2r} \left(\int_A f(\cdot, a) \otimes_r f(\cdot, a) \nu(da) \right)$$

$$= q \sum_{r=0}^{q-1} r! \binom{q-1}{r}^2 I_{2q-2-2r}(f \otimes_{r+1} f)$$

$$= q \sum_{r=1}^{q} (r-1)! \binom{q-1}{r-1}^2 I_{2q-2r}(f \otimes_r f).$$

$$= q! \|f\|_{\mathfrak{H}^{\otimes q}}^2 + q \sum_{r=1}^{q-1} (r-1)! \binom{q-1}{r-1}^2 I_{2q-2r}(f \otimes_r f).$$

Since $E(Z^2) = q! \|f\|_{\mathfrak{H}^{\otimes q}}^2$, the proof of (3.7) is finished. The identity (3.8) follows from (3.7) and the orthogonality properties of multiple stochastic integrals. Using (in order) formula (2.13) and the relation $D(Z^3) = 3Z^2 DZ$, we infer that

$$E(Z^4) = \frac{1}{q} E(\delta DZ \times Z^3) = \frac{1}{q} E(\langle DZ, D(Z^3) \rangle_{\mathfrak{H}}) = \frac{3}{q} E(Z^2 \|DZ\|_{\mathfrak{H}}^2). \quad (3.11)$$

Moreover, the multiplication formula (2.8) yields

$$Z^2 = I_q(f)^2 = \sum_{s=0}^{q} s! \binom{q}{s}^2 I_{2q-2s}(f \otimes_s f). \quad (3.12)$$

By combining this last identity with (3.7) and (3.11), we obtain (3.9) and finally (3.10). □

As a consequence of Lemma 3.2, we deduce the following bound on the Kolmogorov distance – first proved in [22].

Theorem 3.2. *(See [22]) Let Z belong to the qth chaos $\mathcal{H}_q$ of X, for some $q \geq 2$. Suppose moreover that* $\mathrm{Var}(Z) = E(Z^2) = 1$. *Then*

$$d_{Kol}(Z, N) \leq \sqrt{\frac{q-1}{3q}(E(Z^4) - 3)}. \quad (3.13)$$

Proof. Since $L^{-1}Z = -\frac{1}{q}Z$, we have $\langle DZ, -DL^{-1}Z \rangle_{\mathfrak{H}} = \frac{1}{q}\|DZ\|_{\mathfrak{H}}^2$. So, we only need to apply Theorem 3.1 and formula (3.10). □

The estimate (3.13) allows to deduce the following characterization of CLTs on Wiener chaos. Note that the equivalence of Point (i) and Point (ii) in the next statement was first proved by Nualart and Peccati in [27] (by completely different techniques based on stochastic time-changes), whereas the equivalence of Point (iii) was first obtained by Nualart and Ortiz-Latorre in [26] (by means of Malliavin calculus, but not of Stein's method).

Theorem 3.3. *(see [26,27]) Let (Z_n) be a sequence of random variables belonging to the qth chaos $\mathcal{H}_q$ of X, for some fixed $q \geq 2$. Assume that $\mathrm{Var}(Z_n) = E(Z_n^2) = 1$ for all n. Then, as $n \to \infty$, the following three assertions are equivalent:*

(i) $Z_n \xrightarrow{Law} N \sim \mathcal{N}(0,1)$;
(ii) $E(Z_n^4) \to E(N^4) = 3$;
(iii) $\mathrm{Var}\left(\frac{1}{q}\|DZ_n\|_{\mathfrak{H}}^2\right) \to 0$.

Proof. For every n, write $Z_n = I_q(f_n)$ with $f_n \in \mathfrak{H}^{\odot q}$ uniquely determined. The implication (iii) $\to$ (i) is a direct application of Theorem 3.2, and of the fact that the topology of the Kolmogorov distance is stronger than the topology of the convergence in law. The implication (i) $\to$ (ii) comes from a bounded convergence argument (observe that $\sup_{n \geq 1} E(Z_n^4) < \infty$ by the hypercontractivity relation (2.5)). Finally, let us prove the implication (ii) $\to$ (iii). Suppose that (ii) is in order. Then, by virtue of (3.9), we have that $\|f_n \widetilde{\otimes}_r f_n\|_{\mathfrak{H}^{\otimes 2q-2r}}$ tends to zero, as $n \to \infty$, for all (fixed) $r \in \{1, \ldots, q-1\}$. Hence, (3.8) allows to conclude that (iii) is in order. The proof of Theorem 3.3 is thus complete.

Remark 3.5. Theorem 3.3 has been applied to a variety of situations: see, e.g., (but the list is by no means exhaustive) Barndorff-Nielsen et al. [1], Corcuera et al. [8], Marinucci and Peccati [14], Neuenkirch and Nourdin [15], Nourdin and Peccati [17] and Tudor and Viens [39], and the references therein. See Peccati and Taqqu [30] for several combinatorial interpretations of these results.

By combining Theorem 3.2 and Theorem 3.3, we obtain the following result.

Corollary 14.1. *Let the assumptions of Corollary 3.3 prevail. As $n \to \infty$, the following assertions are equivalent:*

(a) $Z_n \xrightarrow{Law} N \sim \mathcal{N}(0,1)$;
(b) $d_{Kol}(Z_n, N) \to 0$.

Proof. Of course, only the implication (a) $\to$ (b) has to be proved. Assume that (a) is in order. By Corollary 3.3, we have that $\mathrm{Var}\left(\frac{1}{q}\|DZ_n\|_{\mathfrak{H}}^2\right) \to 0$. Using Theorem 3.2, we get that (b) holds, and the proof is done.

3.4. Quadratic variation of the fractional Brownian motion, part one

In this section, we use Theorem 3.1 in order to derive an explicit bound for the second-order approximation of the quadratic variation of a fractional Brownian motion.

Let $B = \{B_t : t \geq 0\}$ be a fractional Brownian motion with Hurst index $H \in (0,1)$. This means that B is a centered Gaussian process, started from zero and with covariance function $E(B_s B_t) = R(s,t)$ given by

$$R(s,t) = \frac{1}{2}\left(t^{2H} + s^{2H} - |t-s|^{2H}\right), \quad s,t \geq 0.$$

The fractional Brownian motion of index H is the *only* centered Gaussian processes normalized in such a way that $\text{Var}(B_1) = 1$, and such that B is selfsimilar with index H and has stationary increments. If $H = 1/2$ then $R(s,t) = \min(s,t)$ and B is simply a standard Brownian motion. If $H \neq 1/2$, then B is neither a (semi)martingale nor a Markov process (see, e.g., [25] for more details).

As already explained in the Introduction (see Example 2.3), for any choice of the Hurst parameter $H \in (0,1)$ the Gaussian space generated by B can be identified with an isonormal Gaussian process $X = \{X(h) : h \in \mathfrak{H}\}$, where the real and separable Hilbert space $\mathfrak{H}$ is defined as follows: (i) denote by $\mathscr{E}$ the set of all $\mathbb{R}$-valued step functions on $[0,\infty)$, (ii) define $\mathfrak{H}$ as the Hilbert space obtained by closing $\mathscr{E}$ with respect to the scalar product

$$\langle \mathbf{1}_{[0,t]}, \mathbf{1}_{[0,s]} \rangle_{\mathfrak{H}} = R(t,s).$$

In particular, with such a notation, one has that $B_t = X(\mathbf{1}_{[0,t]})$.

Set

$$Z_n = \frac{1}{\sigma_n} \sum_{k=0}^{n-1} \left[(B_{k+1} - B_k)^2 - 1\right] \stackrel{\text{Law}}{=} \frac{n^{2H}}{\sigma_n} \sum_{k=0}^{n-1} \left[(B_{(k+1)/n} - B_{k/n})^2 - n^{-2H}\right]$$

where $\sigma_n > 0$ is chosen so that $E(Z_n^2) = 1$. It is well-known (see, e.g., [5]) that, for every $H \leq 3/4$ and for $n \to \infty$, one has that Z_n converges in law to $N \sim \mathcal{N}(0,1)$. The following result uses Stein's method in order to obtain an explicit bound for the Kolmogorov distance between Z_n and N. It was first proved in [18] (for the case $H < 3/4$) and [3] (for $H = 3/4$).

Theorem 3.4. *Let $N \sim \mathcal{N}(0,1)$ and assume that $H \leq 3/4$. Then, there exists a constant $c_H > 0$ (depending only on H) such that, for every $n \geq 1$,*

$$d_{Kol}(Z_n, N) \leq c_H \times \begin{cases} \frac{1}{\sqrt{n}} & \text{if } H \in (0, \tfrac{1}{2}] \\ n^{2H - \tfrac{3}{2}} & \text{if } H \in [\tfrac{1}{2}, \tfrac{3}{4}) \\ \frac{1}{\sqrt{\log n}} & \text{if } H = \tfrac{3}{4} \end{cases} \quad (3.14)$$

Remark 3.6.

(1) By inspection of the forthcoming proof of Theorem 3.4, one sees that $\lim_{n\to\infty} \frac{\sigma_n^2}{n} = 2\sum_{r\in\mathbb{Z}} \rho^2(r)$ if $H \in (0, 3/4)$, with ρ given by (3.15), and $\lim_{n\to\infty} \frac{\sigma_n^2}{n \log n} = 9/16$ if $H = 3/4$.

(2) When $H > 3/4$, the sequence (Z_n) does not converge in law to $\mathcal{N}(0,1)$. Actually, $Z_n \xrightarrow[n\to\infty]{\text{Law}} Z_\infty \sim$ 'Hermite random variable' and, using a result by Davydov and Martynova [9], one can also associate a bound to this convergence. See [3] for details on this result.

(3) More generally, and using the analogous computations, one can associate bounds with the convergence of sequence

$$Z_n^{(q)} = \frac{1}{\sigma_n^{(q)}} \sum_{k=0}^{n-1} H_q(B_{k+1} - B_k) \stackrel{\text{Law}}{=} \frac{1}{\sigma_n^{(q)}} \sum_{k=0}^{n-1} H_q(n^H (B_{(k+1)/n} - B_{k/n}))$$

towards $N \sim \mathcal{N}(0,1)$, where H_q ($q \geq 3$) denotes the qth Hermite polynomial (as defined in (2.3)), and $\sigma_n^{(q)}$ is some appropriate normalizing constant. In this case, the critical value is $H = 1 - 1/(2q)$ instead of $H = 3/4$. See [18] for details.

In order to show Theorem 3.4, we will need the following ancillary result, whose proof is obvious and left to the reader.

Lemma 3.3.

1. For $r \in \mathbb{Z}$, let

$$\rho(r) = \frac{1}{2}(|r+1|^{2H} + |r-1|^{2H} - 2|r|^{2H}). \tag{3.15}$$

If $H \neq \frac{1}{2}$, one has $\rho(r) \sim H(2H-1)|r|^{2H-2}$ as $|r| \to \infty$. If $H = \frac{1}{2}$ and $|r| \geq 1$, one has $\rho(r) = 0$. Consequently, $\sum_{r\in\mathbb{Z}} \rho^2(r) < \infty$ if and only if $H < 3/4$.

2. For all $\alpha > -1$, we have $\sum_{r=1}^{n-1} r^\alpha \sim n^{\alpha+1}/(\alpha+1)$ as $n \to \infty$.

We are now ready to prove the main result of this section.

Proof of Theorem 3.4. Since $\|1_{[k,k+1]}\|_{\mathfrak{H}}^2 = E((B_{k+1} - B_k)^2) = 1$, we have, by (2.4),

$$(B_{k+1} - B_k)^2 - 1 = I_2(1_{[k,k+1]}^{\otimes 2})$$

so that $Z_n = I_2(f_n)$ with $f_n = \frac{1}{\sigma_n} \sum_{k=0}^{n-1} 1_{[k,k+1]}^{\otimes 2} \in \mathfrak{H}^{\odot 2}$. Let us compute the exact value of σ_n. Observe that $\langle 1_{[k,k+1]}, 1_{[l,l+1]} \rangle_{\mathfrak{H}} = E((B_{k+1} - B_k)(B_{l+1} - B_l)) = \rho(k-l)$

with ρ given by (3.15). Hence

$$E\left[\left(\sum_{k=0}^{n-1}[(B_{k+1}-B_k)^2-1]\right)^2\right]$$

$$=E\left[\left(\sum_{k=0}^{n-1}I_2(1^{\otimes 2}_{[k,k+1]})\right)^2\right]=\sum_{k,l=0}^{n-1}E\left[I_2(1^{\otimes 2}_{[k,k+1]})I_2(1^{\otimes 2}_{[l,l+1]})\right]$$

$$=2\sum_{k,l=0}^{n-1}\langle 1_{[k,k+1]},1_{[l,l+1]}\rangle^2_{\mathfrak{H}}=2\sum_{k,l=0}^{n-1}\rho^2(k-l).$$

That is,

$$\sigma_n^2=2\sum_{k,l=0}^{n-1}\rho^2(k-l)=2\sum_{l=0}^{n-1}\sum_{r=-l}^{n-1-l}\rho^2(r)=2\left(n\sum_{|r|<n}\rho^2(r)-\sum_{|r|<n}(|r|+1)\rho^2(r)\right).$$

Assume that $H<3/4$. Then, we have

$$\frac{\sigma_n^2}{n}=2\sum_{r\in\mathbb{Z}}\rho^2(r)\left(1-\frac{|r|+1}{n}\right)1_{\{|r|<n\}}.$$

Since $\sum_{r\in\mathbb{Z}}\rho^2(r)<\infty$, we obtain, by bounded Lebesgue convergence:

$$\lim_{n\to\infty}\frac{\sigma_n^2}{n}=2\sum_{r\in\mathbb{Z}}\rho^2(r). \qquad (3.16)$$

Assume that $H=3/4$. We have $\rho^2(r)\sim\frac{9}{64|r|}$ as $|r|\to\infty$. Therefore, as $n\to\infty$,

$$n\sum_{|r|<n}\rho^2(r)\sim\frac{9n}{64}\sum_{0<|r|<n}\frac{1}{|r|}\sim\frac{9n\log n}{32}$$

and

$$\sum_{|r|<n}(|r|+1)\rho^2(r)\sim\frac{9}{64}\sum_{|r|<n}1\sim\frac{9n}{32}.$$

We deduce that

$$\lim_{n\to\infty}\frac{\sigma_n^2}{n\log n}=\frac{9}{16}. \qquad (3.17)$$

Now, we have, see (3.8) for the first equality,

$$\mathrm{Var}(\tfrac{1}{2}\|DZ_n\|_{\mathfrak{H}}^2) = \tfrac{1}{2}\|f_n \otimes_1 f_n\|_{\mathfrak{H}^{\otimes 2}}^2 = \frac{1}{2\sigma_n^4}\left\|\sum_{k,l=0}^{n-1} \mathbf{1}_{[k,k+1]}^{\otimes 2} \otimes_1 \mathbf{1}_{[l,l+1]}^{\otimes 2}\right\|_{\mathfrak{H}}^2$$

$$= \frac{1}{2\sigma_n^4}\left\|\sum_{k,l=0}^{n-1} \rho(k-l)\mathbf{1}_{[k,k+1]} \otimes \mathbf{1}_{[l,l+1]}\right\|_{\mathfrak{H}}^2$$

$$= \frac{1}{2\sigma_n^4}\sum_{i,j,k,l=0}^{n-1} \rho(k-l)\rho(i-j)\rho(k-i)\rho(l-j)$$

$$\leq \frac{1}{4\sigma_n^4}\sum_{i,j,k,l=0}^{n-1} |\rho(k-i)||\rho(i-j)|\big(\rho^2(k-l) + \rho^2(l-j)\big)$$

$$\leq \frac{1}{2\sigma_n^4}\sum_{i,j,k=0}^{n-1} |\rho(k-i)||\rho(i-j)|\sum_{r=-n+1}^{n-1} \rho^2(r)$$

$$\leq \frac{n}{2\sigma_n^4}\left(\sum_{s=-n+1}^{n-1}|\rho(s)|\right)^2 \sum_{r=-n+1}^{n-1} \rho^2(r).$$

If $H \leq 1/2$ then $\sum_{s\in\mathbb{Z}}|\rho(s)| < \infty$ and $\sum_{r\in\mathbb{Z}}\rho^2(r) < \infty$ so that, in view of (3.16), $\mathrm{Var}(\tfrac{1}{2}\|DZ_n\|_{\mathfrak{H}}^2) = O(n^{-1})$. If $1/2 < H < 3/4$ then $\sum_{s=-n+1}^{n-1}|\rho(s)| = O(n^{2H-1})$ (see Lemma 3.3) and $\sum_{r\in\mathbb{Z}}\rho^2(r) < \infty$ so that, in view of (3.16), one has $\mathrm{Var}(\tfrac{1}{2}\|DZ_n\|_{\mathfrak{H}}^2) = O(n^{4H-3})$. If $H = 3/4$ then $\sum_{s=-n+1}^{n-1}|\rho(s)| = O(\sqrt{n})$ and $\sum_{r=-n+1}^{n-1}\rho^2(r) = O(\log n)$ (indeed, by Lemma 3.3, $\rho^2(r) \sim \frac{cst}{|r|}$ as $|r| \to \infty$) so that, in view of (3.17), $\mathrm{Var}(\tfrac{1}{2}\|DZ_n\|_{\mathfrak{H}}^2) = O(1/\log n)$. Finally, the desired conclusion follows from Theorem 3.2. □

3.5. The method of (fourth) moments: explicit estimates via interpolation

It is clear that the combination of Theorem 3.2 and Theorem 3.3 provides a remarkable simplification of the method of moments and cumulants, as applied to the derivation of CLTs on a fixed Wiener chaos (further generalizations of these results, concerning in particular multi-dimensional CLTs, are discussed in the forthcoming Section 4). In particular, one deduces from (3.13) that, for a sequence of chaotic random variables with unit variance, the speed of convergence to zero of the fourth cumulants $E(Z_n^4) - 3$ also determines the speed of convergence in the Kolmogorov distance.

In this section, we shall state and prove a new upper bound, showing that, for a normalized chaotic sequence $\{Z_n : n \geq 1\}$ converging in distribution to $N \sim \mathscr{N}(0,1)$, the convergence to zero of $E(Z_n^k) - E(N^k)$ is always dominated by the speed of convergence of the square root of $E(Z_n^4) - E(N^4) = E(Z_n^4) - 3$. To do this, we shall apply a well-known Gaussian interpolation technique, which has been essentially introduced by Talagrand (see, e.g., [38]); note that a similar approach has recently been adopted in [22], in order to deduce a universal characterization of CLTs for sequences of homogeneous sums.

Remark 3.7.

1. In principle, one could deduce from the results of this section that, for every $k \geq 3$, the speed of convergence to zero of kth cumulant of Z_n is always dominated by the speed of convergence of the fourth cumulant $E(Z_n^4) - 3$.
2. We recall that the explicit computation of moments and cumulants of chaotic random variables is often performed by means of a class of combinatorial devices, known as *diagram formulae*. This tools are not needed in our analysis, as we rather rely on multiplication formulae and integration by parts techniques from Malliavin calculus. See section 3 in [30] for a recent and self-contained introduction to moments, cumulants and diagram formulae.

Proposition 3.1. *Let $q \geq 2$ be an integer, and let Z be an element of the qth chaos $\mathcal{H}_q$ of X. Assume that $\mathrm{Var}(Z) = E(Z^2) = 1$, and let $N \sim \mathcal{N}(0,1)$. Then, for all integer $k \geq 3$,*

$$\left|E(Z^k) - E(N^k)\right| \leq c_{k,q}\sqrt{E(Z^4) - E(N^4)}, \qquad (3.18)$$

where the constant $c_{k,q}$ is given by

$$c_{k,q} = (k-1)2^{k-\frac{5}{2}}\sqrt{\frac{q-1}{3q}}\left(\sqrt{\frac{(2k-4)!}{2^{k-2}(k-2)!}} + (2k-5)^{\frac{kq}{2}-q}\right).$$

Proof. Without loss of generality, we can assume that N is independent of the underlying isonormal Gaussian process X. Fix an integer $k \geq 3$. By denoting $\Psi(t) = E[(\sqrt{1-t}Z + \sqrt{t}N)^k]$, $t \in [0,1]$, we have

$$\left|E(Z^k) - E(N^k)\right| = |\Psi(1) - \Psi(0)| \leq \int_0^1 |\Psi'(t)|\,dt,$$

where the derivative Ψ' is easily seen to exist on $(0,1)$, and moreover one has

$$\Psi'(t) = \frac{k}{2\sqrt{t}}E[(\sqrt{1-t}Z + \sqrt{t}N)^{k-1}N] - \frac{k}{2\sqrt{1-t}}E[(\sqrt{1-t}Z + \sqrt{t}N)^{k-1}Z].$$

By integrating by parts and by using the explicit expression of the Gaussian density, one infers that

$$E[(\sqrt{1-t}Z + \sqrt{t}N)^{k-1}N] = E\left[E[(\sqrt{1-t}z + \sqrt{t}N)^{k-1}N]_{|z=Z}\right]$$

$$= (k-1)\sqrt{t}\,E\left[E[(\sqrt{1-t}z + \sqrt{t}N)^{k-2}]_{|z=Z}\right]$$

$$= (k-1)\sqrt{t}\,E[(\sqrt{1-t}Z + \sqrt{t}N)^{k-2}].$$

Similarly, using this time (2.15) in order to perform the integration by parts and taking into account that $\langle DZ, -DL^{-1}Z\rangle_{\mathfrak{H}} = \frac{1}{q}\|DZ\|_{\mathfrak{H}}^2$ because $Z \in \mathcal{H}_q$, we can write

$$E[(\sqrt{1-t}Z + \sqrt{t}N)^{k-1}Z]$$
$$= E\left[E[(\sqrt{1-t}Z + \sqrt{t}x)^{k-1}Z]_{|x=N}\right]$$
$$= (k-1)\sqrt{1-t}\, E\left[E[(\sqrt{1-t}Z + \sqrt{t}x)^{k-2}\frac{1}{q}\|DZ\|_{\mathfrak{H}}^2]_{|x=N}\right]$$
$$= (k-1)\sqrt{1-t}\, E\left[(\sqrt{1-t}Z + \sqrt{t}N)^{k-2}\frac{1}{q}\|DZ\|_{\mathfrak{H}}^2\right].$$

Hence,

$$\Psi'(t) = \frac{k(k-1)}{2}E\left[\left(1 - \frac{1}{q}\|DZ\|_{\mathfrak{H}}^2\right)(\sqrt{1-t}Z + \sqrt{t}N)^{k-2}\right],$$

and consequently

$$|\Psi'(t)| \leq \frac{k(k-1)}{2}\sqrt{E\left[(\sqrt{1-t}Z + \sqrt{t}N)^{2k-4}\right]} \times \sqrt{E\left[\left(1 - \frac{1}{q}\|DZ\|_{\mathfrak{H}}^2\right)^2\right]}.$$

By (3.5) and (3.10), we have

$$E\left[\left(1 - \frac{1}{q}\|DZ\|_{\mathfrak{H}}^2\right)^2\right] = \operatorname{Var}\left(\frac{1}{q}\|DZ\|_{\mathfrak{H}}^2\right) \leq \frac{q-1}{3q}(E(Z^4) - 3).$$

Using succesively $(x+y)^{2k-4} \leq 2^{2k-5}(x^{2k-4} + y^{2k-4})$, $\sqrt{x+y} \leq \sqrt{x} + \sqrt{y}$, inequality (2.5) and $E(N^{2k-4}) = (2k-4)!/(2^{k-2}(k-2)!)$, we can write

$$\sqrt{E\left[(\sqrt{1-t}Z + \sqrt{t}N)^{2k-4}\right]}$$
$$\leq 2^{k-\frac{5}{2}}(1-t)^{\frac{k}{2}-1}\sqrt{E(Z^{2k-4})} + 2^{k-\frac{5}{2}}t^{\frac{k}{2}-1}\sqrt{E(N^{2k-4})}$$
$$\leq 2^{k-\frac{5}{2}}(1-t)^{\frac{k}{2}-1}(2k-5)^{\frac{kq}{2}-q} + 2^{k-\frac{5}{2}}t^{\frac{k}{2}-1}\sqrt{\frac{(2k-4)!}{2^{k-2}(k-2)!}}$$

so that

$$\int_0^1 \sqrt{E\left[(\sqrt{1-t}Z + \sqrt{t}N)^{2k-4}\right]}\, dt \leq \frac{2^{k-\frac{3}{2}}}{k}\left[(2k-5)^{\frac{kq}{2}-q} + \sqrt{\frac{(2k-4)!}{2^{k-2}(k-2)!}}\right].$$

Putting all these bounds together, one deduces the desired conclusion.

4. Multidimensional case

Here and for the rest of the section, we consider as given an isonormal Gaussian process $\{X(h) : h \in \mathfrak{H}\}$, over some real separable Hilbert space $\mathfrak{H}$.

4.1. Main bounds

We shall now present (without proof) a result taken from [23], concerning the Gaussian approximation of vectors of random variables that are differentiable in the Malliavin sense. We recall that the *Wasserstein distance* between the laws of two $\mathbb{R}^d$-valued random vectors X and Y, noted $d_W(X, Y)$, is given by

$$d_W(X, Y) := \sup_{g \in \mathscr{H}; \|g\|_{Lip} \leq 1} |E[g(X)] - E[g(Y)]|,$$

where $\mathscr{H}$ indicates the class of all Lipschitz functions, that is, the collection of all functions $g : \mathbb{R}^d \to \mathbb{R}$ such that

$$\|g\|_{Lip} := \sup_{x \neq y} \frac{|g(x) - g(y)|}{\|x - y\|_{\mathbb{R}^d}} < \infty$$

(with $\|\cdot\|_{\mathbb{R}^d}$ the usual Euclidian norm on $\mathbb{R}^d$). Also, we recall that the *operator norm* of a $d \times d$ matrix A over $\mathbb{R}$ is given by $\|A\|_{op} := \sup_{\|x\|_{\mathbb{R}^d}=1} \|Ax\|_{\mathbb{R}^d}$.

Note that, in the following statement, we require that the approximating Gaussian vector has a positive definite covariance matrix.

Theorem 4.1. *(See [23]) Fix $d \geq 2$ and let $C = (C_{ij})_{1 \leq i,j \leq d}$ be a $d \times d$ positive definite matrix. Suppose that $N \sim \mathcal{N}_d(0, C)$, and assume that $Z = (Z_1, \ldots, Z_d)$ is a $\mathbb{R}^d$-valued random vector such that $E[Z_i] = 0$ and $Z_i \in \mathbb{D}^{1,2}$ for every $i = 1, \ldots, d$. Then,*

$$d_W(Z, N) \leq \|C^{-1}\|_{op} \|C\|_{op}^{1/2} \sqrt{\sum_{i,j=1}^d E[(C_{ij} - \langle DZ_i, -DL^{-1}Z_j \rangle_{\mathfrak{H}})^2]}.$$

In what follows, we shall use once again interpolation techniques in order to partially generalize Theorem 4.1 to the case where the approximating covariance matrix C is not necessarily positive definite. This additional difficulty forces us to work with functions that are smoother than the ones involved in the definition of the Wasserstein distance. To this end, we will adopt the following simplified notation: for every $\varphi : \mathbb{R}^d \to \mathbb{R}$ of class C^2, we set

$$\|\varphi''\|_\infty = \max_{i,j=1,\ldots,d} \sup_{z \in \mathbb{R}^d} \left|\frac{\partial^2 \varphi}{\partial x_i \partial x_j}(z)\right|.$$

Theorem 4.2. *(See [22]) Fix $d \geq 2$, and let $C = (C_{ij})_{1 \leq i,j \leq d}$ be a $d \times d$ covariance matrix. Suppose that $N \sim \mathcal{N}_d(0, C)$ and that $Z = (Z_1, \ldots, Z_d)$ is a $\mathbb{R}^d$-valued random vector such that $E[Z_i] = 0$ and $Z_i \in \mathbb{D}^{1,2}$ for every $i = 1, \ldots, d$. Then, for every $\varphi : \mathbb{R}^d \to \mathbb{R}$ belonging to C^2 such that $\|\varphi''\|_\infty < \infty$, we have*

$$|E[\varphi(Z)] - E[\varphi(N)]| \leq \frac{1}{2} \|\varphi''\|_\infty \sum_{i,j=1}^d E\left[|C_{i,j} - \langle DZ_j, -DL^{-1}Z_i \rangle_{\mathfrak{H}}|\right]. \tag{4.1}$$

Proof. Without loss of generality, we assume that N is independent of the underlying isonormal Gaussian process X. Let $\varphi : \mathbb{R}^d \to \mathbb{R}$ be a C^2-function such that $\|\varphi''\|_\infty < \infty$. For any $t \in [0,1]$, set $\Psi(t) = E[\varphi(\sqrt{1-t}Z + \sqrt{t}N)]$, so that

$$\left|E[\varphi(Z)] - E[\varphi(N)]\right| = |\Psi(1) - \Psi(0)| \leq \int_0^1 |\Psi'(t)| dt.$$

We easily see that Ψ is differentiable on $(0,1)$ with

$$\Psi'(t) = \sum_{i=1}^d E\left[\frac{\partial \varphi}{\partial x_i}(\sqrt{1-t}Z + \sqrt{t}N)\left(\frac{1}{2\sqrt{t}}N_i - \frac{1}{2\sqrt{1-t}}Z_i\right)\right].$$

By integrating by parts, we can write

$$E\left[\frac{\partial \varphi}{\partial x_i}(\sqrt{1-t}Z + \sqrt{t}N)N_i\right]$$

$$= E\left\{E\left[\frac{\partial \varphi}{\partial x_i}(\sqrt{1-t}z + \sqrt{t}N)N_i\right]_{|z=Z}\right\}$$

$$= \sqrt{t} \sum_{j=1}^d C_{i,j} E\left\{E\left[\frac{\partial^2 \varphi}{\partial x_i \partial x_j}(\sqrt{1-t}z + \sqrt{t}N)\right]_{|z=Z}\right\}$$

$$= \sqrt{t} \sum_{j=1}^d C_{i,j} E\left[\frac{\partial^2 \varphi}{\partial x_i \partial x_j}(\sqrt{1-t}Z + \sqrt{t}N)\right].$$

By using (2.15) in order to perform the integration by parts, we can also write

$$E\left[\frac{\partial \varphi}{\partial x_i}(\sqrt{1-t}Z + \sqrt{t}N)Z_i\right]$$

$$= E\left\{E\left[\frac{\partial \varphi}{\partial x_i}(\sqrt{1-t}Z + \sqrt{t}x)Z_i\right]_{|x=N}\right\}$$

$$= \sqrt{1-t} \sum_{j=1}^d E\left\{E\left[\frac{\partial^2 \varphi}{\partial x_i \partial x_j}(\sqrt{1-t}Z + \sqrt{t}x)\langle DZ_j, -DL^{-1}Z_i\rangle_{\mathfrak{H}}\right]_{|x=N}\right\}$$

$$= \sqrt{1-t} \sum_{j=1}^d E\left[\frac{\partial^2 \varphi}{\partial x_i \partial x_j}(\sqrt{1-t}Z + \sqrt{t}N)\langle DZ_j, -DL^{-1}Z_i\rangle_{\mathfrak{H}}\right].$$

Hence

$$\Psi'(t) = \frac{1}{2} \sum_{i,j=1}^d E\left[\frac{\partial^2 \varphi}{\partial x_i \partial x_j}(\sqrt{1-t}Z + \sqrt{t}N)\left(C_{i,j} - \langle DZ_j, -DL^{-1}Z_j\rangle_{\mathfrak{H}}\right)\right],$$

so that

$$\int_0^1 |\Psi'(t)| dt \leq \frac{1}{2}\|\varphi''\|_\infty \sum_{i,j=1}^d E\left[|C_{i,j} - \langle DZ_j, -DL^{-1}Z_i\rangle_{\mathfrak{H}}|\right]$$

and the desired conclusion follows. □

We now aim at applying Theorem 4.2 to vectors of multiple stochastic integrals.

Corollary 14.2. *Fix integers $d \geq 2$ and $1 \leq q_1 \leq \ldots \leq q_d$. Consider a vector $Z = (Z_1, \ldots, Z_d) := (I_{q_1}(f_1), \ldots, I_{q_d}(f_d))$ with $f_i \in \mathfrak{H}^{\odot q_i}$ for any $i = 1, \ldots, d$. Let $N \sim \mathcal{N}_d(0, C)$, with $C = (C_{ij})_{1 \leq i,j \leq d}$ a $d \times d$ covariance matrix. Then, for every $\varphi : \mathbb{R}^d \to \mathbb{R}$ belonging to C^2 such that $\|\varphi''\|_\infty < \infty$, we have*

$$|E[\varphi(Z)] - E[\varphi(N)]| \leq \frac{1}{2}\|\varphi''\|_\infty \sum_{i,j=1}^d E\left[\left|C_{i,j} - \frac{1}{d_i}\langle DZ_j, DZ_i\rangle_{\mathfrak{H}}\right|\right]. \quad (4.2)$$

Proof. We have $-L^{-1}Z_i = \frac{1}{d_i} Z_i$ so that the desired conclusion follows from (4.1). □

When one applies Corollary 14.2 in concrete situations, one can use the following result in order to evaluate the right-hand side of (4.2).

Proposition 4.2. *Let $F = I_p(f)$ and $G = I_q(g)$, with $f \in \mathfrak{H}^{\odot p}$ and $g \in \mathfrak{H}^{\odot q}$ $(p, q \geq 1)$. Let a be a real constant. If $p = q$, one has the estimate:*

$$E\left[\left(a - \frac{1}{p}\langle DF, DG\rangle_{\mathfrak{H}}\right)^2\right] \leq (a - p!\langle f, g\rangle_{\mathfrak{H}^{\otimes p}})^2$$

$$+ \frac{p^2}{2}\sum_{r=1}^{p-1}(r-1)!^2\binom{p-1}{r-1}^4 (2p-2r)!(\|f \otimes_{p-r} f\|^2_{\mathfrak{H}^{\otimes 2r}} + \|g \otimes_{p-r} g\|^2_{\mathfrak{H}^{\otimes 2r}}).$$

On the other hand, if $p < q$, one has that

$$E\left[\left(a - \frac{1}{q}\langle DF, DG\rangle_{\mathfrak{H}}\right)^2\right] \leq a^2 + p!^2\binom{q-1}{p-1}^2(q-p)!\|f\|^2_{\mathfrak{H}^{\otimes p}}\|g \otimes_{q-p} g\|_{\mathfrak{H}^{\otimes 2p}}$$

$$+ \frac{p^2}{2}\sum_{r=1}^{p-1}(r-1)!^2\binom{p-1}{r-1}^2\binom{q-1}{r-1}^2 (p+q-2r)!$$

$$\times (\|f \otimes_{p-r} f\|^2_{\mathfrak{H}^{\otimes 2r}} + \|g \otimes_{q-r} g\|^2_{\mathfrak{H}^{\otimes 2r}}).$$

Remark 4.8. When bounding the right-hand side of (4.2), we see that it is sufficient to asses the quantities $\|f_i \otimes_r f_i\|_{\mathfrak{H}^{\otimes 2(q_i - r)}}$ for all $i = 1, \ldots, d$ and $r = 1, \ldots, q_i - 1$ on the one hand, and $E(Z_i Z_j) = q_i!\langle f_i, f_j\rangle_{\mathfrak{H}^{\otimes q_i}}$ for all $i, j = 1, \ldots, d$ such that $q_i = q_j$ on the other hand. In particular, this fact allows to recover a result first proved by Peccati and Tudor in [31], namely that, for vectors of multiple stochastic integrals whose covariance matrix is converging, *the componentwise convergence to a Gaussian distribution always implies joint convergence*.

Proof of Proposition 4.2. Without loss of generality, we can assume that $\mathfrak{H} = L^2(A, \mathscr{A}, \mu)$, where $(A, \mathscr{A})$ is a measurable space, and μ is a σ-finite and non-atomic measure. Thus, we can write

$$\langle DF, DG \rangle_{\mathfrak{H}} = pq \langle I_{p-1}(f), I_{q-1}(g) \rangle_{\mathfrak{H}} = pq \int_A I_{p-1}(f(\cdot,t)) I_{q-1}(g(\cdot,t)) \mu(dt)$$

$$= pq \int_A \sum_{r=0}^{p \wedge q - 1} r! \binom{p-1}{r} \binom{q-1}{r} I_{p+q-2-2r}(f(\cdot,t) \widetilde{\otimes}_r g(\cdot,t)) \mu(dt)$$

$$= pq \sum_{r=0}^{p \wedge q - 1} r! \binom{p-1}{r} \binom{q-1}{r} I_{p+q-2-2r}(f \widetilde{\otimes}_{r+1} g)$$

$$= pq \sum_{r=1}^{p \wedge q} (r-1)! \binom{p-1}{r-1} \binom{q-1}{r-1} I_{p+q-2r}(f \widetilde{\otimes}_r g).$$

It follows that

$$E\left[\left(a - \frac{1}{q}\langle DF, DG\rangle_{\mathfrak{H}}\right)^2\right] \tag{4.3}$$

$$= \begin{cases} a^2 + p^2 \sum_{r=1}^{p}(r-1)!^2 \binom{p-1}{r-1}^2 \binom{q-1}{r-1}^2 (p+q-2r)! \|f\widetilde{\otimes}_r g\|^2_{\mathfrak{H}^{\otimes(p+q-2r)}} & \text{if } p < q, \\ (a - p!\langle f, g\rangle_{\mathfrak{H}^{\otimes p}})^2 + p^2 \sum_{r=1}^{p-1}(r-1)!^2 \binom{p-1}{r-1}^4 (2p-2r)! \|f\widetilde{\otimes}_r g\|^2_{\mathfrak{H}^{\otimes(2p-2r)}} & \text{if } p = q. \end{cases}$$

If $r < p \leq q$ then

$$\|f\widetilde{\otimes}_r g\|^2_{\mathfrak{H}^{\otimes(p+q-2r)}} \leq \|f \otimes_r g\|^2_{\mathfrak{H}^{\otimes(p+q-2r)}} = \langle f \otimes_{p-r} f, g \otimes_{q-r} g \rangle_{\mathfrak{H}^{\otimes 2r}}$$

$$\leq \|f \otimes_{p-r} f\|_{\mathfrak{H}^{\otimes 2r}} \|g \otimes_{q-r} g\|_{\mathfrak{H}^{\otimes 2r}}$$

$$\leq \frac{1}{2}\left(\|f \otimes_{p-r} f\|^2_{\mathfrak{H}^{\otimes 2r}} + \|g \otimes_{q-r} g\|^2_{\mathfrak{H}^{\otimes 2r}}\right).$$

If $r = p < q$, then

$$\|f\widetilde{\otimes}_p g\|^2_{\mathfrak{H}^{\otimes(q-p)}} \leq \|f \otimes_p g\|^2_{\mathfrak{H}^{\otimes(q-p)}} \leq \|f\|^2_{\mathfrak{H}^{\otimes p}} \|g \otimes_{q-p} g\|_{\mathfrak{H}^{\otimes 2p}}.$$

If $r = p = q$, then $f\widetilde{\otimes}_p g = \langle f, g \rangle_{\mathfrak{H}^{\otimes p}}$. By plugging these last expressions into (4.3), we deduce immediately the desired conclusion.

4.2. Quadratic variation of fractional Brownian motion, continued

In this section, we continue the example of Section 3.4. We still denote by B a fractional Brownian motion with Hurst index $H \in (0, 3/4]$. We set

$$Z_n(t) = \frac{1}{\sigma_n} \sum_{k=0}^{\lfloor nt \rfloor - 1} [(B_{k+1} - B_k)^2 - 1], \quad t \geq 0,$$

where $\sigma_n > 0$ is such that $E(Z_n(1)^2) = 1$. The following statement contains the multidimensional counterpart of Theorem 3.4, namely a bound associated with the convergence of the finite dimensional distributions of $\{Z_n(t) : t \geq 0\}$ towards a standard Brownian motion. A similar result can be of course recovered from Theorem 4.1 – see again [23].

Theorem 4.3. *Fix $d \geq 1$, and consider $0 = t_0 < t_1 < \ldots < t_d$. Let $N \sim \mathcal{N}_d(0, I_d)$. There exists a constant c (depending only on d, H and $t_1, \ldots, t_d$) such that, for every $n \geq 1$:*

$$\sup \left| E\left[\varphi\left(\frac{Z_n(t_i) - Z_n(t_{i-1})}{\sqrt{t_i - t_{i-1}}} \right)_{1 \leq i \leq d} \right] - E[\varphi(N)] \right| \leq c \times \begin{cases} \frac{1}{\sqrt{n}} & \text{if } H \in (0, \frac{1}{2}] \\ n^{2H - \frac{3}{2}} & \text{if } H \in [\frac{1}{2}, \frac{3}{4}) \\ \frac{1}{\sqrt{\log n}} & \text{if } H = \frac{3}{4} \end{cases}$$

where the supremum is taken over all C^2-function $\varphi : \mathbb{R}^d \to \mathbb{R}$ such that $\|\varphi''\|_\infty \leq 1$.

Proof. We only make the proof for $H < 3/4$, the proof for $H = 3/4$ being similar. Fix $d \geq 1$ and $t_0 = 0 < t_1 < \ldots < t_d$. In the sequel, c will denote a constant independent of n, which can differ from one line to another. First, see e.g. the proof of Theorem 3.4, observe that

$$\frac{Z_n(t_i) - Z_n(t_{i-1})}{\sqrt{t_i - t_{i-1}}} = I_2(f_i^{(n)})$$

with

$$f_n^{(i)} = \frac{1}{\sigma_n \sqrt{t_i - t_{i-1}}} \sum_{k=\lfloor nt_{i-1} \rfloor}^{\lfloor nt_i \rfloor - 1} 1_{[k,k+1]}^{\otimes 2}.$$

In the proof of Theorem 3.4, it is shown that, for any fixed $i \in \{1, \ldots, d\}$ and $r \in \{1, \ldots, q_i - 1\}$:

$$\|f_n^{(i)} \otimes_1 f_n^{(i)}\|_{\mathfrak{H}^{\otimes 2}} \leq c \times \begin{cases} \frac{1}{\sqrt{n}} & \text{if } H \in (0, \frac{1}{2}] \\ n^{2H - \frac{3}{2}} & \text{if } H \in [\frac{1}{2}, \frac{3}{4}) \end{cases}. \quad (4.4)$$

Moreover, when $1 \leq i < j \leq d$, we have, with ρ defined in (3.15),

$$|\langle f_n^{(i)}, f_n^{(j)} \rangle_{\mathfrak{H}^{\otimes 2}}| = \left| \frac{1}{\sigma_n^2 \sqrt{t_i - t_{i-1}} \sqrt{t_j - t_{j-1}}} \sum_{k=\lfloor nt_{i-1} \rfloor}^{\lfloor nt_i \rfloor - 1} \sum_{l=\lfloor nt_{j-1} \rfloor}^{\lfloor nt_j \rfloor - 1} \rho^2(l-k) \right|$$

$$= \frac{c}{\sigma_n^2} \left| \sum_{|r| = \lfloor nt_{j-1} \rfloor - \lfloor nt_i \rfloor + 1}^{\lfloor nt_j \rfloor - \lfloor nt_{i-1} \rfloor - 1} g_{i,j,n}(r) \rho^2(r) \right|$$

$$\leq c \frac{\lfloor nt_i \rfloor - \lfloor nt_{i-1} \rfloor - 1}{\sigma_n^2} \sum_{|r| \geq \lfloor nt_{j-1} \rfloor - \lfloor nt_i \rfloor + 1} \rho^2(r) \quad (4.5)$$

$$= O(n^{4H-3}), \quad \text{as } n \to \infty,$$

where

$$g_{i,j,n}(r) = [(\lfloor nt_j \rfloor - 1 - r) \wedge (\lfloor nt_i \rfloor - 1) - (\lfloor nt_{j-1} \rfloor - r) \vee (\lfloor nt_{i-1} \rfloor)],$$

the last equality coming from (3.16) and

$$\sum_{|r| \geq N} \rho^2(r) = O(\sum_{|r| \geq N} |r|^{4H-4}) = O(N^{4H-3}), \quad \text{as } N \to \infty.$$

Finally, by combining (4.4), (4.5), Corollary 14.2 and Proposition 4.2, we obtain the desired conclusion.

References

1. O. Barndorff-Nielsen, J. Corcuera, M. Podolskij and J. Woerner (2009). Bipower variations for Gaussian processes with stationary increments. *J. Appl. Probab.* **46**, no. 1, 132-150.
2. B. Bercu, I. Nourdin and M.S. Taqqu (2009). A multiple stochastic integral criterion for almost sure limit theorems. Preprint.
3. J.-C. Breton and I. Nourdin (2008). Error bounds on the non-normal approximation of Hermite power variations of fractional Brownian motion. *Electron. Comm. Probab.* **13**, 482-493.
4. J.-C. Breton, I. Nourdin and G. Peccati (2009). Exact confidence intervals for the Hurst parameter of a fractional Brownian motion. *Electron. J. Statist.* **3**, 416-425 (Electronic)
5. P. Breuer et P. Major (1983). Central limit theorems for non-linear functionals of Gaussian fields. *J. Mult. Anal.* **13**, 425-441.
6. D. Chambers et E. Slud (1989). Central limit theorems for nonlinear functionals of stationary Gaussian processes. *Probab. Theory Rel. Fields* **80**, 323-349.
7. L.H.Y. Chen and Q.-M. Shao (2005). Stein's method for normal approximation. In: *An Introduction to Stein's Method* (A.D. Barbour and L.H.Y. Chen, eds), Lecture Notes Series No.4, Institute for Mathematical Sciences, National University of Singapore, Singapore University Press and World Scientific 2005, 1-59.
8. J.M. Corcuera, D. Nualart et J.H.C. Woerner (2006). Power variation of some integral long memory process. *Bernoulli* **12**, no. 4, 713-735.
9. Y.A. Davydov and G.V. Martynova (1987). Limit behavior of multiple stochastic integral. *Preila, Nauka, Moscow* 55-57 (in Russian).
10. R.M. Dudley (1967). The sizes of compact subsets of Hilbert space and continuity of Gaussian processes. *J. Funct. Anal.* **1**, 290-330.
11. L. Giraitis and D. Surgailis (1985). CLT and other limit theorems for functionals of Gaussian processes. *Zeitschrift für Wahrsch. verw. Gebiete* **70**, 191-212.
12. S. Janson (1997). *Gaussian Hilbert Spaces.* Cambridge University Press, Cambridge.
13. P. Major (1981). *Multiple Wiener-Itô integrals.* LNM **849**. Springer-Verlag, Berlin Heidelberg New York.
14. D. Marinucci and G. Peccati (2007). High-frequency asymptotics for subordinated stationary fields on an Abelian compact group. *Stochastic Process. Appl.* **118**, no. 4, 585-613.
15. A. Neuenkirch and I. Nourdin (2007). Exact rate of convergence of some approximation schemes associated to SDEs driven by a fractional Brownian motion. *J. Theoret. Probab.* **20**, no. 4, 871-899.

16. I. Nourdin and G. Peccati (2009). Non-central convergence of multiple integrals. *Ann. Probab.* **37**, no. 4, 1412-1426.
17. I. Nourdin and G. Peccati (2008). Weighted power variations of iterated Brownian motion. *Electron. J. Probab.* **13**, no. 43, 1229-1256 (Electronic).
18. I. Nourdin and G. Peccati (2009). Stein's method on Wiener chaos. *Probab. Theory Rel. Fields* **145**, no. 1, 75-118.
19. I. Nourdin and G. Peccati (2008). Stein's method and exact Berry-Esséen asymptotics for functionals of Gaussian fields. *Ann. Probab.*, to appear.
20. I. Nourdin, G. Peccati and G. Reinert (2009). Second order Poincaré inequalities and CLTs on Wiener space. *J. Func. Anal.* **257**, 593-609.
21. I. Nourdin, G. Peccati and G. Reinert (2008). Stein's method and stochastic analysis of Rademacher functionals. Preprint.
22. I. Nourdin, G. Peccati and G. Reinert (2009). Invariance principles for homogeneous sums: universality of Gaussian Wiener chaos Preprint.
23. I. Nourdin, G. Peccati and A. Réveillac (2008). Multivariate normal approximation using Stein's method and Malliavin calculus. *Ann. Inst. H. Poincaré Probab. Statist.*, to appear.
24. I. Nourdin and F. Viens (2008). Density estimates and concentration inequalities with Malliavin calculus. *Electron. J. Probab.*, to appear.
25. D. Nualart (2006). *The Malliavin calculus and related topics of Probability and Its Applications.* Springer Verlag, Berlin, Second edition, 2006.
26. D. Nualart and S. Ortiz-Latorre (2008). Central limit theorems for multiple stochastic integrals and Malliavin calculus. *Stochastic Process. Appl.* **118** (4), 614-628.
27. D. Nualart and G. Peccati (2005). Central limit theorems for sequences of multiple stochastic integrals. *Ann. Probab.* **33** (1), 177-193.
28. D. Nualart and J. Vives (1990). Anticipative calculus for the Poisson space based on the Fock space. *Séminaire de Probabilités XXIV*, LNM **1426**. Springer-Verlag, Berlin Heidelberg New York, pp. 154-165.
29. G. Peccati, J.-L. Solé, F. Utzet and M.S. Taqqu (2008). Stein's method and normal approximation of Poisson functionals. *Ann. Probab.*, to appear.
30. G. Peccati and M.S. Taqqu (2008). Moments, cumulants and diagram formulae for non-linear functionals of random measures (Survey). Preprint.
31. G. Peccati and C.A. Tudor (2005). Gaussian limits for vector-valued multiple stochastic integrals. *Séminaire de Probabilités XXXVIII*, LNM **1857**. Springer-Verlag, Berlin Heidelberg New York, pp. 247-262.
32. N. Privault (2008). Stochastic analysis of Bernoulli processes. Probability Surveys **5**, 435-483.
33. G. Reinert (2005). Three general approaches to Stein's method. In: *An introduction to Stein's method*, 183-221. Lect. Notes Ser. Inst. Math. Sci. Natl. Univ. Singap. **4**, Singapore Univ. Press, Singapore.
34. B. Rider and B. Virág (2007). The noise in the circular law and the Gaussian free field. *Int. Math. Res. Not.* 2, Art. ID rnm006.
35. S. Sheffield (1997). Gaussian free field for mathematicians. *Probab. Theory Rel. Fields* **139**(3-4), 521-541
36. Ch. Stein (1972). A bound for the error in the normal approximation to the distribution of a sum of dependent random variables. In: *Proceedings of the Sixth Berkeley Symposium on Mathematical Statistics and Probability, Vol. II: Probability theory,* 583-602. Univ. California Press, Berkeley, CA.
37. Ch. Stein (1986). *Approximate computation of expectations.* Institute of Mathematical Statistics Lecture Notes - Monograph Series, **7**. Institute of Mathematical Statistics,

Hayward, CA.
38. M. Talagrand (2003). *Spin Glasses: A Challenge for Mathematicians. Cavity and Mean Fields Models.* Springer, Berlin.
39. C.A. Tudor and F. Viens (2008). Variations and estimators for the selfsimilarity order through Malliavin calculus. *Ann. Probab.*, to appear.
40. F. Viens (2009). Stein's lemma, Malliavin calculus and tail bounds, with applications to polymer fluctuation exponents. *Stochastic Process. Appl.*, to appear.

Chapter 15

On Stochastic Integrals with Respect to an Infinite Number of Poisson Point Process and Its Applications

Guanglin Rang[1], Qing Li[2] and Sheng You[2] *

1. School of Mathematics and Statistics
Wuhan University
Wuhan 430072, China
E-mail: glrang.math@whu.edu.cn

2. Faculty of Mathematics and Computer Science
Hubei University
Wuhan 430062, China
E-mail: 63594643@qq.com

This chapter investigates stochastic integrals with respect to an infinite number of Poisson point processes and its' corresponding martingale representations. Furthermore, with non-Markovian and non-Lipschitz coefficients, stochastic differential equations driven by sequence of Poisson point processes are discussed, where the results are extensions of the linear continuous cases.

Keywords: stochastic integral, infinite number of Poisson point processes, non-Markovian coefficients, integral inequalities

2000 AMS Subject Classification: 60H05, 60H10

Contents

1	Introduction	237
2	Stochastic integral with respect to the infinite number of Poisson point processes	238
3	Martingale representation	241
4	Non-Markovian SDE driven by countably many Poisson point processes	246
References		250

1. Introduction

Since financial market contains infinite number of assets, where each asset price is driven by an idiosyncratic random source as well as by a systematic noise term, it is not enough to describe this market by finite number of stochastic processes. So it necessitates to develop stochastic calculus with respect to a sequence of stochastic

---
*The authors gratefully acknowledge the support by Natural Science Foundation of China (No.10871153).

processes, semi-martingales [5], or slight restrictively speaking, independent increment processes. Such a calculus, in fact, a special case of the theory of cylindrical stochastic calculus, was presented first by M. Hitsuda and H. Watanabe in [6], there stochastic integral, Ito formula and Girsanov transformation with respect to an infinite number of Brownian motions were given systematically, and then applied conveniently to a causal and causally invertible representation of equivalent Gaussian processes. Cao and He in [2] obtained the existence and uniqueness of solution of stochastic differential equation (SDE in short) driven by a sequence of Brownian motion under non-Lipschtiz conditions by a successive approximation (also see [3]). Using a similar method they also got the same results of H-valued SDE and backward SDE in a general setup, i.e., driven by cylindrical Brownian motion with Poisson point process.

All considerations above are based one Poisson process, which as a source of uncertainty are a standard tool for modeling rare and randomly occurring events. These processes can be found, among others, in quality-ladder models of growth, in the endogenous fluctuations and growth literature with uncertainty, in the labor market matching literature, in monetary economics (see [1,13] and references therein). So stochastic calculus related to one jump source might be inappropriate for the use in economic modeling. Our aim, in this chapter, is to establish systematically the theory of stochastic calculus with respect to a sequence of Poisson processes (for simplicity we exclude the term driven by Brownian motion). In most cases Poisson processes affect the concerned variables through a stochastic differential equation, we shall explore the existence and uniqueness of SDE driven by countably many Poisson point processes under non-Markovin coefficients conditions, i.e., the coefficients satisfy integral non-Lipschtiz conditions. The results is obtained by a generalized Biharyi type inequality, so far as we know, this is not discussed yet.

2. Stochastic integral with respect to the infinite number of Poisson point processes

Let $(\Omega, \mathcal{F}, P)$ be the underlying space, with a filtration $\{\mathcal{F}_t, t \geq 0\}$ satisfying the usual conditions. Let $(U_i, \mathcal{B}_i, n_i)$, $i = 1, 2, \cdots$ be a sequence of σ-finite measure spaces (not necessary common). For $i \in N, q_i = (q_i(t_k^i), t_k^i \in D_i, k = 1, 2, \cdots)$ is a Poisson point process defined on $(\Omega, \mathcal{F}, P)$ taking values in U_i with domain D_i. $N_i(dtdx)$ is the counting measure associated with p_i, that is, for $U \in U_i$,

$$N_i([0,t], U) = \sum_{t_k^i \in D_i} 1_U(p_i(t_k^i)) 1_{\{t_k^i \leq t\}},$$

where $1_U(\cdot)$ is the indicator of set U. Furthermore assume p_i stationary, hence $N_i(dtdx)$ admits a compensator $\hat{N}_i(dtdx) = dt\, n_i(dx)$, such that, for every $U \in U_i$, with $n(U_i) < \infty$, $\tilde{N}_i\big((0,t], U\big) = N_i\big((0,t], U\big) - \hat{N}_i\big((0,t), U\big)$ is $\mathcal{F}_t$-square integrable martingale. A real-valued predictable function $f(t,x,w): R^+ \times U_i \times \Omega \to R$ means $\mathcal{P} \otimes \mathcal{B}_i$-measurable, where $\mathcal{P}$ is a σ-algebra generated by all left limited adapted processes. Let

$$F_i^1 = \left\{ f : f \text{ predictable with } \int_0^{t+} \int_{U_i} |f(s,z,w)| \hat{N}_i(ds,dz) < \infty \right\},$$

$$F_i^2 = \left\{ f : f \text{ predictable with } \int_0^{t+} \int_{U_i} f^2(s,z,w) ds \, n_i(dz) < \infty \right\}.$$

For $f \in F_i^1$, define stochastic integrals of f with respect to $\widetilde{N}_i$ by

$$\int_0^{t+} \int_{U_i} f(s-,z,w) \widetilde{N}(ds,dz)$$

$$= \sum_{t_k^i \leq t} f(t_k^i, p(t_k^i), w) - \int_0^t \int_{U_i} f(s,z,w) ds \, n_i(dz),$$

it is an $\mathcal{F}_t$-martingale. For $f \in F_i^2$, we can find predictable functions sequences $\{f_n\} \subset F_i^1 \cap F_i^2$, with $\int_0^t \int_{U_i} |f_n - f_m|^2 ds \, n_i(dz) \to 0$ (as $n,m \to \infty$), by a familiar procedure, stochastic integral of such an f w. r. t. $\widetilde{N}_i$ is obtained, denoted also by

$$\int_0^{t+} \int_{U_i} f(s,z,\omega) \widetilde{N}(ds,dz) = \int_0^{t+} \int_{U_i} f(s,z,\omega)[N(dsdz) - ds \, n_i(dz)],$$

which is an $\mathcal{F}_t$-square integrable martingale. Notice that, in this case, because of the possibility of both integrals being divergent, it is nonsense to take $\int_0^{t+} \int_{U_i} f(t,z,\omega) \widetilde{N}_i(ds,dz)$ as the difference of $\int_0^{t+} \int_{U_i} f(t,z,\omega) N_i(ds,dz)$ and $\int_0^{t+} \int_{U_i} f(s,z,\omega) ds \, n_i(dz)$.

For $U \subset U_i$, with $n_i(U) < \infty$. $N_i = \left\{ N_i(t,U) = N_i((0,t],U), \; t \geq 0 \right\}$ is a pure jump $\mathcal{F}_t$-adapted increasing processes. Let $[N_i] = \left\{ [N_i]_t, \; t \geq 0 \right\}$ be the quadratic variation processes of N_i, then

$$[N_i]_t = \sum_{0 < s \leq t} (\Delta N_i(t))^2 = N_i(t).$$

Actually, if $f \in F_i^2$, denoted by M_t the stochastic integral of f w.r.t. $\widetilde{N}(dt,ds)$ then

$$[M]_t = \sum_{0 < s \leq t} (\Delta M_t)^2 = \sum_{0 < s \leq t} f^2(s,p(s),\omega),$$

with predictable dual projection $\langle M \rangle_t = \int_0^t \int_{U_i} f^2(s,z) ds \, n_i(dz)$.

Next we present the stochastic integral with respect to a sequence q of independent Poisson Point processes $\{q_i\}_{i=1}^\infty$. Let

$$\mathbb{H}_q^2 = \left\{ Y = (y_1, y_2, \cdots) : y_i(s, z_i, \omega) \in F_i^2. \; i = 1, 2, \cdots \right.$$

with $\sum_{i=1}^{\infty} \int_0^t \int_{U_i} \mathbf{E}|y_i(s, z_i, \omega)|^2 ds\, n_i(dz_i) < \infty$. for any $0 \leqslant t \leqslant T \Big\}$

Now, We put $Y^n = (y_1, y_2, \cdots, y_n, 0, 0, \cdots)$ for $Y = (y_1, y_2, \cdots, y_n, \cdots) \in \mathbb{H}_q^2$. Then, we can define a sequence of square integrable martingale.

$$I(t, \omega, Y^n) = \sum_{i=1}^n \int_0^t \int_{U_i} y_i(s, z_i, \omega) \widetilde{N}_i(ds\, dz_i).$$

By independence hypothesis, for any $A_i \subset U_i$ with $n_i(A_i) < \infty$, $i = 1, 2, \cdots, n$. We know $\langle N_i(\cdot, A_i), N_j(\cdot, A_j) \rangle_t = 0$, and

$$\mathbf{E}\Big[\sup_{0 \leq t \leq T} |I(t, \omega, Y_n) - I(t, \omega, Y_m)|^2\Big]$$
$$\leqslant 4\mathbf{E}\Big[|I(T, \omega, Y^n) - I(T, \omega, Y^m)|^2\Big]$$
$$= 4 \sum_{i=n+1}^m \mathbf{E} \int_0^T \int_{U_i} y_i^2(s, z_i, \omega) ds\, n_i(dz_i)$$
$$\to 0,$$

as $m, n \to +\infty$. Therefore, we see easily that $I(t, \omega, Y^n)$ converges uniformly in t in L^2, thus, we can define the stochastic integral for Y with respect to the countably many orthogonal martingale measure related to Poisson point processes $\{q_i\}_{i=1}^{\infty}$, denoted by $I(t, \omega, Y)$, as the limit of $\{I(t, \omega, Y^n)\}_{n=1}^{\infty}$. And write

$$I(t, \omega, Y) = \sum_{i=1}^{\infty} \int_0^t \int_{U_i} y_i^2(s, z_i, \omega) \widetilde{N}_i(dsdz_i) \qquad 0 \leqslant t \leqslant T. \qquad (2.1)$$

Obviously, $I = \{I(t, \omega, Y), t \geq 0\}$ is an $\mathcal{F}_t$ square integrable martingale with $Y \in \mathbb{H}_q^2$ fixed.

Theorem 2.1. *Let $F(t, x_1, x_2, \cdots, x_n)$ be a continuous function defined in $[0,T] \times \mathbb{R}^n$ such that partial derivatives $F'_t = \frac{\partial F}{\partial t}, F'_{x_i} = \frac{\partial F}{\partial x_i}, F''_{x_i x_j} = \frac{\partial^2 F}{\partial x_i \partial x_j}$ are all continuous. Then, the differential of $X(t, \omega) = F(t, I_1, \cdots I_n) = F(t, I)$ is given by*

$$F(t, I(t, \omega, y)) - F(0, I(0))$$
$$= \int_0^t \frac{\partial F}{\partial s}(I(s, \omega, y)) ds + \sum_{i=1}^{\infty} \int_0^{t+} \frac{\partial F}{\partial x_i}(I(s-, \omega, y)) dI_i(s)$$
$$+ \sum_{0 < s \leqslant t} \Big\{ F(I(s)) - F(I(s-)) - \sum_{i=1}^{\infty} \frac{\partial F}{\partial x_i}(I(s)) \Delta I_i(s) \Big\},$$

where, $I_i = I(\cdot, \omega, Y_i)$ with $Y_i \in \mathbb{H}_q^2, i = 1, 2, \ldots, n$, are given by equation (2.1).

Especially,

$$I^2(t,\omega,y)$$
$$= I^2(0) + \sum_{i=1}^{\infty} \int_0^{t^+} 2(I(s^-,\omega,y)dI_i(s) + \sum_{i=1}^{\infty}\sum_{1\leqslant t}(\Delta I_i(s))^2$$
$$= I^2(0) + \sum_{i=1}^{\infty} \int_0^{t^+} 2(I(s^-,\omega,y)dI_i(s)$$
$$+ \sum_{i=1}^{\infty} \int_0^T \int_{U_i} Y_i^2(s,z_i,\omega)ds\, n_i(dz_i).$$

Note that the continuous part of quadratic covariation $[\tilde{N}_i,\tilde{N}_j]_t$ of $\tilde{N}_i$ and $\tilde{N}_j$ is zero when $i \neq j$, i.e., $[\tilde{N}_i,\tilde{N}_j]_t^c = 0$.

Proof. Combining the Itô formula for semimartingale (see [11]) and limit procedure yields the result above. □

3. Martingale representation

Since an important role is played in stochastic analysis and its applications, such as in BSDE, in this section, martingale presentation properties will be presented, i.e., square integrable martingale with the predictable representation properties of this point processes, that is,

Theorem 3.1. *Let $M = \{M_t, t \geqslant 0\}$ be an $\mathcal{F}_t^q$ adapted square integrable martingale, where $\{\mathcal{F}_t^q, t \geqslant 0\}$ is the increasing family of σ-algebras generated by Poisson point processes $q = (q_1, q_2, \cdots)$ up to time t. Then there exists a unique sequence of predictable process*

$$Y = \{Y(t,z,\omega) = (y_1(t,z_1,\omega), y_2(t,z_2,\omega), \cdots), y_i \in F_i^2, i = 1, \cdots\}$$

satisfying $\sum_{i=1}^{\infty} \int_0^t \int_{U_i} y_i^2(s,x_i,\omega)ds\, n_i(dx_i) < \infty\}$, such that

$$M_t = M_0 + \sum_{i=1}^{\infty} \int_0^{t^+} \int_{U_i} Y_i(s,z_i,\omega)\tilde{N}_i(ds dz_i).$$

To prove this theorem we adopt the method in [7] (also see [9]), there the continuous case is considered only (see [11] for semi-martingale case). First we formulate an Itô formula for exponential functions of Poisson point processes. Let $f(\xi_1, \xi_2, \cdots, \xi_n) = \exp\{-\sum_{i=1}^n \theta_i \xi_i\}$, $\theta_i, \xi_i > 0$, $N_i(t) = N_i(t, A_i) = N_i\big((0,t] \times A_i\big)$, $A_i \subset U_i$ with $n_i(A_i) < \infty$, $i = 1, 2, \cdots, n$. Then, by Itô formula for semimartingale, we have

$$f(N_1(t), N_2(t), \cdots, N_n(t))$$
$$= \sum_{s \leq t} \Big[f\big(N_1(s), N_2(s), \cdots, N_n(s)\big)$$
$$\qquad - f\big(N_1(s-), N_2(s-) \cdots, N_n(s-)\big) \Big]$$
$$= \sum_{s \leq t} e^{-\sum_{i=1}^{n} \theta_i N_i(s-)} \cdot \Big[e^{-\sum_{i=1}^{n} \theta_i \triangle N_i(s)} - 1 \Big]$$

Since $\triangle N_i(s)$ takes values 1 or 0, we can show, by induction, that

$$f(N_1(t), N_2(t), \cdots, N_n(t))$$
$$= \sum_{i=1}^{n} (e^{-\theta i} - 1) \sum_{s \leq t} \Gamma(s-) \triangle N_i(s)$$
$$+ \sum_{i<j} (e^{-\theta i} - 1)(e^{-\theta j} - 1) \sum_{s \leq t} \Gamma(s-) \triangle N_i(s) \triangle N_j(s)$$
$$+ \cdots + \prod_{i=1}^{n}(e^{-\theta i} - 1)\big[\sum_{s \leq t} \Gamma(s-) \prod_{i=1}^{n} \triangle N_i(s) \big]$$

where, $\Gamma(s-) = e^{-\sum_{i=1}^{n} \theta_i N_i(s-)}$. If put $[N_{i_1}, N_{i_2}, \cdots, N_{i_l}]_t = \sum_{s \leq t} \triangle N_{i_1}(s) \cdots N_{i_l}(s)$, then

$$f(N_1(t), N_2(t), \cdots, N_n(t))$$
$$= \sum_{i=1}^{n} (e^{-\theta i} - 1) \int_0^{t^+} \Gamma(s-) N_i(s)(\mathrm{d}s, U_i)$$
$$+ \sum_{i<j} (e^{-\theta i} - 1)(e^{-\theta j} - 1) \int_0^{t^+} \Gamma(s-) \mathrm{d}[N_i, N_j]_s + \cdots$$
$$+ \prod_{i=1}^{n} (e^{-\theta i} - 1) \int_0^{t^+} \Gamma(s-) \mathrm{d}[N_1, N_2, \cdots, N_n]_t.$$

Proof. [Proof of Theorem 1.2:] Without loss of generality suppose $M_0 = 0$. First we prove by induction on m, that there exist predictable processes $y_1, y_2, \ldots, y_m$ with $y_i \in F_i^2$, such that

$$Z_t^m \triangleq M_t - \sum_{i=1}^{m} \int_0^{t^+} \int_{U_i} y_i(s, z_i, \omega) \widetilde{N}_i(\mathrm{d}s, \mathrm{d}z_i)$$

is orthogonal to every martingale of the from

$$\sum_{i=1}^{m} \int_0^{t^+} \int_{U_i} r_i(s, x_i, \omega) \widetilde{N}_i(\mathrm{d}s, \mathrm{d}x_i)$$

with $r_i \in F_i^2, i = 1, 2, \ldots, m$. If $m = 1$, this is a direct consequence of Lemma 2.3 in [14]. Suppose such processes exist for $m - 1$, that is

$$Z_t^{m-1} \triangleq M_t - \sum_{i=1}^{m-1} \int_0^{t+} \int_{U_i} y_i(s, z_i, \omega) \widetilde{N}_i(\mathrm{d}s, \mathrm{d}z_i) \qquad 0 \leqslant t \leqslant T$$

is orthogonal to $\sum_{i=1}^{m-1} \int_0^{t+} \int_{U_i} r_i(s, z_i, \omega) \widetilde{N}_i(\mathrm{d}s, \mathrm{d}z_i)$, for $r_i \in F_i^2, i = 1, 2, \ldots, m-1$. Here $Z^{m-1} = \{Z_t^{m-1}, t \geqslant 0\}$ is still a square integrable Martingale, by Lemma 2.3 in [14] again, we get $y_m \in F_m^2$, such that

$$M_t - \sum_{i=1}^{m-1} \int_0^{t+} \int_{U_i} y_i(s, z_i, \omega) \widetilde{N}_i(\mathrm{d}s, \mathrm{d}z_i)$$
$$= \int_0^{t+} \int_{U_m} y_m(s, z_m, \omega) \widetilde{N}_m(\mathrm{d}s, \mathrm{d}z_m) + Z_t,$$

where Z_t is orthogonal to all martingale

$$\int_0^{t+} \int_{U_m} r_m(s, z_m, \omega) \widetilde{N}_m(\mathrm{d}s, \mathrm{d}z_m)$$

for any $r_m \in F_m^2$, and $\int_0^{t+} \int_{U_m} y_m(s, z_m, \omega) \widetilde{N}_m(\mathrm{d}s, \mathrm{d}z_m)$ is orthogonal to the martingale $\sum_{i=1}^{m-1} \int_0^{t+} \int_{U_i} y_i(s, z_i, \omega) \widetilde{N}_i(\mathrm{d}s, \mathrm{d}z_i)$.

Furthermore, by orthogonality, we have

$$\left\| \sum_{i=1}^{m} \int_0^{\cdot +} \int_{U_i} Y_i(s, x_i, \omega) \widetilde{N}_i(\mathrm{d}s, \mathrm{d}x_i) \right\|_{\mathcal{M}^2}$$
$$= \sum_{i=1}^{m} \int_0^{T} \int_{U_i} \mathbf{E} Y_i^2(s, x_i, \omega) \mathrm{d}s \, n_i(\mathrm{d}x_i) \leqslant \mathbf{E}[M_T^2],$$

so $M^m = \left\{ \sum_{i=1}^{m} \int_0^{t+} \int_{U_i} y_i(s, z_i, \omega) \widetilde{N}_i(\mathrm{d}s, \mathrm{d}z_i); \ t \geqslant 0 \right\}$ converges to a limit, then let $m \to \infty$, it justify the following equality:

$$M_t = \sum_{i=1}^{\infty} \int_0^{t+} \int_{U_i} y_i(s, z_i, \omega) \widetilde{N}_i(\mathrm{d}s, \mathrm{d}z_i) + Z_t^{\infty}$$

for any $t \in [0, T]$ with $Z^{\infty} \in \mathfrak{M}^2$. In addition to this, we have also

$$\langle M., \int_0^{\cdot +} \int_{U_i} y_i(s, z_i, \omega) N_i(\mathrm{d}s, \mathrm{d}z_i) \rangle_t = \int_0^{t} \int_{U_i} y_i(s, z_i, \omega) \mathrm{d}s \, n_i(\mathrm{d}z_i) \qquad (3.1)$$

Next we show $Z^{\infty} \equiv 0$ P a.s.. Since Z^{∞} is right continuous left limit (RCLL in short), it is sufficient to prove $Z_t = 0$ P a.s. for any $t \in [0, T]$, therefore we show for any bounded measurable functions $f_k : \mathbb{R}^\infty \to \mathbb{C}$, $0 \leq k \leq n$, such that

$$\mathbf{E}\left[Z_t \cdot \prod_{k=0}^{n} f_k(q_{s_k})\right] = 0, \tag{3.2}$$

where $0 = s_0 \leq s_1 \leq \cdots \leq s_n \leq t$ is a partition of interval $[0,t]$, and $q_{s_k} = (q_1(s_k), \cdots q_n(s_k), \cdots)$. In fact we only take into account the case where f_i ($i = 1, 2, \cdots, k$) depends on arbitrary finite coordinates. Without loss of generality, assume $f_i = f_i(x_1, x_2, \cdots, x_{i_k})$ ($i_k = 1, 2, \cdots$).

We prove equation (3.3) true by induction on n. When $n = 0$ this is an obvious result. Suppose equation (3.3) holds for $n-1$, and let $f_n = f(x_1, x_2, \cdots, x_{i_n}) = \exp\{-\sum_{j=1}^{i_n} \theta_j x_j\}$ with $\theta_j > 0$, $x_j > 0$, $j = 1, 2, \cdots, i_n$.

Let $A_i \subset U_i$ with $n_i(A_i) < \infty$, then by Itô formula (3.1) we have

$$e^{-\sum_{j=1}^{i_k} \theta_j N_j\left([0,s_n], A_i\right)}$$

$$= \exp\{-\sum_{j=1}^{i_k} \theta_j N_j([0, s_{n-1}], A_i)\}$$

$$+ \sum_{j=1}^{i_k}(e^{-\theta_j} - 1) \int_{s_{n-1}}^{s_n+} \Gamma(s-) N_j(\mathrm{d}s, \mathrm{d}x_j)$$

$$+ \sum_{i<j}(e^{-\theta_i} - 1)(e^{-\theta_j} - 1) \int_{s_{n-1}}^{s_n+} \Gamma(s-)\mathrm{d}[N_i, N_j]_s,$$

where $\Gamma(s-)$ was givend in equation (3.1). Put

$$\varphi(s_n) = \mathbf{E}\left[Z_t^\infty \prod_{i=1}^{n-1} f_i(q_{s_i}) \exp\{-\sum_{j=1}^{i_k} \theta_j N_j([0, s_n], A_j)\}\right]$$

$$= \mathbf{E}\left[Z_{s_n}^\infty \prod_{i=1}^{n-1} f_i(q_{s_i}) \exp\{-\sum_{j=1}^{i_k} \theta_j N_j([0, s_n], A_j)\}\right]$$

$$= \mathbf{E}\left[\mathbf{E}\left[Z_{s_n}^\infty \prod_{i=1}^{n-1} f_i(q_{s_i}) \exp\{-\sum_{j=1}^{i_k} \theta_j N_j([0, s_n], A_j)\}\Big|\mathcal{F}_{s_{n-1}}\right]\right]$$

$$= \mathbf{E}\left[\prod_{i=1}^{n-1} f_i(q_{s_i}) \mathbf{E}\left[Z_{s_n}^\infty \exp\{-\sum_{j=1}^{i_k} \theta_j N_j([0, s_n], A_j)\}\Big|\mathcal{F}_{s_{n-1}}\right]\right]$$

$$= \mathbf{E}\left[Z_{s_{n-1}}^\infty \prod_{i=1}^{n-1} f_i(q_{s_i}) \exp\{-\sum_{j=1}^{i_k} \theta_j N_j([0, s_{n-1}], A_j)\}\right]$$

$$+ \mathbf{E}\left[\prod_{i=1}^{n-1} f_i(q_{s_i}) \mathbf{E}\left[Z_{s_n}^\infty \sum_{j=1}^{i_k}(e^{-\theta_l}-1)\int_{s_{n-1}}^{s_n}\Gamma(s-)N_j(ds,A_j)\bigg|\mathcal{F}_{s_{n-1}}\right]\right]$$

$$+ \mathbf{E}\left[\prod_{i=1}^{n-1} f_i(q_{s_i}) \mathbf{E}\left[Z_{s_n}^\infty \sum_{l<m}(e^{-\theta_l}-1)(e^{-\theta_m}-1)\int_{s_{n-1}}^{s_n}\Gamma(s-)d[N_l,N_m]_s\bigg|\mathcal{F}_{s_{n-1}}\right]\right]$$

$$= I_1 + I_2 + I_3.$$

By hypothesis, the first term I_1 above is zero. Since

$$\mathbf{E}\left[Z_{S_n}^\infty \sum_{j=1}^{i_k}(e^{-\theta_j}-1)\int_{s_{n-1}}^{s_n}\Gamma(s-)N_j(ds,A_j)\bigg|\mathcal{F}_{s_{n-1}}\right]$$

$$= \sum_{j=1}^{i_k}(e^{-\theta_j}-1)\mathbf{E}\left[Z_{s_n}^\infty \int_{s_{n-1}}^{S_n}\Gamma(s-)(\widetilde{N}_j(ds,A_j)+n_j(A_j)ds)\bigg|\mathcal{F}_{s_{n-1}}\right]$$

$$= \sum_{j=1}^{i_k}(e^{-\theta_j}-1)\mathbf{E}\left[Z_{s_n}^\infty \int_{s_{n-1}}^{s_n}\Gamma(s-)n_j(A_j)ds\bigg|\mathcal{F}_{s_{n-1}}\right],$$

we have the identity

$$I_2 = \mathbf{E}\left[Z_{s_{n-1}}^\infty \prod_{i=1}^{n-1} f_i(q_{s_i}) \sum_{j=1}^{i_k}(e^{-\theta_j}-1)\int_{s_{n-1}}^{s_n}\Gamma(s-)n_j(A_j)ds\bigg|\mathcal{F}_{s_{n-1}}\right]$$

$$= \sum_{j=1}^{i_k}(e^{-\theta_j}-1)n_j(A_j)\int_{s_{n-1}}^{s_n}\mathbf{E}\left[Z_{s_{n-1}}^\infty \prod_{i=1}^{n-1} f_i(q_{s_i})\Gamma(s)\right]ds$$

$$= \sum_{j=1}^{i_k}(e^{-\theta_j}-1)n_j(A_j)\int_{s_{n-1}}^{s_n}\varphi(s)ds.$$

Therefore the following integral equation is obtained:

$$\varphi(s_n) = \sum_{j=1}^{i_k}(e^{-\theta_j}-1)n_j(A_j)\int_{S_{n-1}}^{S_n}\varphi(s)ds,$$

from which we know $\varphi \equiv 0$.

Then a familiar arguments about Laplacian transformation leads to equation (3.3) holding for all bounded functions f_n. Thus we complete the proof. □

Remark 3.1. We can use such a martingale presentation property to get the existence and uniqueness theory of BSDE driven by countably many Brownian motion with jump, e.g. [12].

4. Non-Markovian SDE driven by countably many Poisson point processes

In this section we shall focus on Non-Markovian SDE driven by countably many Poisson point processes of the form

$$x_t = x_0 + \int_0^t b(s,x)ds + \sum_{i=1}^\infty \int_0^t \int_{U_i} \sigma_i(s,x-,z_i)\tilde{N}_i(dsdz_i). \qquad (4.1)$$

Here we only consider this equation in R^1 for simplicity and follow the notations in the previous sections. We also use the notation $\mathcal{D} = \mathcal{D}([0,T])$ denoting the space of all RCLL functions on $[0,T]$ with sup norm $|x|_T = \sup_{0\leq t\leq T}|x_t|$, and $\mathcal{L}^2 = \mathcal{L}^2(\Omega,\mathcal{D})$ denoting the space of square integral functionals from Ω to $\mathcal{D}$ with norm $\|X\| = (\mathbf{E}|X|_T^2)^{1/2}$. For $x \in \mathcal{D}$, $x-$ is the left limit of x.

Now we will give some assumptions on coefficients b and $\sigma_i, i = 1,2,\cdots$

(A1): $b(t,x) : [0,T] \times \mathcal{D} \longrightarrow R^1$ and $\sigma_i(t,x,z_i) : [0,T] \times \mathcal{D} \times U_i \longrightarrow R^1, i = 1,2,\cdots$, are deterministic measurable functions.

(A2): $b(t,x)$ and $\sigma_i(t,x,z_i), i = 1,2,\cdots$, satisfy integral non-Lipschitz conditions: for any $x,y \in \mathcal{D}, t \in [0,T]$,

$$|b(t,x) - b(t,y)|^2 \leq L_1|x_t - y_t|^2 + L_2 \int_0^t \rho(|x_s - y_s|^2)dA_s,$$

$$\sum_{i=1}^\infty \int_{U_i} |\sigma_i(t,x-,z_i) - \sigma(t,y-,z_i)|^2 n_i(dz_i)$$

$$\leq L_1|x_t - y_t|^2 + L_2 \int_0^t \rho(|x_s - y_s|^2)dA_s, \qquad (4.2)$$

for some RCLL increasing functions $A(s)$ on $[0,T]$ and some positive constants L_1, L_2. ρ is an increasing concave functions satisfying $\int_{0+} \rho^{-1}(r)dr = +\infty$ with $\rho(0) = 0$.

Because of concavity of ρ, integral non-Lipschitz condition (4.2) implies the following linear growth condition (maybe some modifications for constants occur), note that $\sigma_i, i = 1,2,\ldots$, are compelled to be subject to this condition.

$$|b(t,x)|^2 + \sum_{i=1}^\infty \int_{U_i} \sigma_i^2(s,x-,z_i)n_i(dz_i)$$

$$\leq L_1(1 + x_t^2) + L_2 \int_0^t (1 + x_s^2)dA_s. \qquad (4.3)$$

Non-Markovian type SDEs can be used in the theory of transmission of messages in noise channel, such as coding and decoding (see [8] and references thererin), also used in the theory of stochastic optimal control. The corresponding equations are all, usually, with feedback, i.e., the input at time t may include all the past information of the output up to t. The SDEs or BSDEs of this type have been discussed in [3] in a general setup—Hilbert-valued processes driven by cylindrical Brownian motion with jump (also see [9] in real-valued Brownian motion case). In spite of generality in [3], one may be dazzled by the stack of conditions. Here we give a result in a relatively concise form.

First, we formulate the definition of the existence and uniqueness for equation (4.1).

Definition 4.1. There exists a RCLL $\mathcal{F}_t$-adapted process $x(t,\omega)$ satisfying equation (4.1) and if two RCLL $\mathcal{F}_t$-adapted processes x^1, x^2 satisfy equation (4.1) with $\mathbf{E}\sup_{0\le t\le T}(|x^1(t)-x^2(t)|)^2=0$, we say that equation (4.1) has a unique solution.

Theorem 4.1. Suppose $x_0 \in L^2(\Omega)$, conditions (A1) and (A2) hold. Then equation (4.1) admits a unique solution.

Proof. We first show uniqueness. To this end, let x, y be two solutions of eq.(4.1), then we have

$$|x_t - y_t|^2 \le 2T\int_0^t [b(s,x) - b(s,y)]^2 ds$$

$$+ 2\left[\sum_{i=1}^\infty \int_0^{t+}\int_{U_i}(\sigma_i(s,x-,z_i) - \sigma_i(s,y-,z_i))\tilde{N}_i(dsdz_i)\right]^2,$$

thus, by Doob's inequalities

$$\mathbf{E}\sup_{0\le s\le t}|x_s - y_s|^2$$

$$\le 2T\mathbf{E}\int_0^t [b(s,x) - b(s,y)]^2 ds$$

$$+ 2T\mathbf{E}\sum_{i=1}^\infty \int_0^{t+}\int_{U_i}[\sigma_i(s,x^1-,z_i) - \sigma_i(s,x^2-,z_i)]^2 dsn_i(dz_i). \quad (4.4)$$

By integral Lipschitz condition (2.1) plus the concave properties of ρ, we get the following inequality:

$$\mathbf{E}\sup_{0\le s\le t}|x_s - y_s|^2 \quad (4.5)$$

$$\le L_1\int_0^t \mathbf{E}\sup_{0\le u\le s}|x_u - y_u|^2 ds + L_2\int_0^t\int_0^s \rho(\mathbf{E}\sup_{0\le v\le u}|x_v - y_v|^2)dA_u ds$$

$$+ L_1\int_0^t \mathbf{E}\sup_{0\le u\le s}|x_u - y_u|^2 ds + L_2\int_0^t\int_0^s \mathbf{E}\sup_{0\le v\le u}|x_v - y_v|^2 dA_u ds.$$

Put $H(t) = \mathbf{E}\sup_{0\le s\le t}|x_s - y_s|^2$, we have the integral equation

$$H(t) \le L_1\int_0^t H(s)ds + L_2\int_0^t\int_0^s \rho(H(u))dA_u ds$$

$$+ L_2\int_0^t\int_0^s H(u)dA_u ds. \quad (4.6)$$

Therefore, by Lemma 1.1 below we know that $H(t) = 0$ for all $0 \leq t \leq T$, i.e., $P(x = y) = 1$. The proof for uniqueness is complete.

As far as existence is concerned, we will proceed by a Picard iteration procedure as follows: for $0 \leq t \leq T$

$$x^0 = x_0$$

$$x^{n+1}(t) = x_0 + \int_0^t b(s, x^n) ds + \sum_{i=1}^{\infty} \int_0^t \int_{U_i} \sigma_i(s, x^n-, z_i) \tilde{N}_i(dsdz_i). \quad (4.7)$$

Since ρ can be dominated by $ax + b$ in R for some a, b, and by linear growth, we know definition (4.7) reasonable, and we can estimate $\mathbf{E} \sup_{0 \leq s \leq t} |x_s^n|^2 \leq C(x^2, T, L)$ independent of n and $0 \leq t \leq T$ (see [9]).

Also, for $n, m \geq 1, 0 \leq t \leq T$, we have as equation (4.4)

$$\mathbf{E} \sup_{0 \leq s \leq t} |x_s^n - x_s^m|^2 \leq 2L_1 \int_0^t \mathbf{E} \sup_{0 \leq u \leq s} |x_u^n - x_u^m|^2 ds$$

$$+ L_2 \int_0^t \int_0^s \rho(\mathbf{E} \sup_{0 \leq v \leq u} |x_v^n - x_v^m|^2) dA_u ds$$

$$+ L_2 \int_0^t \int_0^s \mathbf{E} \sup_{0 \leq v \leq u} |x_v^n - x_v^m|^2 dA_u ds.$$

If we set $G^{n,m}(t) = \mathbf{E} \sup_{0 \leq s \leq t} |x_s^n - x_s^m|^2$, then by remarks above we know $\limsup_{n,m \to \infty} G^{n,m}(t) = \mathbf{E} \sup_{0 \leq s \leq t} |x_s^n - x_s^m|^2$ exists and denote it by $G(t)$, hence we have

$$G(t) \leq 2L_1 \int_0^t G(s) ds$$

$$+ L_2 \int_0^t \int_0^s \rho(G(u)) dA_u ds + L_2 \int_0^t \int_0^s G(u) dA_u ds \quad (4.8)$$

$$\leq 2L_1 \int_0^t G(s) ds + L_2 \int_0^t \int_0^s \rho_1(G(u)) dA_u ds,$$

where $\rho_1(z) = \rho(z) + z$.

Again by Lemma 1.1 we have $G \equiv 0$, which implies that $\{x^n\}_{n=1}^{\infty}$ is a Cauchy sequence in $L^2(\Omega, \mathcal{D})$ with respect to the norm $(\mathbf{E} \sup_{0 \leq s \leq t} |\cdot|^2)^{1/2}$. We denote by x the limit, then by a limit procedure we have the desired result. □

Lemma 4.1. *Let $f(t)$ be Borel measurable bounded left limit and nonnegative function on $[0, T]$, $H(z)$ be a continuous increasing function with the property that $u^{-1}H(z) \leq H(u^{-1}z)$ for $z \geq 0, u > 0$. If*

$$f(t) \leq \int_0^t m(s) f(s) ds + \int_0^t n(s) \int_0^s l(\tau) H(f(\tau)) dA_\tau ds, \quad (4.9)$$

and m, n are two continuous functions and can be comparable each other. A_t is a nondecreasing function on $[0, T]$. Then $f(t) \equiv 0$ for $t \in [0, T]$.

Proof. First let $c_0 > 0$, and consider the following inequality:

$$f(t) \leq c_0 + \int_0^t m(s)f(s)\mathrm{d}s + \int_0^t n(s)\int_0^s l(\tau)H(f(\tau))\mathrm{d}A_\tau \mathrm{d}s, \qquad (4.10)$$

or

$$g(t) \leq 1 + \int_0^t m(s)g(s)\mathrm{d}s + \int_0^t n(s)\int_0^s l(\tau)H(g(\tau))\mathrm{d}A_\tau \mathrm{d}s, \qquad (4.11)$$

with $g = f/c_0$. Denote by $v(t)$ the right hand of equation (4.11) and differentiate, we find $g(t) \leq v(t)$ and

$$\begin{aligned} v'(t) &\leq m(t)g(t) + n(t)\int_0^t l(\tau)H(g(\tau))\mathrm{d}A_\tau \\ &\leq m(t)v(t) + n(t)\int_0^t l(\tau)H(v(\tau))\mathrm{d}A_\tau. \end{aligned} \qquad (4.12)$$

Put $W(t) = v(t) + \int_0^t l(\tau)H(v(\tau))\mathrm{d}A_\tau$, then $v(t) \leq W(t)$ and

$$W(t) \leq v(t) + \int_0^t l(\tau)H(W(\tau))\mathrm{d}A_\tau,$$

i.e.,

$$\frac{W(t)}{v(t)} \leq 1 + \int_0^t l(\tau)H(W(\tau)/v(\tau))\mathrm{d}A_\tau,$$

An extension of Bihari's inequality to Lebesgue-Stieltjes integral (see [10]) justifies the following inequality,

$$W(t) \leq v(t)\Psi^{-1}(\Psi(1) + \int_0^t l(\tau)\mathrm{d}A_\tau),$$

where $\Psi(t) = \int_{t_0}^t \frac{1}{H(s)}\mathrm{d}s$, Ψ^{-1} is the inverse of Ψ. If assume $m(t) \leq n(t)$, then we have, by inequality (4.12), $v'(t) \leq g(t)W(t)$, and

$$v'(t) \leq m(t)v(t)\Psi^{-1}(\Psi(1) + \int_0^t l(\tau)\mathrm{d}A_\tau).$$

Thus, since $v(0) = 1$, we have

$$v(t) \leq \exp \int_0^t m(s)\Psi^{-1}(\Psi(1) + \int_0^s l(s)\mathrm{d}A_s),$$

that is,

$$f(t) \leq c_0 \exp \int_0^t m(s)\Psi^{-1}(\Psi(1) + \int_0^s l(s)\mathrm{d}A_s).$$

Finally, let $c_0 \to 0$, the desired result is followed. $\square$

Remark 4.1. This lemma is an extension of theorem 1 in [15]. Here Lebesgue-Stieljes integral is involved, the corresponding difficulty is that the differentiation can not be implemented.

Remark 4.2. If f in the first term of the right hand of inequality (4.9) is replaced by $\gamma(f)$ for some nonlinear functions γ, the conclusion may not hold (see [4]).

References

1. T. Björk and B. Näslund, Diversified portfolios in continuous time. European Financial Review, 1(1998), 361-387.
2. G. Cao and K. He, On a type of stochastic differential equations driven by countably many Brownian motions. Journal of Functional Analysis, **203**(2003),262-285.
3. G. Cao, K. He and X. Zhang, Successive approximations of infinite dimensional SDEs with jump. Stoch. Dyn. **5** (4)(2005), 609-619.
4. M. Dannan, Integral Inequalities of Gronwall-Bellman-Bihari Type and Asymptotic Behavior of Certain Second Order Nonlinear Differential Equations. J. Math. Anal. Appl. **108**(1985), 151-164.
5. M. Donno and M. Pratelli, Stochastic Integration with Respect to a sequence of semimartingales. LNM, Springer-Verlag, 2006.
6. M. Hitsuda and H. Watanabe, On Stochastic Integrals with Respect to an Infinite Nunber of Brownian Motions and Its Applications. Proc.of Ineqn.Symp.SDE, **57-74**.Kyoto:John Wiley&Sons,Inc, 1976.
7. I. Karatzas and E. Shreve, Brownian Motion and Stochastic Calculus. Springer, 1988.
8. P. Katyshev, Uniform Optimal Transmission of Gaussian Messages, From Stochastic Calculus to Mathematical Finance –The Shiryaev Festschrift, Yuri Kabanov, Robert Liptser and Jordan Stoyanov (eds), Springer 369-384,2006.
9. R. Lipster and A. Shiryaev, Statistics of Stochastic Processes. Moscow, "Naukau" (1974).
10. X. Mao, Lebesgue-Stieljes integral inequlities in several variables with retardation. Proc. Indian. Acad. Sci. (Math. Sci.), **100**(3)(1990), 231-243.
11. E. Protter, Stochastic Integration and Differential Equation. Second Edition, Springing, 2004.
12. G. Rang and H. Jiao, The existence and uniquness of the solution of backward stochastic differential equation driven by countably many Brownian motions with jump. J. Hubei University, 2009.
13. K. Sennewald and K. Wälde, "Ito's Lemma" and the Bellman Equation for Poisson Processes: An Applied View. Journal of Economics, **89**(1)(2006),1-36.
14. S. Tang and X. Li, Necessary conditions for optimal control of stochastic systems with random jumps. SIMA J. Control Optim, **32**(1994), 1447-1475.
15. C. Young, ON BELLMAN-BIHARI INTEGRAL INEQUALITIES. Intern. J. Math. & Math. Sci., **5**(1)(1982),97-103.

Chapter 16

Lévy White Noise, Elliptic SPDEs and Euclidean Random Fields

Jiang-Lun Wu

Department of Mathematics, Swansea University, Singleton Park, Swansea SA2 8PP, United Kingdom

In this article, we start with a briefly survey of the recent development on Euclidean random fields along with constructive, relativistic quantum field theory. We then present a unified account of Lévy white noise and related elliptic SPDEs driven by Lévy white noise. We explicate the link, via analytic continuation, from the Euclidean random fields obtained as solutions of the elliptic SPDEs to local, relativistic quantum fields with indefinite metric. By comparing the derived vector and scalar relativistic quantum field models, we reformulate the elliptic SPDE for the scalar model with a new Gaussian white noise term so that the associated Euclidean random field possesses a feature of avoiding the re-definition of the two point Schwinger functions needed for nontrivial scattering in the relativistic model, which then leads to a scalar model of local, relativistic quantum field theory in indefinite metric with nontrivial scattering behavior. Finally, we demonstrate a lattice approximation for the induced Euclidean random field.

Contents

1 Introduction 251
2 Lévy white noise 253
3 Lévy white noise and random fields 256
4 Comparison of vector and scalar models 258
5 New formulation of elliptic SPDEs and the lattice approximation 263
References 265

1. Introduction

Since the pioneer works of Symanzik[51] and Nelson,[46,47] the construction of local, relativistic quantum fields via analytic continuation from Euclidean random fields has been the most vital and productive paradigm in constructive quantum field theory (QFT), see e.g. the by now classical expositions in Ref. 31. This Euclidean strategy has been completed successfully in $d = 2$ space-time dimensions[49] (cf. also Ref. 1) and partial results have been obtained for $d = 3$ (see Refs. 21 and 30). In the physical space-time dimension $d = 4$, however, the standard approach to the definition of local potentials via renormalization up to now is plagued by seemingly incurable ultra-violet divergences, and no construction of a non-trivial (interacting) quantum field is known within that approach.

In the series of papers by Albeverio and Høegh–Krohn[12–14] (see also Ref. 11), a different approach for $d = 4$ was started of construction Euclidean covariant vector Markov random fields as solutions of stochastic quaternionic Cauchy-Riemann equations with multiplicative white noise. This has been associated, in Ref. 15, with some mass zero local relativistic quantum filed models of gauge type by obtaining their Wightman functions via performing the analytic continuation of the corresponding Schwinger functions (moments) of these Euclidean covariant random fields. However, the peculiarities of mass zero and gauge invariance in the relativistic vector models rise difficulties of physical interpretation, especially concerning to the construction of the "physical Hilbert space", in QFT. But remarkably in Ref. 5, it has been proved that these Wightman functions satisfy the so-called "Hilbert space structure condition" which permits the construction of (non unique) physical Hilbert spaces associated to the set of Wightman functions and hence leads, by Refs. 44 and 50, to local relativistic vector field models for indefinite metric QFT. Furthermore, in Ref. 7, explicit formulae for the (gauge invariant) scattering amplitudes for these local relativistic vector fields with indefinite metric have been carried out, which shows that such models have nontrivial scattering behaviour.

Moreover, Euclidean covariant (elliptic) SPDEs with multiplicative white noise have been systematically studied in Refs. 22, 8 and 9, leading to local relativistic vector models which include massive quantum vector fields. Further in Ref. 8, necessary and sufficient conditions on the mass spectrum of the given covariantly differential operators (with constant coefficients) have been presented. Such conditions imply nontrivial scattering behaviour of the relativistic models. In Ref. 9, in order to avoid renormalising the two point Schwinger functions and to keep the induced relativistic fields with nontrivial scattering, a revised formulation of the covariant SPDEs with a newly added Gaussian white noise term is proposed. Furthermore, asymptotic states and the S-matrix are constructed. The scattering amplitudes can be explicitly calculated out and the masses of particles are then determined by the mass spectrum. Thus one can have a nice particle interpretation picture of the obtained vector models (Ref. 35). In Ref. 23, however, a no-go theorem for Euclidean random vector fields has been established, showing why Euclidean field theory has been less successful in the vector case than in the scalar case. For example, as a consequence of the no-go theorem, it follows that there is no simple vector analogue of scalar $P(\phi)_2$ -theory.

On the other hand, by considering stochastic pseudo-differential equations with generalized white noise, scalar models have been started in Ref. 17 (massless) and in Ref. 4 (massive). In the latter case, a better approach of anayltic continuation has been developed via truncation techniques. Again, in Ref. 5, it has been proved that the obtained Wightman functions satisfy the "Hilbert space structure condition" which then permits the construction of (non unique) physical Hilbert spaces and hence leads to local relativistic scalar field models for indefinite metric QFT, as in the vector case. However, there is no scattering theory (neither Lehmann-Symanzik-Zimmerman theory nor Haag-Ruelle theory) to the scalar case. In Ref. 6, the spectral condition on the translation group for the scalar models has been proved, which is an important step towards Haag-Ruelle scattering theory in the usual standard (namely, positive definite metric) QFT, cf. e.g. Ref. 36. In Ref. 2, axiomatically scattering theory for local relativisitic QFT with indefinite metric has been established. Let us also mention Refs. 10, 20, 24–28, 32–34, 37–40, 43, 45, 48 for further investigations of the scalar models.

The rest of the paper is organised as follows. In the next section, we shall briefly introduce Lévy white noise with several concrete examples of Lévy type noise in SPDEs. In Section 3, we establish a link of Lévy white noise with generalized random fields (i.e., multiplicative white noise). In Section 4, starting with elliptic SPDEs driven by Lévy white noise, we explicate the link, via analytic continuation, from the Euclidean random fields obtained as solutions of the elliptic SPDEs to local, relativistic quantum fields with indefinite metric. By comparing the derived vector and scalar relativistic quantum field models, we then reformulate the elliptic SPDE for the scalar model with a new Gaussian white noise term so that the associated Euclidean random field possesses a feature of avoiding the re-definition of the two point Schwinger function needed for nontrivial scattering in the relativistic model, which then leads to a scalar model of local, relativistic quantum field theory in indefinite metric with nontrivial scattering behavior. Section 5, the final section, is devoted to demonstrate a lattice approximation for the induced Euclidean random field.

2. Lévy white noise

This section is devoted to a brief account to Lévy white noise. We start with Poisson white noise. Let $(\Omega, \mathcal{F}, P)$ be a given complete probability space with a filtration $\{\mathcal{F}_t\}_{t \in [0,\infty)}$ and let $(U, \mathcal{B}(U), \nu)$ be an arbitrary σ-finite measure space.

Definition 2.1. Let $(E, \mathcal{E}, \mu)$ be a σ-finite measure space. By a Poisson white noise on $(E, \mathcal{E}, \mu)$ we mean an integer-valued random measure

$$N : (E, \mathcal{E}, \mu) \times (U, \mathcal{B}(U), \nu) \times (\Omega, \mathcal{F}, P) \to \mathbf{N} \cup \{0\} \cup \{\infty\} =: \tilde{\mathbf{N}}$$

with the following properties:

(i) for $A \in \mathcal{E}$ and $B \in \mathcal{B}(U)$, $N(A, B, \cdot) : (\Omega, \mathcal{F}, P) \to \tilde{\mathbf{N}}$ is a Poisson distributed random variable with

$$P\{\omega \in \Omega : N(A, B, \omega) = n\} = \frac{e^{-\mu(A)\nu(B)}[\mu(A)\nu(B)]^n}{n!}$$

for each $n \in \tilde{\mathbf{N}}$. (Here we take the convention that when $\mu(A) = \infty$ or $\nu(B) = \infty$, $N(A, B, \cdot) = \infty$, P-a.s.);

(ii) for any fixed $B \in \mathcal{B}(U)$ and any $n \geq 2$, if $A_1, \ldots, A_n \in \mathcal{E}$ are pairwise disjoint, then $N(A_1, B, \cdot), \ldots, N(A_n, B, \cdot)$ are mutually independent random variables such that

$$N(\cup_{j=1}^n A_j, B, \cdot) = \sum_{j=1}^n N(A_j, B, \cdot), \quad P-a.s.$$

Clearly, the mean measure of N is

$$\mathbf{E}[N(A, B, \cdot)] = \mu(A)\nu(B), \quad A \in \mathcal{E}, B \in \mathcal{B}(U).$$

N is nothing but a Poisson random measure on the product measure space $(E \times U, \mathcal{E} \times \mathcal{B}(U), \mu \otimes \nu)$ and can be constructed canonically as

$$N(A, B, \omega) := \sum_{n \in \mathbf{N}} \sum_{j=1}^{\eta_n(\omega)} 1_{(A \cap E_n) \times (B \cap U_n)}(\xi_j^{(n)}(\omega)) 1_{\{\omega \in \Omega : \eta_n(\omega) \geq 1\}}(\omega) \quad (2.1)$$

for $A \in \mathcal{E}, B \in \mathcal{B}(U)$ and $\omega \in \Omega$, where
(a) $\{E_n\}_{n \in \mathbf{N}} \subset \mathcal{E}$ is a partition of E with $0 < \mu(E_n) < \infty, n \in \mathbf{N}$, and $\{U_n\}_{n \in \mathbf{N}} \subset \mathcal{B}(U)$ is a partition of U with $0 < \nu(U_n) < \infty, n \in \mathbf{N}$;
(b) $\forall n, j \in \mathbf{N}, \xi_j^{(n)} : \Omega \to E_n \times U_n$ is $\mathcal{F}/\mathcal{E}_n \times \mathcal{B}(U_n)$-measurable with

$$P\{\omega \in \Omega : \xi_j^{(n)}(\omega) \in A \times B\} = \frac{\mu(A)\nu(B)}{\mu(E_n)\nu(U_n)}, \quad A \in \mathcal{E}_n, B \in \mathcal{B}(U_n),$$

where $\mathcal{E}_n := \mathcal{E} \cap E_n$ and $\mathcal{B}(U_n) := \mathcal{B}(U) \cap U_n$;
(c) $\forall n \in \mathbf{N}, \eta_n : \Omega \to \tilde{\mathbf{N}}$ is Poisson distributed with

$$P\{\omega \in \Omega : \eta_n(\omega) = k\} = \frac{e^{-\mu(E_n)\nu(U_n)}[\mu(E_n)\nu(U_n)]^k}{k!}, k \in \tilde{\mathbf{N}};$$

(d) $\xi_j^{(n)}$ and η_n are mutually independent for all $n, j \in \mathbf{N}$.

Next, let us briefly recall the notion of Gaussian white noise on $(E, \mathcal{E}, \mu)$. It is a random measure

$$W : (E, \mathcal{E}, \mu) \times (\Omega, \mathcal{F}, P) \to [0, \infty)$$

such that $\{W(A, \cdot)\}_{A \in \mathcal{E}_F}$ is a Gaussian family of random variables with $\mathbf{E}[W(A, \cdot)] = \mu(A)$ and $\mathbf{E}[W(A_1, \cdot)W(A_2, \cdot)] = \mu(A_1 \cap A_2)$, where $\mathcal{E}_F := \{A \in \mathcal{E} : \mu(A) < \infty\}$.

Definition 2.2. By a Lévy white noise on $(E, \mathcal{E}, \mu)$, we mean a random measure $L : (E, \mathcal{E}, \mu) \times (\Omega, \mathcal{F}, P) \to [0, \infty)$ having the following expression

$$L(A, \omega) := W(A, \omega) + \int_A \int_U a(x, y) N(dx, dy, \omega)$$
$$+ \int_A \int_U b(x, y) \mu(dx) \nu(dy) \quad (2.2)$$

for $(A, \omega) \in \mathcal{E} \times \Omega$, where $a, b : E \times U \to \mathbf{R}$ are measurable.

Let us end this section with some examples of Lévy white noise used in SPDEs.

Example 1. Take $(E, \mathcal{E}) = ([0, \infty), \mathcal{B}([0, \infty)))$ and μ to be Lebesgue measure on the Borel σ-algebra $\mathcal{B}([0, \infty))$. Let the Poisson white noise N on $([0, \infty), \mathcal{B}([0, \infty)), \mu)$ be constructed canonically by (2.1). Such a random measure N is called an extended Poisson measure on $[0, \infty) \times U$ in Ref. 42. Alternatively, N can be constructed as the Poisson random measure on $[0, \infty) \times U$ associated with an $\{\mathcal{F}_t\}$-Poisson point process as in Ref. 41. In the sense of Schwartz distributions, we then define

$$N_t(B, \omega) := \frac{N(dt, B, \omega)}{dt}(t), \quad (t, B, \omega) \in [0, \infty) \times \mathcal{B}(U) \times \Omega.$$

We call N_t *Poisson time white noise*.

Example 2. Take $(E, \mathcal{E}, \mu) = ([0, \infty) \times \mathbf{R}^d, \mathcal{B}([0, \infty)) \times \mathcal{B}(\mathbf{R}^d), dt \otimes dx)$ and let the Poisson white noise N be constructed canonically by (2.1). Define (again in the sense of Schwartz distributions) the Radon-Nikodym derivative

$$N_{t,x}(B, \omega) := \frac{N(dt, dx, B, \omega)}{dt\,dx}(t, x), \quad (B, \omega) \in \mathcal{B}(U) \times \Omega$$

for $t \in [0, \infty)$ and $x \in \mathbf{R}^d$. We call $N_{t,x}$ *Poisson space-time white noise*.

Accordingly, we can define the compensating martingale measure

$$M(t, A, B, \omega) := N([0, t], A, B, \omega) - t|A|\nu(B)$$

for any $(t, A, B) \in [0, \infty) \times \mathcal{B}(\mathbf{R}^d) \times \mathcal{B}(U)$ with $|A|\nu(B) < \infty$ (where $|A|$ stands for the Lebesgue measure of A). Then we have

$$\mathbf{E}[M(t, A, B, \cdot)] = 0, \quad \mathbf{E}([M(t, A, B, \cdot)]^2) = t|A|\nu(B).$$

Moreover, we can define the Radon-Nikodym derivative

$$M_{t,x}(dy, \omega) := \frac{M(dt, dx, dy, \omega)}{dt\,dx}(t, x).$$

For a Brownian sheet $\{W(t, x, \omega)\}_{(t,x,\omega) \in [0,\infty) \times \mathbf{R}^d \times \Omega}$ on $[0, \infty) \times \mathbf{R}^d$ (cf. e.g. Ref. 52), we define

$$W_{t,x}(\omega) := \frac{\partial^{d+1} W(t, x, \omega)}{\partial t\, \partial x_1 \ldots \partial x_d}.$$

Let the given N and W be independent and let $U_0 \in \mathcal{B}(U)$ with $\nu(U \setminus U_0) < \infty$ be arbitrarily given. Set

$$L_{t,x}(\omega) = W_{t,x}(\omega) + \int_{U_0} c_1(t, x; y) M_{t,x}(dy, \omega)$$

$$+ \int_{U \setminus U_0} c_2(t, x; y) N_{t,x}(dy, \omega)$$

for $(t, x, \omega) \in [0, \infty) \times \mathbf{R}^d \times \Omega$, where $c_1, c_2 : [0, \infty) \times [0, L] \times U \to \mathbf{R}$ are measurable. $L_{t,x}$ is called *Lévy space-time white noise*.

Example 3. Take $(E, \mathcal{E}, \mu) = ([0, \infty), \mathcal{B}([0, \infty)), dt)$ and $(U, \mathcal{B}(U), \nu) = (\mathbf{R}^d \setminus \{0\}, \mathcal{B}(\mathbf{R}^d \setminus \{0\}), \frac{dx}{|x|^{d+p}}), p \in (0, 2)$. Let the Poisson random measure N and the compensating martingale measure M be defined in the same manner as in previous examples. Let us restrict ourselves to the nonnegative valued M. Namely, we consider those M whose Laplace transform is given by

$$\mathbf{E} e^{\theta M([0,t], A, \cdot)} = e^{\theta^p t |A|}$$

for $t \geq 0, \theta \geq 0, A \in \mathcal{B}(\mathbf{R}^d \setminus \{0\})$ with $|A| < \infty$. Notice that in this case, $M([0, t], A, \omega)$ is nonnegative valued and its Fourier transform is given by

$$\mathbf{E} e^{i\theta M([0,t], A, \cdot)} = e^{(i\theta)^p t |A|}$$

for $t \geq 0, \theta \in \mathbf{R}, A \in \mathcal{B}(\mathbf{R}^d \setminus \{0\})$ with $|A| < \infty$. From this it is easy to see that when $p = 2$, M then determines a Brownian sheet. We call the (distributional) Radon-Nikodym derivative

$$M_{t,x}(\omega) = \frac{M(dt, dx, \omega)}{dtdx}(t, x)$$

a *p-stable space-time white noise*.

3. Lévy white noise and random fields

Here we want to give a link between Lévy white noise and random fields. First of all, we have the following definition for random fields

Definition 3.1. Let $(\Omega, \mathcal{F}, P)$ be a probability space and $(V, \mathcal{T})$ be a (real) topological vector space. By a random field X on $(\Omega, \mathcal{F}, P)$ with parameter space V, we mean a system $\{X(f, \omega), \omega \in \Omega\}_{f \in V}$ of random variables on $(\Omega, \mathcal{F}, P)$ having the following properties:

(1) $P\{\omega \in \Omega : X(c_1 f_1 + c_2 f_2, \omega) = c_1 X(f_1, \omega) + c_2 X(f_2, \omega)\} = 1$,
 for $c_1, c_2 \in \mathbf{R}, f_1, f_2 \in V$;
(2) $f_n \xrightarrow{(V, \mathcal{T})} f \Rightarrow X(f_n, \cdot) \xrightarrow{\text{in law}} X(f, \cdot)$.

Now let $\mathcal{S}(\mathbf{R}^d)$ ($d \in \mathbf{N}$) be the Schwartz space of rapidly decreasing (real) C^∞-functions on $\mathbf{R}^d$ and $\mathcal{S}'(\mathbf{R}^d)$ the topological dual of $\mathcal{S}(\mathbf{R}^d)$. Denote by $<\cdot, \cdot>$ the natural dual pairing between $\mathcal{S}(\mathbf{R}^d)$ and $\mathcal{S}'(\mathbf{R}^d)$. Let $\mathcal{B}$ be the σ-algebra generated by cylinder sets of $\mathcal{S}'(\mathbf{R}^d)$. Then $(\mathcal{S}'(\mathbf{R}^d), \mathcal{B})$ is a measurable space.

Let $\psi : \mathbf{R} \to \mathbb{C}$ be a continuous, negative definite function having the following Lévy-Khinchine representation

$$\psi(t) = iat - \frac{\sigma^2 t^2}{2} + \int_{\mathbf{R} \setminus \{0\}} \left(e^{ist} - 1 - \frac{ist}{1+s^2} \right) dM(s), \quad t \in \mathbf{R} \quad (3.1)$$

where $a, \sigma \in \mathbf{R}$ and the measure M (Lévy measure) satisfies
$$\int_{\mathbf{R}\setminus\{0\}} \min(1, s^2) dM(s) < \infty.$$
Hereafter, we call ψ a Lévy-Khinchine function.

From Gelfand and Vilenkin,[29] the following functional
$$C(f) = \exp\left\{\int_{\mathbf{R}^d} \psi(f(x)) dx\right\}, \quad f \in \mathcal{S}(\mathbf{R}^d)$$
is a characteristic functional on $\mathcal{S}(\mathbf{R}^d)$. Thus by Bochner-Minlos theorem (Ref. 29) there exists a unique probability measure P_ψ on $(\mathcal{S}'(\mathbf{R}^d), \mathcal{B})$ such that
$$\int_{\mathcal{S}'(\mathbf{R}^d)} e^{i<f,\omega>} dP_\psi(\omega) = \exp\left\{\int_{\mathbf{R}^d} \psi(f(x)) dx\right\}, \quad f \in \mathcal{S}(\mathbf{R}^d).$$

We call P_ψ the Lévy white noise measure and $(\mathcal{S}'(\mathbf{R}^d), \mathcal{B}, P_\psi)$ the Lévy white noise space associated with ψ. The associated (coordinate) canonical process
$$F : \mathcal{S}(\mathbf{R}^d) \times (\mathcal{S}'(\mathbf{R}^d), \mathcal{B}, P_\psi) \to \mathbf{R}$$
defined by
$$F(f, \omega) = <f, \omega>, \quad f \in \mathcal{S}(\mathbf{R}^d), \omega \in \mathcal{S}'(\mathbf{R}^d)$$
is a random field on $(\mathcal{S}'(\mathbf{R}^d), \mathcal{B}, P_\psi)$ with parameter space $\mathcal{S}(\mathbf{R}^d)$. Such kind of (Euclidean) random fields has been used substantially in recent years in constructive (indefinite metric) quantum field theory.

Remark 3.1. In the terminology of Gelfand and Vilenkin, F is called a generalized random process with independent value at every point, namely, the random variables $<f_1, \cdot>$ and $<f_2, \cdot>$ are independent whenever $f_1(x) f_2(x) = 0$ for $f_1, f_2 \in \mathcal{S}(\mathbf{R}^d)$. Moreover, F is also named multiplicative white noise or generalized white noise in the literature.

In order to present the relation between Lévy white noise L and the random field F, we need to define $F(1_A, \omega)$ for $A \in \mathcal{B}(\mathbf{R}^d)$. Notice that $1_A \in L^2(\mathbf{R}^d)$, thus we need to extend F to the parameter space $L^2(\mathbf{R}^d)$ from $\mathcal{S}(\mathbf{R}^d)$. This can be done by the following argument (cf. Ref. 17). Remarking that $\mathcal{S}(\mathbf{R}^d)$ is dense in $L^2(\mathbf{R}^d)$, for any $f \in L^2(\mathbf{R}^d)$, there exists a sequence $\{f_n\}_{n \in \mathbf{N}} \subset \mathcal{S}(\mathbf{R}^d)$ converging to f in $L^2(\mathbf{R}^d)$. Since $F(f_n, \omega)$ is well-defined for each $n \in \mathbf{N}$, one can define
$$F(f, \omega) := \lim_{n \to \infty} F(f_n, \omega)$$
where the limit is understood in law. Then the linear operator
$$F(\cdot, \omega) : \mathcal{S}(\mathbf{R}^d) \to L(\mathcal{S}'(\mathbf{R}^d), \mathcal{B}, P_\psi)$$
can be extended uniquely to a continuous linear operator
$$F(\cdot, \omega) : L^2(\mathbf{R}^d) \to L(\mathcal{S}'(\mathbf{R}^d), \mathcal{B}, P_\psi),$$

where $L(\mathcal{S}'(\mathbf{R}^d), \mathcal{B}, P_\psi)$ is the Fréchet space of real random variables on $(\mathcal{S}'(\mathbf{R}^d), \mathcal{B}, P_\psi)$ with quasi-norm $||\xi||_0 := \mathbf{E}_{P_\psi}(|\xi| \wedge 1)$. Now in Definition 2.2 by taking $(E, \mathcal{E}, \mu) = (\mathbf{R}^d, \mathcal{B}(\mathbf{R}^d), dx)$ and $(\Omega, \mathcal{F}, P) = (\mathcal{S}'(\mathbf{R}^d), \mathcal{B}, P_\psi)$, we have Lévy white noise $L(A, \omega)$, for $(A, \omega) \in \mathcal{B}(\mathbf{R}^d) \times \Omega$. By virtue of the Fourier transform,

$$F(1_A, \omega) \stackrel{\text{in law}}{=} L(A, \omega), \quad (A, \omega) \in \mathcal{B}(\mathbf{R}^d) \times \Omega.$$

4. Comparison of vector and scalar models

In this section, let us briefly introduce both the vector and scalar models and then give a comparison between them. We start with the construction of covariant Euclidean random fields via the quaternionic Cauchy-Riemann operator. To this end, we begin with quaternions and the quaternionic Cauchy-Riemann operator.

Let $\mathbf{Q}$ be the skew field of all quaternions and let $\{1, e_1, e_2, e_3\}$ be its canonical basis with multiplication rules

$$e_1 e_1 = e_2 e_2 = e_3 e_3 = -1 \quad \text{and} \quad e_1 e_2 = -e_2 e_1 = e_3.$$

A quaternion $x \in \mathbf{Q}$ is represented by

$$x = x_0 1 + x_1 e_2 + x_2 e_2 + e_3 x_3, \quad (x_0, x_1, x_2, x_3) \in \mathbf{R}^4.$$

Thus, $\mathbf{Q}$ is isomorphic to $\mathbf{R}^4$, the isomorphism being given by $x \in \mathbf{Q} \mapsto (x_0, x_1, x_2, x_3) \in \mathbf{R}^4$, regarding $\mathbf{Q}$ as a real vector space. Furthermore, $\mathbf{R}$ can be imbedded into $\mathbf{Q}$ by identifying $x_0 \in \mathbf{R}$ with $x_0 1 \in \mathbf{Q}$. Hence, $\mathbf{Q}$ forms a real associative algebra with identity 1 under the multiplication rules of the canonical basis. In fact, $\mathbf{Q}$ is a Clifford algebra. There is a distinct automorphism of $\mathbf{Q}$ called conjugation which is defined by

$$\bar{x} := x_0 1 - x_1 e_2 - x_2 e_2 - e_3 x_3, \quad x \in \mathbf{Q}.$$

One can then define a norm $|| \cdot ||$ and a scalar inner product $(\cdot, \cdot)_\mathbf{Q}$ on $\mathbf{Q}$:

$$||x|| := (x\bar{x})^{\frac{1}{2}} = (x_0^2 + x_1^2 + x_2^2 + x_3^2)^{\frac{1}{2}}, \quad x \in \mathbf{Q}.$$

and

$$(x, y)_\mathbf{Q} := \frac{1}{4}(||x + y||^2 - ||x - y||^2)^{\frac{1}{2}} = xy$$
$$= x_0 y_0 + x_1 y_1 + x_2 y_2 + x_3 y_3, \quad x, y \in \mathbf{Q}.$$

The inverse x^{-1} of a quaternion $x \in \mathbf{Q} \setminus \{0\}$ with respect to the multiplication is given by $x^{-1} = \frac{\bar{x}}{||x||}$. Setting $Sp(1) := \{x \in \mathbf{Q} : ||x|| = 1\}$, we notice that $Sp(1)$ is a subgroup of the multiplicative group $\mathbf{Q} \setminus \{0\}$ and is isomorphic to $SU(2)$. By $x \in \mathbf{Q} \mapsto uxv^{-1} \in \mathbf{Q}$ for $u, v \in Sp(1)$, we have a surjective homomorphism $Sp(1) \times Sp(1) \to SO(4)$, whose kernel is given by $\{(1, 1), (-1, -1)\} \cong \mathbb{Z}_2$, and hence $[Sp(1) \times Sp(1)]/\mathbb{Z}_2 \cong SO(4)$. Now by identifying $\mathbf{Q}$ with $\mathbf{R}^4$, we have the following two distinct $Sp(1) \times Sp(1)$ actions on the collection $\mathcal{X}(\mathbf{R}^4, \mathbf{Q})$ of all 4-vector fields on $\mathbf{R}^4$: the first one is given by

$$A(x) \longrightarrow uA(u^{-1}(x-y)v)v^{-1},$$

where

$$x, y \in \mathbf{R}^4, \quad (u,v) \in Sp(1) \times Sp(1), \quad A \in \mathcal{X}(\mathbf{R}^4, \mathbf{Q}),$$

and A obeying this rule is called a covariant 4-vector field; the second one is given by

$$A(x) \longrightarrow vA(u^{-1}(x-y)v)v^{-1},$$

where

$$x, y \in \mathbf{R}^4, \quad (u,v) \in Sp(1) \times Sp(1), \quad A \in \mathcal{X}(\mathbf{R}^4, \mathbf{Q}),$$

and A obeying this rule is called a covariant scalar 3-vector field.

Let $C_0^\infty(\mathbf{R}^4, \mathbf{Q})$ be the space of all $\mathbf{Q}$-valued smooth functions with compact supports (where the smoothness of $\mathbf{Q}$-valued functions is defined in terms of the four real components). We define a bilinear form, for $f \in C_0^\infty(\mathbf{R}^4, \mathbf{Q})$ and $A \in \mathcal{X}(\mathbf{R}^4, \mathbf{Q})$, via

$$<f, A> := \int_{\mathbf{R}^4} (f(x), A(x))_\mathbf{Q} dx$$

and then extend this relation to a distributional pairing in the natural manner. We remark that $<\cdot, \cdot>$ is invariant under the above two $Sp(1) \times Sp(1)$ actions.

We can now define the quaternionic differential operators $\partial, \bar{\partial}$ and Δ as

$$\partial := 1\partial_0 - e_1\partial_1 - e_2\partial_2 - e_3\partial_3,$$

$$\bar{\partial} := 1\partial_0 + e_1\partial_1 + e_2\partial_2 + e_3\partial_3,$$

$$\Delta := \partial\bar{\partial} = \bar{\partial}\partial = \partial_0^2 + \partial_1^2\partial_2^2 + \partial_3^2,$$

where $\partial_k := \frac{\partial}{\partial x_k}, k = 0, 1, 2, 3$. The operator ∂ is called the quaternionic Cauchy-Riemann operator. Moreover, we consider the two variable transformation $x \to x' = u^{-1}xv$ for some $(u,v) \in Sp(1) \times Sp(1)$ and define the corresponding quaternionic Cauchy-Riemann operator ∂' and its conjugate $\bar{\partial}'$ in the same manner as ∂ and $\bar{\partial}$, then $\partial' = v^{-1}\partial u$ and $\bar{\partial}' = u^{-1}\bar{\partial}v$. Let

$$\sigma := A_0 dx_0 + A_1 dx_1 + A_2 dx_2 + A_3 dx_3$$

be a 1-form with the orientation adapted to the canonical basis $\{1, e_1.e_2, e_3\}$ and $*\sigma$ be the associated Hodge dual. Further by identifying anti-selfdual 2-forms with 3-vector fields, we get

$$<{}^*d^*\sigma, d\sigma - {}^*\sigma> = <F_0, F_1 e_1 + F_2 e_2 + F_3 e_3>,$$

where

$$F_0 := \partial_0 A_0 + \partial_1 A_1 + \partial_2 A_2 + \partial_3 A_3,$$

$$F_1 := -\partial_1 A_0 + \partial_0 A_1 + \partial_3 A_2 - \partial_2 A_3,$$

$$F_2 := -\partial_2 A_0 - \partial_3 A_1 + \partial_0 A_2 + \partial_1 A_3,$$

$$F_3 := -\partial_3 A_0 + \partial_2 A_1 - \partial_1 A_2 + \partial_0 A_3.$$

Therefore, if A is a covariant 4-vector field, then ∂A is a covariant scalar 3-vector filed. However, ∂A is not covariant under reflections since $F_1 e_1 + F_2 e_2 + F_3 e_3$ corresponds to an anti-selfdual 2-form. Remarking that the Green's function for the operator $-\Delta$ is given by

$$g(x) := \frac{1}{2\pi^2 ||x||^2}, \quad x \in \mathbf{Q} \setminus \{0\}.$$

Thus, we define $S(x) := -\bar{\partial} g(x) = \frac{x}{\pi^2 ||x||^4}$ and $\bar{S}(x) := \frac{\bar{x}}{\pi^2 ||x||^4}$ for $x \in \mathbf{Q} \setminus \{0\}$. We then see that

$$\partial S(x) = -\partial \bar{\partial} g(x) = -\Delta g(x) = \delta(x), \quad x \in \mathbf{Q}$$

where δ stands for the Dirac distribution with support at the origin.

The Euclidean vector (massless) field models considered in Ref. 5, 7, 15 are the solutions of the following covariant SPDEs:

$$\partial X = F \tag{4.1}$$

where $F : \mathcal{S}(\mathbf{R}^4, \mathbf{Q}) \times (\mathcal{S}'(\mathbf{R}^4, \mathbf{Q}), P) \to \mathbf{R}$ is a Lévy white noise whose Fourier transform is given by

$$\int_{\mathcal{S}'(\mathbf{R}^4,\mathbf{Q})} e^{iF(f,\omega)} P(d\omega) = \exp\left\{\int_{\mathbf{R}^4} \psi(f(x)) dx\right\}, \, f \in \mathcal{S}(\mathbf{R}^4, \mathbf{Q})$$

with ψ being a Lévy-Khinchine function on $\mathbf{Q}$ with the following expression

$$\psi(x) := i\beta x_0 - \frac{a_0}{2} x_0^2 - \frac{a}{2} ||x - x_0||^2$$

$$+ \int_{\mathbf{Q}\setminus\{0\}} [e^{ixy} - 1 - ixy 1_{(0,1)}(||y||)] \lambda(dy), \quad x \in \mathbf{Q} \tag{4.2}$$

where $\beta \in \mathbf{R}, a_0, a \in (0, \infty)$ and λ is a $Sp(1)$ adjoint invariant Lévy measure on $\mathbf{Q} \setminus \{0\}$, namely, $\lambda(udyu^{-1}) = \lambda(dy)$ for $u \in Sp(1)$. The function ψ on $\mathbf{Q}$ is invariant under $SO(\mathbf{Q}) = SO(\mathbf{R}^4)$. The SPDEs can be solved by the convolution $X = (-\bar{\partial} g) * F$. Thus, the solution to the above SPDE is clearly related to mass zero.

The Schwinger functions associated to X can be calculated explicitly and the corresponding Wightman functions are obtained explicitly by performing analytic continuation. The explicit formulae for the truncated Wightman functions can be found in Ref. 7.

It is proved in Ref. 5 that the Wightman functions of this model fulfil the modified Wightman axioms (i.e., Poincaré invariance, locality, hermiticity, spectral condition and Hilbert space structure condition) of Morchio and Strocchi.[44] Thus, these Wightman functions are associated with a local relativistic quantum field theory with indefinite metric (this is different from the free electromagnetic potential field unless the noise F is purely Gaussian). Furthermore, explicit formulae for scattering amplitudes have been obtained in Ref. 7 which show a convergence to free asymptotic in- and out fields of mass zero and nontrivial scattering behaviour.

Moreover, the following generalized covariant SPDEs (in any space-time dimension d) have been considered in Ref. 22 and Ref. 8:

$$DX = F \qquad (4.3)$$

where F is a Lévy white noise transforming covariantly under a representation τ of $SO(\mathbf{R}^d)$ acting on the spin components of F, i.e. $\tau(\Lambda)F(\Lambda^{-1}x) = F(x)$ in law, and D is a τ-covariant differential operator with constant coefficients whose Fourier transformed Green's function has the form

$$\hat{D}^{-1}(k) = \frac{Q_E(k)}{\prod_{l=1}^{N}(|k|^2 + m_l^2)^{\nu_l}}$$

with $m_l \in \mathbb{C}$ (the complex mass parameters), $\nu_l \in \mathbf{N}, j \in \mathbf{N}$, and $m_j \neq m_l$ if $j \neq l$. $Q_E(k)$ is certain matrix with polynomial entries of order less or equal to $\kappa := 2(\sum_{l=1}^{N} \nu_l - 1)$ which fulfils the Euclidean transformation law

$$\tau(\Lambda)Q_E(k)\tau(\Lambda^{-1}) = Q_E(\Lambda k), \quad \forall \Lambda \in SO(d).$$

The covariant SPDE is solved by setting $X = D^{-1} * F$. Again, as has been carried out in Ref. 8, one can compute explicitly the Schwinger functions of X and the associated Wightman functions which determine a massive local relativistic quantum vector field model with indefinite metric. Furthermore, necessary and sufficient conditions in terms of the mass spectrum of D are obtained for the massive vector model having nontrivial scattering behaviour (cf. Theorem 2 in Ref. 8).

Thus, the vector models have an advantage that they possess nontrivial scattering behaviour in which the masses of the asymptotic fields can be obtained from the mass-spectrum of the Green's function D^{-1}. While their disadvantage is that the two point Wightman function in general is ill-defined and has to be "renormalized" (cf. Ref. 15 for $m = 0$), otherwise the two point function cannot be associated to free field and the scattering behaviour is unclear, cf. Formula (8) in Ref. 8. This procedure of "renormalisation", however, does not have a clear interpretation in terms of the models. This makes it unclear whether the physical insight, gained e.g. by the calculation of lattice action functionals on the Euclidean domain, is still meaningful in the "renormalised" models.

Now let us turn to the scalar case. The stochastic (elliptic) pseudo-differential equations for local scalar models with sharp mass (in any space-time dimension d) are considered in the following form (Ref. 4)

$$(-\Delta + m^2)^\alpha X = F, \quad \alpha \in (0, 1/2] \tag{4.4}$$

where Δ is the Laplace operator on $\mathbf{R}^d$, and the mass $m > 0$ if $d = 1, 2$ and $m \geq 0$ if $d \geq 3$, $F : \mathcal{S}(\mathbf{R}^d) \times (\mathcal{S}'(\mathbf{R}^d), P) \to \mathbf{R}$ is a Lévy white noise determined by

$$\int_{\mathcal{S}'(\mathbf{R}^d)} e^{iF(f,\omega)} P(d\omega) = \exp\left\{\int_{\mathbf{R}^d} \psi(f(x))dx\right\}, \quad f \in \mathcal{S}(\mathbf{R}^d)$$

with ψ being a Lévy-Khinchine function on $\mathbf{R}$ as given by (3.1) in Section 3 having the following representation

$$\psi(t) = iat - \frac{\sigma^2 t^2}{2} + \int_{\mathbf{R}\setminus\{0\}} \left(e^{ist} - 1 - \frac{ist}{1+s^2}\right) dM(s), \quad t \in \mathbf{R}$$

for $a \in \mathbf{R}, \sigma \geq 0$ and M satisfying $\int_{\mathbf{R}\setminus\{0\}} \min(1, s^2) dM(s) < \infty$.

The truncated Wightman functions are derived whose Fourier transforms $\hat{W}_n^T, n \in \mathbf{N}$, are given by the following explicit formulae:

$$\hat{W}_n^T = c_n \int_{(\mathbf{R}_+)^n} \hat{W}_{\underline{m},n}^T \rho^{\otimes n}(d\underline{m}^2), \quad n \geq 3, \tag{4.5}$$

where $c_n, n \in \mathbf{N}$, are constants depending on the probability law of the noise $F, \rho(d\mu) = 2\sin(\pi\alpha)1_{\{\mu > m^2\}}/(\mu - m^2)^\alpha d\mu$ is a Borel measure on $\mathbf{R}_+$ with $\mathrm{supp}\rho \subset [m^2, \infty), \underline{m} = (m_1, \ldots, m_n) \in (\mathbf{R}_+)^n$, and

$$\hat{W}_{\underline{m},n}^T = (2\pi)^{d-1-\frac{dn}{2}} \left\{\sum_{j=1}^n \prod_{l=1}^{j-1} \delta_{m_l}^-(k_l) \frac{(-1)}{k_j^2 - m_j^2} \prod_{l=j+1}^n \delta_{m_l}^+(k_l)\right\} \delta(\sum_{l=1}^n k_l) \tag{4.6}$$

and

$$\hat{W}_2^T = 2(2\pi)^{-d/2} \sin(2\pi\alpha) \frac{1_{\{k_1^0 < 0, k_1^2 > m^2\}}}{(k_1^2 - m^2)^{2\alpha}} \delta(k_1 + k_2) \tag{4.7}$$

Here $\delta_{m_l}^-(k_l) := 1_{\{k_l^0 < 0\}}(k_l)\delta(k_l^2 - m_l^2)$, $\delta_{m_l}^+(k_l) := 1_{\{k_l^0 > 0\}}(k_l)\delta(k_l^2 - m_l^2)$, $k_l^2 := {k_l^0}^2 - \vec{k}_l^2$ for $k_l = (k_l^0, \vec{k}_l) \in \mathbf{R} \times \mathbf{R}^{d-1}$.

From the above formulae, one sees that the two point Wightman function is a regular two point function of a generalized free field and hence no "renormalisation" is required. However, pseudo-differential operators with continuous mass distribution are involved (leading to an "infraparticle"-interpretation of the model), which rise difficulties for scattering theory in the scalar case. In other words, there is no clear picture for particle interpretation.

Comparing both vector and scalar cases, we see that in order to get rid of psuedo-differential operators and to consider scattering theory for the scalar case, it is worthwhile to reform the Euclidean (scalar) random fields by modifying the corresponding elliptic SPDEs.

5. New formulation of elliptic SPDEs and the lattice approximation

First of all, let us consider the following elliptic SPDE

$$(-\Delta + m^2)X = F. \tag{5.1}$$

It is not hard to see that the two point Schwinger function of X is $c_2(-\Delta + m^2)^{-2}$ instead of $c_2(-\Delta + m^2)^{-1}$, where the latter is the right one we expect to have for the two point Schwinger function of a QFT-model of mass m. In order to have that, we propose to consider the following elliptic SPDE

$$(-\Delta + m^2)A = F + G \tag{5.2}$$

where $G : \mathcal{S}(\mathbf{R}^d) \times (\mathcal{S}'(\mathbf{R}^d), P) \to \mathbf{R}$ is a ultralocal Gaussian white noise, which is independent of F, with Fourier transform

$$\int_{\mathcal{S}'(\mathbf{R}^d)} e^{iG(f,\omega)} P(d\omega) = \exp\left\{\frac{c_2}{m^2} \int_{\mathbf{R}^d} f(x)(\Delta f)(x) dx\right\}, \quad f \in \mathcal{S}(\mathbf{R}^d).$$

Obviously, the covariance of G is $\frac{c_2}{m^2}(-\Delta)\delta(x)$. Moreover, the two point Schwinger function of A can be then calculated as follows

$$\begin{aligned}
\text{Cov}(A)(x) &= (-\Delta + m^2)^{-2}[\text{Cov}(F) + \text{Cov}(G)](x) \\
&= c_2[(-\Delta + m^2)^{-2} - (-\Delta + m^2)^{-2}\Delta/m^2](x) \\
&= (c_2/m^2)(-\Delta + m^2)^{-1}(x).
\end{aligned}$$

The truncated n-point Schwinger functions of A remain the same as those of X for $n \geq 3$ and the associated truncated Wightman functions are (up to a constant) given by formula (4.6) for the case $m_l = m, l = 1, \ldots, n$. Thus, Euclidean random field A has the "right" Schwinger functions, and the associated Wightman functions possess a well-defined (nontrivial) scattering behaviour and permit the construction of associated relativistic quantum field to asymptotic in- and out- fields of mass m. Therefore, it is possible to develop a properly nontrivial scattering behaviour of relativistic quantum field just starting from A (cf. e.g. Ref. 2).

In order to understand the "physics" behind the structure functions, it seems thus to be reasonable to study the lattice approximation of A so that one can see its action functional. The lattice approximation for Euclidean random fields has been discussed in Ref. 16 and Ref. 18 for vector models and in Ref. 17 for scalar models. In Ref. 19 (and most recently Ref. 53), a nonstandard lattice formulation has been set up which gives a rigorous functional integration formula for the Euclidean random field measures (namely the inverse Fourier transform formula for the measures).

Let $\varepsilon > 0$ be arbitrarily fixed and define the lattice $\mathcal{L}_\varepsilon$ with spacing ε to be the set

$$\mathcal{L}_\varepsilon := \varepsilon \mathbb{Z}^d = \{\varepsilon z : z \in \mathbb{Z}^d\}.$$

Denote for $z = (z_1, \ldots, z_d) \in \mathbf{Z}^d$

$$[\varepsilon z, \varepsilon(z+1)) := \prod_{j=1}^{d} [\varepsilon z_j, \varepsilon(z_j+1))$$

then the indicator function $1_{[\varepsilon z, \varepsilon(z+1))} \in L^2(\mathbf{R}^d)$ for any $z \in \mathbf{Z}^d$. Remarking that $\mathcal{S}(\mathbf{R}^d)$ is dense in $L^2(\mathbf{R}^d)$, there exists a sequence $\{h_n\}_{n \in \mathbf{N}} \subset \mathcal{S}(\mathbf{R}^d)$ converging to $1_{[\varepsilon z, \varepsilon(z+1))}$ in $L^2(\mathbf{R}^d)$. We can then define

$$F_{\varepsilon z}(\cdot) := F(1_{[\varepsilon z, \varepsilon(z+1))}, \cdot) = P - \lim_{\varepsilon \to 0} F(h_n, \cdot)$$

and

$$G_{\varepsilon z}(\cdot) := G(1_{[\varepsilon z, \varepsilon(z+1))}, \cdot) = P - \lim_{\varepsilon \to 0} G(h_n, \cdot).$$

Clearly the random variables $F_{\varepsilon z}, G_{\varepsilon z}, \varepsilon z \in \mathcal{L}_\varepsilon$, are mutually independent. Let Δ_ε be the discretized Laplace operator on $\mathcal{L}_\varepsilon$ and K_ε the lattice Green's function of $-\Delta_\varepsilon + m^2$:

$$(-\Delta_\varepsilon + m^2)f(\varepsilon z) = \sum_{\varepsilon z' \in \mathcal{L}_\varepsilon} \varepsilon^d K_\varepsilon(\varepsilon z - \varepsilon z')f(\varepsilon z'). \tag{5.3}$$

Setting

$$A_{\varepsilon z}(\cdot) := \sum_{\varepsilon z' \in \mathcal{L}_\varepsilon} \varepsilon^d K_\varepsilon(\varepsilon z - \varepsilon z')[F_{\varepsilon z'}(\cdot) + G_{\varepsilon z'}(\cdot)], \tag{5.4}$$

then following the proofs of Theorems 5.1 and 5.1' in Ref. 17 and Theorem 4.3 in Ref. 18, we can show

Theorem 5.1. $\{A_{\varepsilon z}, \varepsilon z \in \mathcal{L}_\varepsilon\}$ *approximates A in the sense that*

$$\lim_{\varepsilon \to 0} C_{A_{\varepsilon z}}(f) = C_A(f), \quad f \in \mathcal{S}(\mathbf{R}^d)$$

where $C_A(f) := \int_{\mathcal{S}'(\mathbf{R}^d)} e^{iA(f,\omega)} P(d\omega)$ is the characteristic functional of A. Namely the convergence is in the sense of characteristic functionals or in law.

Let $P_F^\varepsilon, P_G^\varepsilon$, and P_A^ε be the probability measures on $\mathbf{R}^{\mathcal{L}_\varepsilon}$ associated with $F_\varepsilon = (F_{\varepsilon z}, \varepsilon z \in \mathcal{L}_\varepsilon), G_\varepsilon = (G_{\varepsilon z}, \varepsilon z \in \mathcal{L}_\varepsilon)$, and $A_\varepsilon = (A_{\varepsilon z}, \varepsilon z \in \mathcal{L}_\varepsilon)$, respectively. Then one can carry out that $\forall q = (q_{\varepsilon z}) \in \mathbf{R}^{\mathcal{L}_\varepsilon} \cap \mathcal{S}(\mathbf{R}^d)$,

$$P_F^\varepsilon(dq) = e^{-\sum_{\varepsilon z \in \mathcal{L}_\varepsilon} \varepsilon^d W(q_{\varepsilon z})} \prod_{\varepsilon z \in \mathcal{L}_\varepsilon} dq_{\varepsilon z} \tag{5.5}$$

and

$$P_G^\varepsilon(dq) = Z^{-1} e^{-\frac{1}{2} \sum_{\varepsilon z, \varepsilon z' \in \mathcal{L}_\varepsilon} B(\varepsilon z - \varepsilon z')q_{\varepsilon z}q_{\varepsilon z'}} \prod_{\varepsilon z \in \mathcal{L}_\varepsilon} dq_{\varepsilon z} \tag{5.6}$$

where

$$W(q_{\varepsilon z}) = -\varepsilon^{-d} \log \left[(-2\pi)^{-d} \int_{\mathbf{R}} e^{\varepsilon^d(-iq_{\varepsilon z}t + \psi(t))} dt\right],$$

$(B(\varepsilon z - \varepsilon z'))_{\varepsilon z, \varepsilon z' \in \mathcal{L}_\varepsilon}$ is a symmetric, positive definite matrix determined by $(-\Delta_\varepsilon + m^2)$ and the lattice Green's function of $-\Delta_\varepsilon$, and Z is the normalization constant. Furthermore,

$$\begin{aligned}
P_A^\varepsilon(dq) &= P_F^\varepsilon * P_G^\varepsilon \left((\varepsilon^d K_\varepsilon(\varepsilon z - \varepsilon z'))_{\varepsilon z, \varepsilon z' \in \mathcal{L}_\varepsilon}^{-1} dq \right) \\
&= Z^{-1} [\int_{\mathbf{R}^{\mathcal{L}_\varepsilon}} \exp\{ -\frac{1}{2} \sum_{\varepsilon z, \varepsilon z' \in \mathcal{L}_\varepsilon} B(\varepsilon z - \varepsilon z')(q_{\varepsilon z} - p_{\varepsilon z'})^2 \\
&\quad - \sum_{\varepsilon z \in \mathcal{L}_\varepsilon} \varepsilon^d W((-\Delta_\varepsilon + m^2) p_{\varepsilon z}) \} \prod_{\varepsilon z \in \mathcal{L}_\varepsilon} dp_{\varepsilon z}] \prod_{\varepsilon z \in \mathcal{L}_\varepsilon} dq_{\varepsilon z}.
\end{aligned}$$
(5.7)

Finally from Theorem 5.1, we get

Corollary 16.1. $P_A^\varepsilon \xrightarrow{w} P_A$ as $\varepsilon \to 0$.

More details and proofs regarding to the lattice approximation of A will appear in our forthcoming work.

Acknowledgement

I would like to thank Sergio Albeverio for introducing me to the wonderful world of constructive quantum field theory and for his deep insight discussion. I also thank Hanno Gottschalk for the joyful collaboration on topics related to this article.

References

1. S. Albeverio, J. E. Fenstad, R. Høegh-Krohn, T. Lindstrøm: *Nonstandard Methods in Stochastic Analysis and Mathematical Physics.* Pure and Applied Mathematics, **122**, Academic Press, Inc., Orlando, FL, 1986; MIR, Moscow, 1988 (in Russian).
2. S. Albeverio, H. Gottschalk: Scattering theory for local relativistic QFT with indefinite metric. *Commun. Math. Phys.* **216** (2001), 491–513.
3. S. Albeverio, H. Gottschalk, J.-L. Wu: Euclidean random fields, pseudodifferential operators, and Wightman functions. *Proc. Gregynog Symposium Stochastic Analysis and Applications* (eds. I.M. Davies, A. Truman and K.D. Elworthy), pp20–37, World Scientific, Singapore, 1996.
4. S. Albeverio, H. Gottschalk, J.-L. Wu: Convoluted generalized white noise, Schwinger functions and their analytic continuation to Wightman functions. *Rev. Math. Phys.* **8** (1996), 763–817.
5. S. Albeverio, H. Gottschalk, J.-L. Wu: Models of local relativistic quantum fields with indefinite metric (in all dimensions). *Commun. Math. Phys.* **184** (1997), 509–531.
6. S. Albeverio, H. Gottschalk, J.-L. Wu: Remarks on some new models of interacting quantum fields with indefinite metric. *Reports Math. Phys.* **40** (1997), 385–394.
7. S. Albeverio, H. Gottschalk, J.-L. Wu: Nontrivial scattering amplitudes for some local relativistic quantum field models with indefinite metric. *Phys. Lett.* **B 405** (1997), 243–248.

8. S. Albeverio, H. Gottschalk, J.-L. Wu: Scattering behaviour of relativistic quantum vector fields obtained from Euclidean covariant SPDEs. *Reports Math. Phys.* **44** (1999), 21–28.
9. S. Albeverio, H. Gottschalk, J.-L. Wu: SPDEs leading to local, relativistic quantum vector fields with indefinite metric and nontrivial S-matrix. *Stochastic Partial Differential Equations and Applications* (eds. G. Da Prato and L. Tubaro.), pp21–38, Lecture Notes in Pure and Appl. Math., **227**, Dekker, New York, 2002.
10. S. Albeverio, H. Gottschalk, M.W. Yoshida: Systems of classical particles in the grand canonical ensemble, scaling limits and quantum field theory. *Rev. Math. Phys.* **17** (2005), 175–226.
11. S. Albeverio, H. Holden, R. Høegh–Krohn, T. Kolsrud: Representation and construction of multiplicative noise. *J. Funct. Anal.* **87** (1989), 250–272.
12. S. Albeverio, R. Høegh–Krohn: Euclidean Markov fields and relativistic quantum fields from stochastic partial differential equations. *Phys. Lett.* **B177** (1986), 175–179.
13. S. Albeverio, R. Høegh–Krohn: Quaternionic non–abelian relativistic quantum fields in four space–time dimensions. *Phys. Lett.* **B189** (1987), 329–336.
14. S. Albeverio, R. Høegh–Krohn: Construction of interacting local relativistic quantum fields in four space–time dimensions. *Phys. Lett.* **B200** (1988), 108–114; with erratum in ibid. **B202** (1988), 621.
15. S. Albeverio, K. Iwata, T. Kolsrud: Random fields as solutions of the inhomogeneous quaternionic Cauchy–Riemann equation.I.Invariance and analytic continuation. *Commun. Math. Phys.* **132** (1990), 555–580.
16. S. ALbeverio, K. Iwata, M. Schmidt: A convergent lattice approximation for nonlinear electromagnetic fields in four dimensions. *J. Math. Phys.* **34** (1993), 3327–3342.
17. S. Albeverio, J.-L. Wu: Euclidean random fields obtained by convolution from generalized white noise. *J. Math. Phys.* **36** (1995), 5217–5245.
18. S. Albeverio, J.-L. Wu: On the lattice approximation for certain generalized vector Markov fields in four space-time dimensions. *Acta Appl. Math.* **47** (1997), 31–48.
19. S. Albeverio, J.-L. Wu: On nonstandard construction of stable type Euclidean random field measures and large deviation. *Reuniting the Antipodes: Constructive and Nonstandard Views of the Continuum* (eds. U. Berger, H. Osswald, P. Schuster), pp1–18, Synthese Library, vol. **306**, Kluwer Academic Publishers, Boston, Dordrecht, London, 2001.
20. S. Albeverio, M.W. Yoshida: H-C^1 maps and elliptic SPDEs with polynomial and exponential perturbations of Nelson's Euclidean free field. *J. Funct. Anal.* **196** (2002), 265–322.
21. T. Balaban, Ultra violet stability in field theory. The ϕ_3^4 model. *Scaling and Self-Similarity in Physics* (ed. J. Fröhlich), pp297–319, Birkhäuser, Boston, Basel, Stuttgart, 1983.
22. C. Becker, R. Gielerak, P. Ługiewicz: Covariant SPDEs and quantum field structures. *J. Phys.* **A 31** (1998), 231–258.
23. C. Becker, H. Gottschalk, J.-L. Wu: Generalized random vector fields and Euclidean quantum vector fields.*Second Seminar on Stochastic Analysis, Random Fields and Applications* (eds. R. Dalang, M. Dozzi, F. Russo), pp15–24, Progr. Probab., **45**, Birkhäuser, Basel, 1999.
24. T. Constantinescu, A. Gheondea: On L. Schwartz's boundedness condition for kernels. *Positivity* **10** (2006), 65–86.
25. T. Constantinescu, A. Gheondea: Invariant Hermitian kernels and their Kolmogorov decompositions. *C. R. Acad. Sci. Paris Sr. I Math.* **331** (2000), 797–802.
26. T. Constantinescu, A. Gheondea: Representations of Hermitian kernels by means of

Krein spaces. II. Invariant kernels. *Commun. Math. Phys.* **216** (2001), 409–430.
27. S.H. Djah, H. Gottschalk, H. Ouerdiane: Feynman graph representation of the perturbation series for general functional measures. *J. Funct. Anal.* **227** (2005), 153–187.
28. C. H. Eab, S. C. Lim, L. P. Teo: Finite temperature Casimir effect for a massless fractional Klein-Gordon field with fractional Neumann conditions. *J. Math. Phys.* **48** (2007), no. 8, 082301, 24 pp.
29. I.M. Gelfand, N.Ya. Vilenkin: *Generalized Functions, IV. Some Applications of Harmonic Analysis.* Academic Press, New York, London, 1964.
30. J. Glimm, A. Jaffe: Positivity of the φ_3^4 Hamiltonian. *Fortschritte der Physik* **21** (1973), 327–376.
31. J. Glimm, A. Jaffe: *Quantum physics: A Functional Integral Point of View.* 2nd ed., Springer-Verlag, New York, Berlin, Heidelberg, 1987.
32. H. Gottschalk, H. Thale: An indefinite metric model for interacting quantum fields on globally hyperbolic space-times. *Ann. Henri Poincar* **4** (2003), 637–659.
33. H. Gottschalk, B. Smii: How to determine the law of the solution to a stochastic partial differential equation driven by a Lévy space-time noise? *J. Math. Phys.* **48** (2007), no. 4, 043303, 22 pp.
34. M. Grothaus, L. Streit: Construction of relativistic quantum fields in the framework of white noise analysis. *J. Math. Phys.* **40** (1999), 5387–5405.
35. R. Haag: Quantum fields with composite particles and asymptotic conditions. *Phys. Rev.* **112** (1958), 669–673.
36. K. Hepp: On the connection between the LSZ and Wightman quantum field theory. *Commun. Math. Phys.* **1** (1965), 95–111.
37. G. Hofmann: The Hilbert space structure condition for quantum field theories with indefinite metric and transformations with linear functionals. *Lett. Math. Phys.* **42** (1997), 281–295.
38. G. Hofmann: On inner characterizations of pseudo-Krein and pre-Krein spaces. *Publ. Res. Inst. Math. Sci.* **38** (2002), 895–922.
39. Z. Huang, C. Li: On fractional stable processes and sheets: white noise approach. *J. Math. Anal. Appl.* **325** (2007), 624–635.
40. Z. Huang, Y. Wu: Interacting Fock expansion of Lévy white noise functionals. *Acta Appl. Math.* **82** (2004), 333–352.
41. N. Ikeda, S. Watanabe: *Stochastic Differential Equations and Diffusion Processes.* 2nd. North-Holland, Kodansha, 1989.
42. J. Jacod, A.N. Shiryaev: *Limit Theorems for Stochastic Processes.* Springer-Verlag, Berlin, 1987.
43. S. C. Lim, L. P. Teo: Sample path properties of fractional Riesz-Bessel field of variable order. *J. Math. Phys.* **49** (2008), no. 1, 013509, 31 pp.
44. G. Morchio, F. Strocchi: Infrared singularities, vacuum structure and pure phases in local quantum field theory. *Ann. Inst. H. Poincaré* **A33** (1980), 251–282.
45. G. Morchio, F. Strocchi: Representation of *-alegbras in indefinite inner product spaces. *Stochastic Processes, Physics and Geometry: New Interplays, II (a volume in honor of Sergio Albeverio)*, pp 491–503, CMS Conf. Proc., **29**, Amer. Math. Soc., Providence, RI, 2000.
46. E. Nelson: Construction of quantum fields from Markoff fields. *J. Funct. Anal.* **12** (1973), 97–112.
47. E. Nelson: The free Markoff field. *J. Funct. Anal.* **12** (1973), 211–227.
48. W.G. Ritter: Description of noncommutative theories and matrix models by Wightman functions. *J. Math. Phys.* **45** (2004), 4980–5002.
49. B. Simon: *The $P(\phi)_2$ Euclidean (Quantum) Field Theory.* Princeton University Press,

Princeton, 1975.
50. F. Strocchi: *Selected Topics on the General Properties of Quantum Field Theory.* Lect. Notes in Physics **51**. World Scientific, Singapore, 1993.
51. K. Symanzik: Euclidian quantum field theory I. Equations for a scalar model. *J. Math. Phys.* **7** (1966), 510–525.
52. J.B. Walsh: An introduction to stochastic partial differential equations. *Ecole d'Été de Probabilités de St. Flour XIV*, pp. 266–439, Lect. Notes in Math. **1180**, Springer-Verlag, Berlin, 1986.
53. J.-L. Wu: A hyperfinite flat integral for generalized random fields. *J. Math. Anal. Appl.* **330** (2007), 133–143.

Chapter 17

A Short Presentation of Choquet Integral

Jia-an Yan*

Institute of Applied Mathematics
Academy of Mathematics and Systems Science
Chinese Academy of Sciences, Beijing 100190, P. R. China
Email: jayan@amt.ac.cn

The paper provides a short representation of Choquet integral. The main content is from the book D. Denneberg (1994) and the paper Zhou (1998).

Contents

1 Introduction 269
2 Integration of Monotone Functions 270
3 Monotone Set Functions, Measurability of Functions 271
4 Comonotonicity of Functions 275
5 The Choquet Integral 277
    5.1 Definition and basic properties 277
    5.2 Example 1: Distorted probability measures 279
    5.3 Example 2: λ-fuzzy measures 280
6 The Subadditivity Theorem 281
7 Representing Functionals as Choquet Integrals 285
References 291

1. Introduction

The Choquet integral was introduced by Choquet (1953), and originally used in statistical mechanics and potential theory. After the work of Dempster (1967), later developed by Shafer (1976), it was applied to uncertainty and the representation of beliefs. The interest of statistics for the subject was started with the work on robust Bayesian inference by Huber (1973). Some years later integral representation theorems based on Choquet integral were established by Greco (1982) and Schmeidler (1986). Yaari (1987) establishes a dual theory for risk and uncertainty, in which the certainty equivalent of a uniformly bounded economic prospect can be represented as a Choquet integral.

---

*This work was supported by the National Natural Science Foundation of China (No. 10571167), the National Basic Research Program of China (973 Program, No.2007CB814902), and the Science Fund for Creative Research Groups (No.10721101).

The purpose of this note is to give a short representation of some basic results about the Choquet integral. The main content is from the book "Non-additive Measure and Integral" by D. Denneberg (1994) and the paper Zhou(1998). In fact, this note has been presented by the author at a seminar of the Department of Systems Engineering and Engineering Management, Chinese University of Hong Kong, during his visit to CUHK in 2005.

2. Integration of Monotone Functions

Let I be an (open, closed or semiclosed)interval of $\overline{\mathbb{R}}$. Let $f : I \to \overline{\mathbb{R}}$ be a decreasing function on I. Put $a = \inf\{x : x \in I\}$, and $J = [\inf_{x \in I} f(x), \sup_{x \in I} f(x)]$. There always exists a decreasing function $g : J \to \overline{\mathbb{R}}$ such that

$$a \vee \sup\{x|f(x) > y\} \leq g(y) \leq a \vee \sup\{x|f(x) \geq y\}.$$

We call such a g a *pseudo-inverse* of f, and denoted by $\check{f}$. Note that $\check{f}$ is unique except on an at most countable set (e.c. for short). We have $(\check{f})\check{}= f$, e.c., and $f \leq g$, e.c. is equivalent $\check{f} \leq \check{g}$, e.c.. If $f(x)$ is a continuity point of $\check{f}$, then $\check{f}(f(x)) = x$.

Proposition 2.1 For a decreasing function $f : \overline{\mathbb{R}}_+ \to \overline{\mathbb{R}}_+$ with $\lim_{x \to \infty} f(x) = 0$ and any pseudo-inverse $\check{f}$ of f, we have

$$\int_0^\infty \check{f}(y) dy = \int_0^\infty f(x) dx,$$

where we extended $\check{f}$ from $[0, f(0)]$ to $\mathbb{R}_+$ by letting $\check{f}(x) = 0$ for $x > f(0)$. For a decreasing function $f : [0, b] \to \overline{\mathbb{R}}$ with $0 < b < \infty$, and any pseudo-inverse $\check{f}$ of f, we have

$$\int_0^b f(x) dx = \int_0^\infty \check{f}(y) dy + \int_{-\infty}^0 (\check{f}(y) - b) dy,$$

where we extended $\check{f}$ from $[f(b), f(0)]$ to $\mathbb{R}$ by letting $\check{f}(x) = 0$ for $x > f(0)$ and $\check{f}(x) = f(b)$ for $x < f(b)$.

Proof. In order to prove the first result we put

$$S_f := \{(x, y) \in \mathbb{R}_+^2 \, | \, 0 \leq y \leq f(x), x \in \mathbb{R}_+\},$$

$$S_{\check{f}} := \{(x, y) \in \mathbb{R}_+^2 \, | \, 0 \leq x \leq \check{f}(y), y \in \mathbb{R}_+\}.$$

Then the closures $\overline{S}_f$ and $\overline{S}_{\check{f}}$ of S_f and $S_{\check{f}}$ in $\mathbb{R}^2$ are the same. However the integrals f and $\check{f}$ are the areas of $\overline{S}_f$ and $\overline{S}_{\check{f}}$, so they are equal.

Now assume $f : [0, b] \to \overline{\mathbb{R}}$ with $0 < b < \infty$ is a decreasing function. There is a point $a \in [0, b]$ such that $f(x) \geq 0$ for $x < a$ and $f(x) \leq 0$ for $x > a$. Define

$$g(x) = f(x) I_{[0,a)}(x), \quad h(x) = -f(b - x) I_{(0, b-a)}(x), x \in [0, \infty].$$

Then $\check{f} = \check{g}, \check{h}(x) = b - \check{f}(-x)$ on $\mathbb{R}_+$, e.c., and applying the above proved result gives

$$\int_0^a f(x)dx = \int_0^\infty g(x)dx = \int_0^\infty \check{g}(y)dy = \int_0^\infty \check{f}(y)dy,$$

and

$$\int_a^b f(x)dx = -\int_0^\infty h(x)dx = -\int_0^\infty \check{h}(x)dx$$
$$= -\int_0^\infty (b - \check{f}(-x))dx = \int_{-\infty}^0 (\check{f}(y) - b)dy.$$

Adding two equalities gives the desired result. □

3. Monotone Set Functions, Measurability of Functions

Let Ω be a non-empty set. We denote by 2^Ω the family of all subsets of Ω. By a *set system*, we mean any sub-family of 2^Ω containing $\emptyset$ and Ω. By a *set function* on a set system $\mathcal{S}$, we mean a function $\mu : \mathcal{S} \to \overline{\mathbb{R}}_+ = [0, \infty]$ with $\mu(\emptyset) = 0$.

Definition 3.1 A set function μ on $\mathcal{S}$ is called *monotone*, if $\mu(A) \leq \mu(B)$ whenever $A \subset B, A, B \in \mathcal{S}$. μ is called *submodular* (resp. *supermodular*), if $A, B \in \mathcal{S}$ such that $A \cup B, A \cap B \in \mathcal{S}$ implies $\mu(A \cup B) + \mu(A \cap B) \leq$ (resp.$\geq$)$\mu(A) + \mu(B)$. μ is called *modular* if it is sub- and supermodular. μ is called *subadditive*(resp., *superadditive*), if $A, B \in \mathcal{S}$ such that $A \cup B \in \mathcal{S}, A \cap B = \emptyset$ implies $\mu(A \cup B) \leq$ (resp.$\geq$)$\mu(A) + \mu(B)$.

If $\mathcal{S}$ is an algebra then μ is modular iff it is additive. If $\mathcal{S}$ is a σ-algebra then μ is σ-additive iff it is additive and continuous from below.

For a system $\mathcal{S}$ the *closure from below* $\overline{\mathcal{S}}$ of $\mathcal{S}$ is defined by

$$\overline{\mathcal{S}} := \left\{ A \subset \Omega \,\bigg|\, \exists \text{ increasing sequence } A_n \in \mathcal{S} \text{ such that } A = \bigcup_{n=1}^\infty A_n \right\}.$$

A set system $\mathcal{S}$ is called *closed from below* if $\mathcal{S} = \overline{\mathcal{S}}$.

Definition 3.2 A set system $\mathcal{S}$ is called a *chain*, if $A, B \in \mathcal{S}$ implies $A \subset B$ or $B \subset A$.

Proposition 3.1 Let $\mathcal{S} \subset 2^\Omega$ be a chain and μ a monotone set function on $\mathcal{S}$. We denote by $\mathcal{A}$ the algebra generated by $\mathcal{S}$. Then μ is a modular and there exists a unique modular, i.e. additive extension $\alpha : \mathcal{A} \to \mathbb{R}_+$ of μ on $\mathcal{A}$.

Proof It is easy to prove that

$$\mathcal{A} = \left\{ \bigcup_{i=1}^n (A_i \setminus B_i) \,\bigg|\, n \in \mathbb{N}, A_i, B_i \in \mathcal{S}, B_i \subset A_i, A_{i+1} \subset B_i, 1 \leq i \leq n \right\}.$$

For a set in $\mathcal{A}$ as above, if we require that $A_i \neq B_i$ and $B_i \neq A_{i+1}$ then this representation is unique, and we define a set function on $\mathcal{A}$ by

$$\alpha(\bigcup_{i=1}^{n}(A_i \setminus B_i)) := \sum_{i=1}^{n}(\mu(A_i) - \mu(B_i)).$$

Here we understand $\infty - \infty$ to be 0. Obviously, α is an additive extension of μ on $\mathcal{A}$. □

Definition 3.3 Let μ be a monotone set function on 2^Ω and $X: \Omega \to \overline{\mathbb{R}}$ be a function on Ω. Put

$$G_{\mu,X}(x) := \mu(X > x).$$

We call $G_{\mu,X}$ the (decreasing) distribution function of X w.r.t. μ, and call the pseudo-inverse function $\check{G}_{\mu,X}$ the *quantile function* of X w.r.t. μ. Since $0 \leq G_{\mu,X} \leq \mu(\Omega)$, $\check{G}_{\mu,X}$ is defined on $[0, \mu(\Omega)]$.

Proposition 3.2 Let μ be a monotone set function on 2^Ω and $X: \Omega \to \overline{\mathbb{R}}$ be a function on Ω. If u is an increasing function and u and $G_{\mu,X}$ have no common discontinuities, then

$$\check{G}_{\mu,u(X)} = u \circ \check{G}_{\mu,X}.$$

Proof. Let $u^{-1}(y) = \inf\{x \mid u(x) > y\}$. Then

$\{x \mid u(x) > y\}$
$= \{x \mid x > u^{-1}(y)\} \cup \{x \mid x = u^{-1}(y), u(x) > y\} \subset \{x \mid x \geq u^{-1}(y)\}.$

Consequently, if $[X = u^{-1}(y), u(X) > y] = \emptyset$, then it holds that $[u(X) > y] = [X > u^{-1}(y)]$; otherwise $u^{-1}(y)$ is a discontinuity point of u so that $G_{\mu,X}$ is continuous at $u^{-1}(y)$. In that case we have $\mu([X > u^{-1}(y)]) = \mu([X \geq u^{-1}(y)])$ which implies $\mu([u(X) > y]) = \mu([X > u^{-1}(y)])$, i.e.,

$$G_{\mu,u(X)} = G_{\mu,X} \circ u^{-1}.$$

In order to prove the proposition, we only need to show that

$\sup\{x \mid G_{\mu,X} \circ u^{-1}(x) > y\}$
$\leq u \circ \check{G}_{\mu,X}(y)$
$\leq \sup\{x \mid G_{\mu,X} \circ u^{-1}(x) \geq y\}.$

We first show the left inequality. Assume $G_{\mu,X} \circ u^{-1}(x) > y$, then $u^{-1}(x) \leq \check{G}_{\mu,X}(y)$. We consider separately two cases: when $u^{-1}(x) < \check{G}_{\mu,X}(y)$, then $x < u \circ \check{G}_{\mu,X}(y)$; when $u^{-1}(x) = \check{G}_{\mu,X}(y)$, then $G_{\mu,X}$ is discontinuous at $\check{G}_{\mu,X}(y)$, so that u is continuous at $\check{G}_{\mu,X}(y)$. In the latter case we have $x = u(u^{-1}(x)) = u \circ \check{G}_{\mu,X}(y)$. This proves the left inequality.

Now we show the right inequality. If $x < u \circ \check{G}_{\mu,X}(y)$, then $u^{-1}(x) \leq \check{G}_{\mu,X}(y)$. We consider separately two cases: when $u^{-1}(x) < \check{G}_{\mu,X}(y)$, then $G_{\mu,X} \circ u^{-1}(x) >$

y; when $u^{-1}(x) = \check{G}_{\mu,X}(y)$, then u is discontinuous at $\check{G}_{\mu,X}(y)$, so that $G_{\mu,X}$ is continuous at $\check{G}_{\mu,X}(y)$. In the latter case we have $G_{\mu,X} \circ u^{-1}(x) = G_{\mu,X} \circ \check{G}_{\mu,X}(y) = y$. This proves the right inequality. □

Now we consider a monotone set function on a set system $\mathcal{S} \subset 2^\Omega$. For any $A \subset \Omega$ we define

$$\mu^*(A) := \inf\{\mu(B) \mid A \subset B, B \in \mathcal{S}\},$$

$$\mu_*(A) := \sup\{\mu(C) \mid C \subset A, C \in \mathcal{S}\}.$$

We call the set functions μ^* and μ_* the outer and inner set function of μ, respectively.

The following two results are typical ones from measure theory. We omit their proofs.

Proposition 3.3 Let μ be a monotone set function on $\mathcal{S} \subset 2^\Omega$.
(i) μ_*, μ^* are monotone.
(ii) Let $\mathcal{S}$ be closed under union and intersection. μ^* is submodular if μ is; μ_* is supermodular (superadditive) if μ is. When $\mathcal{S}$ is an algebra, μ^* is subadditive if μ is.
(iii) Let $\mathcal{S}$ be closed under union and intersection and closed from below. μ^* is submodular and continuous from below if μ is.

Proposition 3.4 Let μ be an arbitrary set function on 2^Ω. Put

$$\mathcal{A}_\mu := \{A \subset \Omega \mid \mu(C) = \mu(A \cap C) + \mu(A^c \cap C) \text{ for all } C \subset \Omega\}.$$

Then $\mathcal{A}_\mu$ is an algebra and

$$\mu(A \cup B) + \mu(A \cap B) = \mu(A) + \mu(B), \quad \text{for all } A \in \mathcal{A}_\mu, B \subset \Omega.$$

In particular, μ is additive on $\mathcal{A}_\mu$.

We call $\mathcal{A}_\mu$ the *Caratheodory algebra* of μ.

Corollary 3.1 If μ is a monotone set function on 2^Ω and is subadditive and continuous from below, then $\mathcal{A}_\mu$ is a σ-algebra and μ is σ-additive on $\mathcal{A}_\mu$.

Definition 3.4 A function $X : \Omega \to \overline{\mathbb{R}}$ is called *upper μ-measurable* if

$$G_{\mu_*,X} = G_{\mu^*,X}, \text{ e.c..}$$

We denote this function, unique e.c., by $G_{\mu,X}$, and call it the *(decreasing) distribution function* of X w.r.t μ on $\mathcal{S}$.

Definition 3.5 A function $X : \Omega \to \overline{\mathbb{R}}$ is called *lower μ-measurable*, if $-X$ is upper μ-measurable. X is called *μ-measurable*, if it is lower and upper μ-measurable. X is called *(upper, lower) $\mathcal{S}$-measurable*, if it is (lower, upper) μ-measurable for any monotone set function μ on $\mathcal{S}$. X is called *strongly $\mathcal{S}$-measurable*, if $\mathcal{M}_X, \mathcal{M}_{-X} \in \mathcal{S}$, where $\mathcal{M}_X$ the so-called *upper set system* of X, i.e.

$$\mathcal{M}_X = \{[X > x], [X \geq x], \ x \in \overline{\mathbb{R}}\}.$$

If $\mathcal{S}$ is a σ-algebra, then X is strongly $\mathcal{S}$-measurable iff it is $\mathcal{S}$-measurable in the usual sense.

The following hereditary properties for measurability are an immediate consequence of Proposition 3.2.

Proposition 3.5 Let μ be a monotone set function on $\mathcal{S} \subset 2^\Omega$ and X a upper μ-measurable function on Ω. If $u : \overline{\mathbb{R}} \to \overline{\mathbb{R}}$ is increasing (decreasing) and continuous, then $u \circ X$ is upper (lower) μ-measurable. In particular, $X + c, X \wedge c, X \vee c$ are upper μ-measurable for $c \in \mathbb{R}$, and cX is upper (lower) μ-measurable for $c > 0$ ($c < 0$).

The following proposition, due to Greco (1982), gives a necessary and sufficient condition for upper $\mathcal{S}$-measurability.

Proposition 3.6 A function $X : \Omega \to \overline{\mathbb{R}}$ is upper $\mathcal{S}$-measurable iff for every pair $a, b \in \mathbb{R}, a < b$, there exists a set $S \in \mathcal{S}$ so that $[X > b] \subset S \subset [X > a]$.

Proof. Sufficiency. Let μ be a monotone on $\mathcal{S}$ and $x \in \mathbb{R}$ a continuous point of $G_{\mu^*,X}$. It is sufficient to show that $G_{\mu_*,X}(x) = G_{\mu^*,X}(x)$. Let $b > x$. The assumption implies

$$G_{\mu^*,X}(b) = \inf\{\mu(A) \,|\, [X > b] \subset A, A \in \mathcal{S}\}$$
$$\leq \sup\{\mu(B) \,|\, B \subset [X > x], B \in \mathcal{S}\}$$
$$= G_{\mu_*,X}(x).$$

Letting $b \to x$ gives $G_{\mu^*,X}(x) \leq G_{\mu_*,X}(x)$. Since $\mu_* \leq \mu^*$ the reversed inequality holds and we are done.

Necessity. Let $a < b$. We put

$$\mu(A) = \inf\{(b - a \vee x)^+ \,|\, A \subset [X > x], x \in \mathbb{R}\}, \quad A \in \mathcal{S}.$$

Here we make a convention that $\inf \emptyset := b - a$. By assumption X is upper μ-measurable. So we can find a real number x, $a < x < b$, such that $G_{\mu_*,X}(x) = G_{\mu^*,X}(x)$. If $[X > x] \in \mathcal{S}$, we can take $S = [X > x]$ so that $[X > b] \subset S \subset [X > a]$. If $[X > x] \notin \mathcal{S}$, then we have

$$(b - a \vee x)^+ \geq \sup\{\mu(A) \,|\, A \in \mathcal{S}, A \subset [X > x]\}$$
$$= \inf\{\mu(B) \,|\, B \in \mathcal{S}, [X > x] \subset B\} \geq (b - a \vee x)^+.$$

Hence equality holds and we can find $S \in \mathcal{S}$ so that

$$(b - a \vee x)^+ \leq \mu(S) < b - a.$$

Consequently, we must have

$$[X > b] \subset [X > x] \subset S \subset [X > a].$$

The proposition is proved. $\square$

Remark 3.1 If $\mathcal{S}$ is a σ-algebra, then by Proposition 3.6 a function $X : \Omega \to \overline{\mathbb{R}}$ is upper $\mathcal{S}$-measurable iff it is $\mathcal{S}$-measurable.

The following important result is also due to Greco (1982).

Proposition 3.7 Let $\mathcal{S} \subset 2^\Omega$ be a set system which is closed under union and intersection. If $X, Y : \Omega \to \mathbb{R}$ are upper $\mathcal{S}$-measurable functions, which are bounded below, then $X + Y$ is upper $\mathcal{S}$-measurable, too.

Proof. Instate of adding a constant to X and Y we may assume $X, Y \geq 0$. Given $a < b$ we have to find a set S so that
$$[X + Y > b] \subset S \subset [X + Y > a].$$
First we select $n \in \mathbb{N}$ so large that $\frac{n-4}{n}b > a$. Let $a_i = (i-1)\frac{b}{n}, i = 0, \ldots, n$. Since $a_{i+1} + a_{n-i} = \frac{n-1}{n}b < b$, it is easy to see that
$$[X + Y > b] \subset \bigcup_{i+j=n} [X > a_i] \cap [Y > a_j].$$
On the other hand, by Proposition 2.10 there exist $S_i, T_j \in \mathcal{S}$ so that
$$[X > a_i] \subset S_i \subset [X > a_{i-1}], \quad [Y > a_j] \subset T_j \subset [Y > a_{j-1}].$$
Consequently,
$$[X > a_i] \cap [Y > a_j] \subset S_i \cap T_j \subset [X > a_{i-1}] \cap [Y > a_{j-1}]$$
$$\subset [X + Y > a_{i-1} + a_{j-1}].$$
Since $a_{i-1} + a_{j-1} = \frac{n-4}{n}b > a$, we have that the last set is a subset of $[X + Y > a]$ if $i + j = n$. Thus $S := \bigcup_{i+j=n} S_i \cap T_j$ is the desired set. $\square$

4. Comonotonicity of Functions

A class $\mathcal{C}$ of functions $\Omega \to \overline{\mathbb{R}}$ is called *comonotonic* if $\bigcup_{X \in \mathcal{C}} \mathcal{M}_X$ is a chain. Clearly, a class $\mathcal{C}$ of functions is comonotonic iff each pair of functions in $\mathcal{C}$ is comonotonic. The following proposition gives equivalent conditions for a pair of functions to be comonotonic.

Proposition 4.1 For two functions $X, Y : \Omega \to \overline{\mathbb{R}}$ the following conditions are equivalent:

(i) X, Y are comonotonic.

(ii) There is no pair $\omega_1, \omega_2 \in \Omega$ such that $X(\omega_1) < X(\omega_2)$ and $Y(\omega_1) > Y(\omega_2)$.

(iii) The set $\{(X(\omega), Y(\omega)) \mid \omega \in \Omega\} \subset \overline{\mathbb{R}}^2$ is a chain w.r.t. the $\leq$-relation in $\overline{\mathbb{R}}^2$.

If X, Y are real valued, the above conditions and the following two conditions are equivalent:

(iv) There exists a function $Z : \Omega \to \mathbb{R}$ and increasing functions u, v on $\mathbb{R}$ such that $X = u(Z), Y = v(Z)$.

(v) There exist continuous, increasing functions u, v on $\mathbb{R}$ such that $u(z)+v(z) = z, z \in \mathbb{R}$, and
$$X = u(X+Y), \quad Y = v(X+Y).$$

Proof. The equivalences (i) $\Leftrightarrow$(ii)$\Leftrightarrow$(iii) are easy to check. For real valued X, Y, the implications (v)$\Rightarrow$(iv)$\Rightarrow$(ii) are trivial. We only need to prove (ii)$\Rightarrow$(v).

Now assume (ii) is valid and X, Y are real valued. Let $Z = X + Y$. Then from (ii) it is easy to see that any $z \in Z(\Omega)$ possesses a unique decomposition $z = x + y$ with $z = Z(\omega), x = X(\omega), y = Y(\omega)$ for some $\omega \in \Omega$. We denote x and y by $u(z)$ and $v(z)$. By (ii) it is easy to check that u and v are increasing on $Z(\Omega)$.

Now we prove that u, v are continuous on $Z(\Omega)$. First notice that for $z, z + h \in Z(\Omega)$ with $h > 0$ we have
$$z + h = u(z+h) + v(z+h)$$
$$\geq u(z+h) + v(z) = u(z+h) + z - u(z).$$

Thus we have
$$u(z) \leq u(z+h) \leq u(z) + h.$$

Similarly, for $z, z - h \in Z(\omega)$ with $h > 0$ we have
$$u(z) - h \leq u(z-h) \leq u(z).$$

These two inequalities together imply the continuity of u. By the symmetry of the roles of u and v, v is continuous, too.

It remains to show that u, v can be extended continuously from $Z(\Omega)$ to $\mathbb{R}$. Fist extend to the closure $\overline{Z(\Omega)}$. If $z \in \partial Z(\Omega)$ is only one sided boundary point, there is no problem, because u, v are increasing functions. If z is two sided limiting point of $Z(\Omega)$, then the above inequalities imply that two sided continuous extensions coincide. Finally, the extension of u, v from $\overline{Z(\Omega)}$ to $\mathbb{R}$ is done linearly on each connected component of $\mathbb{R}\setminus\overline{Z(\Omega)}$ in order to maintain the condition $u(z)+v(z) = z$. □

Corollary 4.1 Let μ be a monotone set function on 2^Ω. If X, Y are real valued comotononic functions on Ω, then
$$\check{G}_{\mu,X+Y} = \check{G}_{\mu,X} + \check{G}_{\mu,Y}, \text{ e.c..}$$

Proof Using the above notations in (v), we have $X = u(X+Y), Y = v(X+Y)$. By Proposition 3.2 we get
$$\check{G}_{\mu,X+Y} = (u+v) \circ \check{G}_{\mu,X+Y} = u \circ \check{G}_{\mu,X+Y} + v \circ \check{G}_{\mu,X+Y} = \check{G}_{\mu,X} + \check{G}_{\mu,Y}, \text{ e.c..}$$
□

5. The Choquet Integral

In this section, we define the Choquet integral of functions w.r.t. a monotone set function, and show their basic properties. As two examples of monotone set functions, the distorted probability measure and the λ-fuzzy measure are studied.

5.1. Definition and basic properties

Let μ be a monotone set function on a set system $\mathcal{S} \subset \mathcal{F}$, and $X : \Omega \to \overline{\mathbb{R}}$ an upper μ-measurable function. If the following Lebesgue integral

$$\int_0^{\mu(\Omega)} \check{G}_{\mu,X}(t) dt$$

exists, where $\check{G}_{\mu,X}$ is the quantile function of X, then we say that X is integrable w.r.t. μ, and define it as the Choquet integral of X w.r.t. μ. We denote it by $\int X d\mu$ or $\mu(X)$.

By Proposition 2.1 and using the fact that $(\check{f})\check{} = f$, e.c., we have

$$\mu(X) = \int_0^\infty G_{\mu,X}(x) dx + \int_{-\infty}^0 (G_{\mu,X}(x) - \mu(\Omega)) dx, \text{ if } \mu(\Omega) < \infty.$$

Recall that if μ is a probability measure and X is a random variable, then the expectation of X w.r.t. μ can be expressed by

$$\mu(X) = \int_0^\infty \mu(X \geq t) dt + \int_{-\infty}^0 (\mu(X \geq t) - 1) dt.$$

So the Choquet integral of a real valued function X w.r.t. a probability measure μ coincide with its expectation.

If X is a simple function of the form $X = \sum_{i=1}^n x_i I_{A_i}$, where $A_1, \cdots, A_n$ are disjoint and enumerated so that (x_i) are in descending order, i.e. $x_1 \geq \cdots \geq x_n$, then

$$\mu(X) = \sum_{i=1}^n (x_i - x_{i+1}) \mu(S_i) = \sum_{i=1}^n x_i (\mu(S_i) - \mu(S_{i-1})),$$

where $S_i = A_1 \cup \cdots \cup A_i, i = 1, \cdots, n, S_0 = \emptyset$, and $x_{n+1} = 0$.

Now we investigate the basic properties of the Choquet integral.

Proposition 5.1 If μ is a monotone set function on $\mathcal{S} \subset 2^\Omega$ and $X, Y : \Omega \to \overline{\mathbb{R}}$ are upper μ-measurable functions, then
(i) $\int I_A d\mu = \mu(A), A \in \mathcal{S}$.
(ii) (*positive homogeneity*) $\int cX d\mu = c \int X d\mu$, if $c \geq 0$.
(iii) (*asymmetry*) If μ is finite then $\int X d\mu = -\int(-X) d\bar{\mu}$, where $\bar{\mu}(A) = \mu(\Omega) - \mu(A)$.
(iv) (*monotonicity*) If $X \leq Y$ then $\int X d\mu \leq \int Y d\mu$.
(v) $\int (X + c) d\mu = \int X d\mu + c\mu(\Omega), c \in \mathbb{R}$.

(vi) (*comonotonic additivity*)　If X, Y are comonotonic and real valued then
$$\int (X+Y)d\mu = \int X d\mu + \int Y d\mu.$$

(vii) (*transformation rule*)　For a $T: \Omega \to \Omega'$ with $T^{-1}(\mathcal{S}') \subset \mathcal{S}$, let $\mu^T(A) = \mu(T^{-1}(A)), A \in \mathcal{S}'$. Then for a function $Z: \Omega' \to \mathbb{R}$, we have $G_{\mu, Z \circ T} = G_{\mu^T, Z}$ and
$$\int Z d\mu^T = \int Z \circ T d\mu.$$

Proof.　Since for upper μ-measurable functions X we have
$$\int X d\mu = \int X d\mu_* = \int X d\mu^*,$$

and μ_*, μ^* are monotone set functions defined on 2^Ω, instate of replacing μ by μ_* or μ^*, we may assume that μ is a monotone set function on 2^Ω.

(i) is trivial. (ii) follows from $\check{G}_{\mu, cX} = c\check{G}_{\mu, X}$ for $c > 0$ (Proposition 3.2). (iii) is due to the fact that $G_{\bar{\mu}, X}(x) = \mu(\Omega) - G_{\mu, -X}(-x)$. (vi) is derived from Corollary 4.1. Other properties are easy to check. □

Proposition 5.2　Let $X: \Omega \to \mathbb{R}$ be a $\mathcal{S}$-measurable function and μ, ν monotone set functions on $\mathcal{S} \subset 2^\Omega$. Then

(i) $G_{c\mu, X} = cG_{\mu, X}$, $\int X d(c\mu) = c \int X d\mu$, if $c > 0$.

(ii) If μ and ν are finite and $\mathcal{S}$ is closed under union and intersection, then
$$G_{\mu+\nu, X} = G_{\mu, X} + G_{\nu, X}, \text{ e.c., } \int X d(\mu + \nu) = \int X d\mu + \int X d\nu.$$

(iii) If $\mu(\Omega) = \nu(\Omega) < \infty$ or $X \geq 0$ then $\mu \leq \nu$ implies
$$G_{\mu, X} \leq G_{\nu, X}, \text{ e.c., } \int X d\mu \leq \int X d\nu.$$

(iv) If μ_n is a sequence of monotone set function on $\mathcal{S}$ with $\mu_n \leq \mu_{n+1}$ and $\lim_{n \to \infty} \mu_n(A) = \mu(A)$, $A \in \mathcal{S}$, then for bounded below X
$$\lim_{n \to \infty} \int X d\mu_n = \int X d\mu.$$

Proof.　(i) and (iii) are trivial. (ii) is trivial, too, if $\mathcal{S} = 2^\Omega$. In the general case, (ii) is also true, because we can show that $(\mu + \nu)^* = \mu^* + \nu^*$. In order to prove (iv), notice that
$$\lim_n (\mu_n)_*(A) = \sup_n (\mu_n)_*(A) = \sup_n \sup_{B \in \mathcal{S}, B \subset A} \mu_n(B)$$
$$= \sup_{B \in \mathcal{S}, B \subset A} \sup_n \mu_n(B) = \sup_{B \in \mathcal{S}, B \subset A} \mu(B) = \mu_*(A), \quad A \in 2^\Omega,$$

we may assume $\mathcal{S} = 2^\Omega$. If $X \geq 0$, then $\int X d\mu_n = \int G_{\mu_n, X}(x) dx$ and the monotone convergence theorem gives the desired assertion. Subtraction a constant shows that the assertion is true for bounded below function X. □

Remark 5.1 If X is μ-measurable then X^+ and $-X^-$ are upper μ-measurable (Proposition 3.5). Since $X^+, -X^-$ are comonotonic functions we have, if X is real valued,

$$\int X d\mu = \int X^+ d\mu + \int (-X^-) d\mu,$$

and

$$\int X d\mu = \int X^+ d\mu - \int X^- d\bar{\mu}, \text{ if } \mu(\Omega) < \infty.$$

Proposition 5.3 Let μ be a monotone set functions on 2^Ω. For any $q, 0 < q < \mu(\Omega)$, define

$$\mu_q(A) := q \wedge \mu(A), \ A \in 2^\Omega.$$

μ_q is monotone and for an arbitrary function $X : \Omega \to \overline{\mathbb{R}}$

$$\lim_{q \to \mu(\Omega)} \int X d\mu_q = \int X d\mu.$$

Proof. Since

$$G_{\mu_q, X}(x) = \mu_q(X > x) = q \wedge \mu(X > x) = q \wedge G_{\mu, X},$$

$\check{G}_{\mu, X}$ and $\check{G}_{\mu_q, X}$ coincide on $[0, q)$. Hence we have

$$\int X d\mu_q = \int_0^q \check{G}_{\mu_q, X}(t) dt = \int_0^q \check{G}_{\mu, X}(t) dt$$

$$\to \int_0^{\mu(\Omega)} \check{G}_{\mu, X}(t) dt = \int X d\mu.$$

$\square$

5.2. Example 1: Distorted probability measures

Let P be a probability measure on a measurable space $(\Omega, \mathcal{F})$ and $\gamma : [0,1] \to [0,1]$ an increasing function with $\gamma(0) = 0, \gamma(1) = 1$. Then $\mu = \gamma \circ P$ is a monotone set function. μ is called a *distorted probability* and γ the corresponding *distortion*.

If γ is a concave (convex) function then $\gamma \circ P$ is a submodular (supermodular) set function. This assertion is also valid for a normalized additive set function on an algebra, instate of a probability measure on a σ-algebra. We only consider the concave case, the convex case being similar. Let $A, B \in \mathcal{S}$. Assume $a := P(A) \leq P(B) =: b$. Denote $c = P(A \cap B), d = P(A \cup B)$. Then $c \leq a \leq b \leq d$. By modularity of P we have $c + d = a + b$. Thus concavity of γ implies $\gamma(c) + \gamma(d) \leq \gamma(a) + \gamma(b)$. This proves submodularity of $\gamma \circ P$.

For a distortion $g \circ P$ the Choquet integral $(g \circ P)(X)$ of X w.r.t. $g \circ P$ can be expressed in the following form:

$$(g \circ P)(X) = \int_0^1 q_X(1-x) dg(x) = \int_0^1 q_X(t) d\gamma(t),$$

where $q_X(t)$ is the right-continuous inverse of the distribution function F_X of X, and $\gamma(t) = 1 - g(1-t)$.

5.3. Example 2: λ-fuzzy measures

Let $\lambda \in (-1, \infty)$. A normalized monotone set function μ_λ defined on an algebra $\mathcal{S} \subset 2^\Omega$ is called a λ-fuzzy measure on $\mathcal{S}$, if for every pair of disjoint subsets A and B of Ω

$$\mu_\lambda(A \cup B) = \mu_\lambda(A) + \mu_\lambda(B) + \lambda \mu_\lambda(A) \mu_\lambda(B).$$

If $\lambda = 0$, μ_0 is additive. For $\lambda \in (-1, \infty)$ and $\lambda \neq 0$, we define

$$\psi_\lambda(r) = \log_{(1+\lambda)}(1 + \lambda r).$$

The inverse of ψ_λ is

$$\psi_\lambda^{-1} = \frac{1}{\lambda}[(1+\lambda)^r - 1].$$

It is easy to check that $\psi_\lambda \circ \mu_\lambda$ is additive. Since ψ_λ^{-1} is a concave (resp. convex) function for $\lambda > 0$ (resp. $\lambda \in (-1,0)$), μ_λ is submodular (resp. supermodular) if $\lambda > 0$ (resp. $\lambda \in (-1,0)$).

For every finite sequence of mutually disjoint subsets $A_1, A_2, \ldots, A_n$ of Ω,

$$\mu_\lambda\left(\bigcup_{i=1}^n A_i\right) = \psi_\lambda^{-1}\left[\sum_{i=1}^n \psi_\lambda(\mu_\lambda(A_i))\right].$$

Thus,

$$\mu_\lambda\left(\bigcup_{i=1}^n A_i\right) = \psi_\lambda^{-1}\left[\sum_{i=1}^n \log_{(1+\lambda)}(1 + \lambda \mu_\lambda(A_i))\right]$$

$$= \frac{1}{\lambda}\left(\prod_{i=1}^n [1 + \lambda \mu_\lambda(A_i)] - 1\right).$$

Let P be a probability measure on a measurable space $(\Omega, \mathcal{F})$. Then the set function $\psi_\lambda^{-1} \circ P$ is a λ-fuzzy measure. For every $\mathcal{F}$-measurable function X on Ω, we define its λ-expectation $E_\lambda(X)$ as

$$E_\lambda[X] = \int X d(\psi_\lambda^{-1} \circ P).$$

The λ-expectation has the following properties:
(i) If $\lambda \leq \lambda'$, then $E_\lambda[X] \geq E_{\lambda'}[X]$.
(ii) $\lim_{\lambda \to -1} E_\lambda[X] = \operatorname{esssup}_{\omega \in \Omega} X(\omega)$.
(iii) $\lim_{\lambda \to \infty} E_\lambda[X] = \operatorname{essinf}_{\omega \in \Omega} X(\omega)$.

For decision problem, the region of λ representing risk proneness is $(-1, 0)$ and the one representing risk aversion is $(0, \infty)$. When $\lambda = 0$, the decision maker is risk neutral.

For $\lambda \in (-1, \infty)$, let $\bar{\lambda} = -\frac{\lambda}{1+\lambda}$, then the two λ-fuzzy measures $\psi_\lambda^{-1} \circ P$ and $\psi_{\bar{\lambda}}^{-1} \circ P$ are conjugate to each other, i.e.,

$$\psi_\lambda^{-1} \circ P(A) = 1 - \psi_{\bar{\lambda}}^{-1} \circ P(A^c),$$

and that

$$E_\lambda(X) = -E_{\bar{\lambda}}(-X).$$

6. The Subadditivity Theorem

Let μ be a monotone set functions on 2^Ω. The Choquet integral w.r.t. μ is called *subadditive* if for upper μ-measurable functions X and Y

$$\int (X + Y) d\mu \leq \int X d\mu + \int Y d\mu.$$

A necessary condition for the Choquet integral w.r.t. μ be subadditive is submodularity of μ, because $I_{A \cup B}$ and $I_{A \cap B}$ are comonotonic, and we have

$$\int (I_A + I_B) d\mu = \int (I_{A \cup B} + I_{A \cap B}) d\mu = \int I_{A \cup B} d\mu + \int I_{A \cap B} d\mu$$
$$= \mu(A \cup B) + \mu(A \cap B), \quad A, B \subset \Omega.$$

We shall prove that submodularity of the set function is also sufficient for subadditivity of the Choquet integral.

The following lemma contains the core of the proof.

Lemma 6.1 Let Ω be the disjoint union of the sets $A_1, \ldots, A_n$. Let $\mathcal{A}$ be the algebra generated by $\{A_1, \ldots, A_n\}$ and $\mu : \mathcal{A} \to [0, 1]$ be a monotone set function with $\mu(\Omega) = 1$. For any permutation π of $\{1, \ldots, n\}$ define

$$S_i^\pi := \bigcup_{j=1}^i A_{\pi_j}, \ i = 1, \ldots, n, \quad S_0^\pi := \emptyset.$$

We define a probability measure P^π on $\mathcal{A}$ through

$$P^\pi(A_{\pi_i}) := \mu(S_i^\pi) - \mu(S_{i-1}^\pi), \ i = 1, \ldots, n.$$

Now let $X : \Omega \to \mathbb{R}$ be $\mathcal{A}$-measurable, i.e. constant on each A_i. If μ is submodular then

$$\int X d\mu \geq \int X dP^\pi,$$

and equality holds if

$$X(A_{\pi_1}) \geq X(A_{\pi_2}) \geq \cdots \geq X(A_{\pi_n}).$$

Proof. It suffices to prove the case $\pi = id$. We denote S_i^{id} by S_i, P^{id} by P, and let $x_i := X(A_i)$. We first prove the assertion on equality. Assume that $x_1 \geq x_2 \geq \cdots \geq x_n$. Since $S_1 \subset S_2 \subset \cdots \subset S_n$, the class $\{I_{S_1}, \cdots I_{S_n}\}$ is comonotonic. Thus we have (letting $x_{i+1} := 0, S_0 := \emptyset$)

$$\int X d\mu = \int \sum_{i=1}^n x_i I_{A_i} d\mu = \int \sum_{i=1}^n (x_i - x_{i+1}) I_{S_i}$$
$$= \sum_{i=1}^n (x_i - x_{i+1}) \mu(S_i) = \sum_{i=1}^n x_i (\mu(S_i) - \mu(S_{i-1})) = \int X dP.$$

Now assume that for some $i < n$ we have $x_i < x_{i+1}$. Let φ be the permutation which just interchanges i and $i+1$. Then $S^\varphi_{i-1} = S_{i-1} = S^\varphi_i \cap S_i, S^\varphi_{i+1} = S_{i+1} = S^\varphi_i \cup S_i$. Submodularity of μ implies

$$P(A_{i+1}) = \mu(S_{i+1}) - \mu(S_i) \leq \mu(S^\varphi_i) - \mu(S^\varphi_{i-1}) = P^\varphi(A_{\varphi_i}) = P^\varphi(A_{i+1}).$$

Multiplying by $x_{i+1} - x_i > 0$ gives

$$(x_{i+1} - x_i) P(A_{i+1}) \leq (x_{i+1} - x_i) P^\varphi(A_{i+1}).$$

On the other hand, we have

$$P(A_i) + P(A_{i+1}) = \mu(S_{i+1}) - \mu(S_{i-1}) = \mu(S^\varphi_{i+1}) - \mu(S^\varphi_{i-1})$$
$$= P^\varphi(A_{i+1}) + P^\varphi(A_i).$$

Multiplying by x_i and adding to the last inequality gives

$$x_i P^\varphi(A_i) + x_{i+1} P^\varphi(A_{i+1}) \geq x_i P(A_i) + x_{i+1} P(A_{i+1}),$$

which implies

$$\int X dP^\varphi \geq \int X dP.$$

By induction, we can construct from finitely many permutation of type φ a permutation θ with

$$X(A_{\theta_1}) \geq X(A_{\theta_2}) \geq \cdots \geq X(A_{\theta_n})$$

and

$$\int X dP^\theta \geq \int X dP.$$

Since we have proved that the left hand side integral is $\int X d\mu$, we conclude the proof of the desired result. □

For convenience we will say that a property of an upper μ-measurable function X holds μ-*essentially* if the same property holds for the quantile function $\check{G}_{\mu,X}$ e.c.. For example, we say X is μ-essentially $> -\infty$, if $\check{G}_{\mu,X}(t) > -\infty$ for all $t \in [0, \mu(\Omega)]$, e.c..

The following is the *subadditivity theorem*.

Theorem 6.1 Let μ be a monotone, submodular set functions on 2^Ω and X, Y upper μ-measurable functions on Ω. If X, Y are μ-essentially $> -\infty$, i.e.,

$$\lim_{x \to -\infty} G_{\mu,X}(x) = \mu(\Omega), \quad \lim_{x \to -\infty} G_{\mu,Y}(x) = \mu(\Omega),$$

then

$$\int (X+Y) d\mu \leq \int X d\mu + \int Y d\mu.$$

If μ is continuous from below the assumption on X, Y can be dropped.

Proof. First of all, we assume $\mu(\Omega) = 1$. If X, Y are simple functions, the $Z := X + Y$ is also a simple functions. Let $A_1, A_2, \cdots, A_n$ be a partition of Ω such that X and Y are constant on each A_i, and $Z(A_1) \geq Z(A_2) \geq \cdots \geq Z(A_n)$. By Lemma 6.1 there is a probability measure on $\mathcal{A}$, the algebra generated by $A_1, A_2, \cdots, A_n$ such that

$$\int Z d\mu = \int Z dP = \int X dP + \int Y dP.$$

Once again, Lemma 6.1 implies

$$\int (X+Y) d\mu \leq \int X d\mu + \int Y d\mu.$$

Now assume that X, Y are bounded. Let $Z := X + Y$ and $X_n := u_n(X), Y_n := u_n(Y), Z_n := u_n(Z)$, where

$$u_n := \inf \left\{ \frac{k}{n} \mid k \in \mathbb{Z}, \frac{k}{n} \geq x \right\}, n \in \mathbb{N}.$$

Then X_n, Y_n, Z_n are sequences of simple functions, and

$$X \leq X_n \leq X + \frac{1}{n}, Y \leq Y_n \leq Y + \frac{1}{n},$$

$$X_n + Y_n - \frac{2}{n} \leq Z_n \leq X_n + Y_n.$$

Proposition 3.1 (iv) and (v) imply

$$\int X d\mu \leq \int X_n d\mu \leq \int X d\mu + \frac{1}{n}.$$

Hence

$$\lim_{n \to \infty} \int X_n d\mu = \int X d\mu.$$

The same is valid for Y and Z. However, monotonicity of the integral the subadditivity for simple functions imply

$$\int Z_n d\mu \leq \int (X_n + Y_n) d\mu \leq \int X_n d\mu + \int Y_n d\mu,$$

from which we get the desired inequality.

Assume $\mu(\Omega) = 1$ and that X, Y are bounded below. By adding a constant we may assume $X, Y \geq 0$. Let $X_n := n \wedge X, Y_n := n \wedge Y$. Since the increasing sequence $G_{\mu, X_n + Y_n}$ converges to $G_{\mu, X+Y}$, we have

$$\int (X_n + Y_n) d\mu = \int_0^\infty G_{\mu, X_n + Y_n}(x) dx \to \int_0^\infty (X + Y) d\mu.$$

On the other hand, monotonicity of the integral the subadditivity for bounded functions imply

$$\int (X_n + Y_n) d\mu \leq \int X_n d\mu + \int Y_n d\mu \leq \int X d\mu + \int Y d\mu,$$

from which we get the desired inequality.

Assume $\mu(\Omega) = 1$ and that $\check{G}_{\mu, X}(t)$ and $\check{G}_{\mu, Y}$ are bounded below e.c.. In this case there is an $a \in \mathbb{R}$ so that $G_{\mu, X}(a) = 1, G_{\mu, Y}(a) = 1$. Define $\overline{X} := a \vee X$, which is bounded below. Then $G_{\mu, \overline{X}} = G_{\mu, X}$, hence $\int X d\mu = \int \overline{X} d\mu$. Doing the same for Y we get

$$int(X + Y) d\mu \leq \int (\overline{X} + \overline{Y}) d\mu$$

$$\leq \int \overline{X} d\mu + \int \overline{Y} d\mu = \int X d\mu + \int Y d\mu.$$

Now we come to the general case. First of all Proposition 5.2 (i) extends the desired inequality from normalized μ to finite μ. We will use $\mu_q = q \wedge \mu, 0 < q < \mu(\Omega)$ to extend the assertion on unbounded X, Y and infinite $\mu(\Omega)$. In fact, since $\lim_{t \to -\infty} G_{\mu, X}(t) = \mu(\Omega) > q$, we can find an $a \in \mathbb{R}$ with $G_{\mu, X}(a) \geq q$ so that $G_{\mu_q, X} = q \wedge G_{\mu, X}(a) = q = \mu_q(\Omega)$. That means $\check{G}_{\mu_q, X}(t)$, and similarly $\check{G}_{\mu_q, Y}$, are bounded below e.c.. From the above proved result we have

$$\int (X + Y) d\mu_q \leq \int X d\mu_q + \int Y d\mu_q.$$

Letting $q \to \mu(\Omega)$, Proposition 5.3 implies

$$\int (X + Y) d\mu \leq \int X d\mu + \int Y d\mu.$$

Finally, we come to the case where μ is continuous below. We treat two cases separately. In case $\mu(X + Y > -\infty) < \mu(\Omega)$, either $\int (X + Y) d\mu$ does not exist or is $-\infty$. So nothing has to be proved or the assertion is trivial. Now assume $\mu(X + Y > -\infty) = \mu(\Omega)$. Since $\{X + Y > -\infty\} = \{X > -\infty\} \cap \{Y > -\infty\}$, monotonicity of μ implies $\mu(X > -\infty) = \mu(\Omega)$ and $\mu(Y > -\infty) = \mu(\Omega)$. Then it is easy to see that $\check{G}_{\mu, X}(t) > -\infty, \check{G}_{\mu, Y}(t) > -\infty$ for all $t \in [0, \mu(\Omega)]$ e.c.. So we are in the situation already proved. □

Corollary 6.1 Let μ be a monotone, submodular set functions on 2^Ω and X, Y upper μ-measurable functions on Ω. Moreover, if $X, Y, X - Y$ and $Y - X$ are

μ-essentially $> -\infty$, then
$$\left|\int X d\mu - \int Y d\mu\right| \leq \int |X - Y| d\mu.$$
Especially, we have
$$\left|\int X d\mu\right| \leq \int |X| d\mu.$$

Proof. We may assume $\int X d\mu \geq \int Y d\mu$. By Theorem 6.1 we have
$$\int X d\mu = \int (X - Y + Y) d\mu \leq \int (X - Y) d\mu + \int Y d\mu$$
and, using $X - Y \leq |X - Y|$,
$$0 \leq \int X d\mu - \int Y d\mu \leq \int (X - Y) d\mu \leq \int |X - Y| d\mu,$$
the latter is the desired inequality. □

7. Representing Functionals as Choquet Integrals

Given a family $\mathcal{F}$ of functions $X : \Omega \to \overline{\mathbb{R}}$ and a functional $\Gamma : \mathcal{F} \to \overline{\mathbb{R}}$, we are interested in conditions under which Γ can be represented as a Choquet integral:
$$\Gamma(X) = \int X d\gamma, \quad X \in \mathcal{F},$$
where γ is a monotone set function on 2^Ω.

To begin with we prepare a lemma.

Lemma 7.1 Let μ be a monotone set function on $\mathcal{S} \subset 2^\Omega$ and $X : \Omega \to \overline{\mathbb{R}}_+$ a upper μ-measurable function. Then at any continuity point $x \geq 0$ of $G_{\mu,X}$ the function $g : x \mapsto \int X \wedge x d\mu$ is differentiable with derivative $G_{\mu,X}(x)$. If $G_{\mu,X}$ is right continuous then $G_{\mu,X}$ is the derivative from right of $\int X \wedge x d\mu$ at all points $x \geq 0$.

Proof. Since for $x, y \geq 0$
$$G_{\mu, X \wedge x}(y) = \mu^*(X \wedge x > y) = G_{\mu,X}(y) I_{[0,x)}(y),$$
we have
$$g(x) := \int X \wedge x d\mu = \int_0^\infty G_{\mu, X \wedge x}(y) dy = \int_0^x G_{\mu,X}(y) dy, \quad x \geq 0,$$
whence the desired assertion. □

The following representation theorem is due to Greco (1982).

Theorem 7.1 Let $\mathcal{F}$ be a family of functions which has the following properties:
a) $X \geq 0$ for all $X \in \mathcal{F}$,

b) $aX, X \wedge a, X - X \wedge a \in \mathcal{F}$, if $X \in \mathcal{F}, a \in \mathbb{R}_+$.

Assume that the functional $\Gamma : \mathcal{F} \to \overline{\mathbb{R}}$ satisfies the following conditions:
  (i) (*positive homogeneity*): $\Gamma(cX) = c\Gamma(X)$ for $X \in \mathcal{F}, a \in \mathbb{R}_+$,
  (ii) (*monotonicity*): $X, Y \in \mathcal{F}, X \leq Y$ imply $\Gamma(X) \leq \Gamma(Y)$,
  (iii) (*comonotonic additivity*): $\Gamma(X+Y) = \Gamma(X) + \Gamma(Y)$ for comonotonic $X, Y \in \mathcal{F}$ with $X + Y \in \mathcal{F}$,
  (iv) (*lower marginal continuity*): $\lim_{a \to 0} \Gamma(X - X \wedge a) = \Gamma(X)$ for $X \in \mathcal{F}$,
  (V) (*upper marginal continuity*): $\lim_{b \to \infty} \Gamma(X \wedge b) = \Gamma(X)$ for $X \in \mathcal{F}$.

Put
$$\alpha(A) := \sup\{\Gamma(X) \mid X \in \mathcal{F}, X \leq I_A\},$$
$$\beta(A) := \inf\{\Gamma(Y) \mid X \in \mathcal{F}, Y \geq I_A\}, \quad A \in 2^\Omega.$$

Then $\alpha \leq \beta$ and α, β are monotone. Let γ be a monotone set function on 2^Ω so that $\alpha \leq \gamma \leq \beta$. Then γ represents Γ.

Proof. Let $X \in \mathcal{F}$. For $n \in \mathbb{N}$ we define
$$u_n(x) := \frac{1}{2^n} \sum_{i=1}^{n2^n - 1} I_{x > \frac{i}{2^n}}.$$

Since
$$2^n \left(x \wedge \frac{i+1}{2^n} - x \wedge \frac{i}{2^n} \right) \leq I_{x > \frac{i}{2^n}} \leq 2^n \left(x \wedge \frac{i}{2^n} - x \wedge \frac{i-1}{2^n} \right),$$
we have
$$2^n \Gamma \left(X \wedge \frac{i+1}{2^n} - X \wedge \frac{i}{2^n} \right) \leq \alpha \left(X > \frac{i}{2^n} \right)$$
$$\leq \gamma \left(X > \frac{i}{2^n} \right) \leq \beta \left(X > \frac{i}{2^n} \right)$$
$$\leq 2^n \Gamma \left(X \wedge \frac{i}{2^n} - X \wedge \frac{i-1}{2^n} \right).$$

Summing up these inequalities with running index i and observing that the functions $X \wedge \frac{i}{2^n} - X \wedge \frac{i-1}{2^n}$ are comonotonic (Proposition 4.1) we get, using comonotonic additivity of Γ,
$$\Gamma \left(X \wedge n - X \wedge \frac{1}{2^n} \right) \leq \int u_n(X) d\gamma \leq \Gamma \left(X \wedge \left(n - \frac{1}{2^n} \right) \right) \leq \Gamma(X).$$

In order to prove that $\lim_{n \to \infty} \int u_n(X) d\gamma = \Gamma(X)$, we only need to show
$$\lim_{n \to \infty} \Gamma \left(X \wedge n - X \wedge \frac{1}{2^n} \right) = \Gamma(X).$$

To this end we rewrite the functions on the left hand side as follows:
$$X \wedge n - X \wedge \frac{1}{2^n} = (X \wedge n - X \wedge 1) + (X \wedge 1 - X \wedge \frac{1}{2^n})$$
$$= (X - X \wedge 1) \wedge (n - 1) + (X \wedge 1 - (X \wedge 1) \wedge \frac{1}{2^n}).$$

The last two functions being comonotonic we derive from (iii)-(v)

$$\lim_{n\to\infty} \Gamma\left(X \wedge n - X \wedge \frac{1}{2^n}\right) = \Gamma(X - X \wedge 1) + \Gamma(X \wedge 1) = \Gamma(X).$$

Finally, to conclude the proof it remains to show

$$\lim_{n\to\infty} \int u_n(X) d\gamma = \int X d\gamma.$$

This can be shown as follows:

$$X \wedge n - X \wedge \frac{1}{2^n} \le u_n(X) \le X$$

implies

$$\int_{\frac{1}{2^n}}^{n} G_{\gamma,X}(x) dx = \int \left(X \wedge n - X \wedge \frac{1}{2^n}\right) d\gamma \le \int u_n(X) d\gamma$$

$$\le \int X d\gamma = \int_0^\infty G_{\gamma,X}(x) dx,$$

from which letting $n \to \infty$ we get the desired result. □

Remark 7.1 1) In the statement of theorem condition (i) is implied by conditions (ii) and (iii). In fact, for a positive rational number c, $\Gamma(cX) = c\Gamma(X)$ is implied by comonotonic additivity. The monotonicity assumption which is also a continuity assumption implies the above equality for all non-negative numbers c.

2) The set functions α, β are the smallest and the largest monotone set functions, respectively, which represent Γ.

3) Conditions (i) through (v) are not only sufficient for representing Γ as an integral but necessary, too.

Corollary 7.1 Let $\mathcal{F}$ be a family of functions which has the properties b), c) (or c)') and d), where
  c) $X + 1 \in \mathcal{F}$ for all $X \in \mathcal{F}$,
  c)' $1 \in \mathcal{F}$ and a) is true,
  d) X is bounded for $X \in \mathcal{F}$.
Given a real functional Γ on $\mathcal{F}$ satisfying properties (i)-(iii), then there is a monotone finite set function γ on 2^Ω representing Γ.

Proof. First we assume c)' and d) are true. d) implies (v), and (iv) follows from the fact that $\Gamma(X - X \wedge a) = \Gamma(X) - \Gamma(X \wedge a)$ and $0 \le \Gamma(X \wedge a) \le \Gamma(a) = a\Gamma(1) \to 0$, as $a \to 0$. Thus all assumptions of Theorem 7.1 are valid.

Now we assume c) and d) are valid. Since b) implies $0 \in \mathcal{F}$, by c) 1 must in $\mathcal{F}$. Let $\mathcal{F}_+ = \{X \in \mathcal{F} \mid X \ge 0\}$. Then according to the above proved result $\mathcal{F}_+$ matches all assumptions of Theorem 7.1. Now for $X \in \mathcal{F}$ there is, according to d), a constant $c > 0$ such that $X + c \ge 0$ and $X + c = c(\frac{1}{c}X + 1) \in \mathcal{F}_+$ and, since the assertion is valid for $X + c$,

$$\Gamma(X) = \Gamma(X + c) - c\Gamma(1) = \int (X + c) d\gamma - c\gamma(\Omega) = \int X d\gamma.$$ □

The following corollary gives sufficient conditions on $\mathcal{F}$ and Γ under which a submodular (resp. supermodular) γ representing Γ exists.

Corollary 7.2 In Theorem 7.1 if $\mathcal{F}$ further has the lattice property

e) $X \wedge Y, X \vee Y \in \mathcal{F}$ if $X, Y \in \mathcal{F}$,

and Γ further has the following property

(vi) (*submodularity*): $\Gamma(X \vee Y) + \Gamma(X \wedge Y) \leq \Gamma(X) + \Gamma(Y)$ if $X, Y \in \mathcal{F}$, or

(vii) (*supermodularity*): $\Gamma(X \vee Y) + \Gamma(X \wedge Y) \geq \Gamma(X) + \Gamma(Y)$ if $X, Y \in \mathcal{F}$,

then β (resp. α) defined in the proof of Theorem 7.1 is a monotone, submodular (resp. supermodular) which represents Γ.

Proof. For any $X : \Omega \to \overline{I\!R}_+$ let

$$S_X := \{(\omega, x) \in \Omega \times I\!R_+ \mid x < X(\omega)\}$$

be the *subgraph* of X. Then by e) the system $\mathcal{S} := \{S_X \mid X \in \mathcal{F}\}$ is closed under union and intersection. We introduce an auxiliary set function ν on $\mathcal{S}$ by

$$\nu(S_X) := \Gamma(X), \ X \in \mathcal{F}.$$

Since $S_X \subset S_Y$ iff $X \leq Y$, ν is monotone. It is easy to see that (vi) implies that ν is submodular, and the outer set function ν^* of ν is submodular, too, by Proposition 5.2.

Now we return from $\Omega \times I\!R_+$ to Ω in defining

$$\gamma(A) := \nu^*(S_{I_A}), \ A \in 2^\Omega.$$

Clearly γ is monotone and submodular since

$$S_{I_{A \cup B}} = S_{I_A} \cup S_{I_B}, \quad S_{I_{A \cap B}} = S_{I_A} \cap S_{I_B}.$$

Now we show that $\gamma = \beta$. In fact, for any $A \in 2^\Omega$,

$$\beta(A) = \inf\{\Gamma(Y) \mid I_A \leq Y \in \mathcal{F}\}$$
$$= \inf\{\nu(S_Y) \mid S_{I_A} \subset S_Y \in \mathcal{S}\}$$
$$= \nu^*(S_{I_A}) = \gamma(A).$$

Similarly, we can treat with the case where (vii) holds. □

A normalized monotone set function is called a *capacity*. If a capacity is continuous from above we call it a *upper-continuous capacity*.

Let $\mathcal{F}$ be a collection of bounded real-valued functions on Ω. $\mathcal{F}$ is called a *Stone vector lattice* if : (i) $\mathcal{F}$ is a vector space; (ii) $\mathcal{F}$ is a lattice, i.e., $X \vee Y, X \wedge Y \in \mathcal{F}$ for all $X, Y \in \mathcal{F}$; and (iii) $\mathcal{F}$ contains all constant functions on Ω. Let I be a (real-valued) function from $\mathcal{F}$ to $I\!R$. I is called a *quasi-integral* if: I is comonotonically additive, monotonic and continuous in the sense that

$$\lim_{n \to \infty} I(X_n) = I(X),$$

if $X, X_n \in \mathcal{F}, n \geq 1$, (X_n) is deccreasing and tends to X.

According to Remark 7.1.1), a quasi-integral I is always positive homogeneous.

The following theorem (due to Zhou(1998)) establishes a one-to-one correspondence between upper-continuous capacities and quasi-integrals.

Theorem 7.2 Assume that I is a quasi-integral on a Stone lattice $\mathcal{F}$ on Ω and $I(1) = 1$. Then there exists a unique upper-continuous capacity μ on $\mathcal{S} := \{[X \geq c], X \in \mathcal{F}, c \in \mathbb{R}\}$ which represents I. On the other hand, for any upper-continuous capacity μ, the functional I defined on $\mathcal{F}$ by the Choquet integral $I(X) := \int X d\mu$ is a quasi-integral.

Before proving Theorem 7.2 we prepare a lemma.

Lemma 7.2 Let I be a functional on a lattice $\mathcal{F}$ that is comonotonically additive and monotonic. Then I is continuous iff for any decreasing sequence (X_n) and X in $\mathcal{F}$ such that for all $\omega \in \Omega$, there is an $n_\omega \in \mathbb{N}$ with $X_n(\omega) \leq X(\omega)$ for all all $n \geq n_\omega$, $\lim_{n \to \infty} I(X_n) \leq I(X)$.

Proof. Necessity. Assume that I is continuous. Take any decreasing sequence (X_n) and X in $\mathcal{F}$ satisfying the required conditions. Since $X_n \vee X$ decreasingly tends to X, we have

$$\lim_{n \to \infty} I(X_n) \leq \lim_{n \to \infty} I(X_n \vee X) = I(X).$$

Sufficiency. Take any decreasing sequence (X_n) and X in $\mathcal{F}$ such that $\lim_{n \to \infty} X_n(\omega) = X(\omega)$ for all ω. Fix an $\epsilon > 0$. Since for all ω there is an n_ω with $X_n(\omega) \leq X(\omega) + \epsilon$, by assumption $\lim_{n \to \infty} I(X_n) \leq I(X + \epsilon) = I(X) + \epsilon I(1)$. Let ϵ go to zero, we have $\lim_{n \to \infty} I(X_n) \leq I(X)$. But the inverse inequality is also true by monotonicity. Hence, $\lim_{n \to \infty} I(X_n) = I(X)$. □

Proof of Theorem 7.2 Suppose that I is a quasi-integral on $\mathcal{F}$. Let $A = [X \geq c] \in \mathcal{S}$. Put $X_n^A = (1 - n(c - X)^+)^+$, then $X_n^A \in \mathcal{F}$ and X_n^A decreasingly tends to I_A. Since I is monotone, we denote by $\mu(A)$ the limit of $I(X_n^A)$. We will show that the definition of $\mu(A)$ in independent of the expression of A. In fact, if $A = [Y \geq b]$ with $Y \in \mathcal{F}$ and let $Y_n^A = (1 - n(b - Y)^+)^+$, by Lemma 5.6 one can show that for any fixed m,

$$\lim_{n \to \infty} I(X_n^A \vee Y_m^A) \leq I(Y_m^A).$$

Consequently we have $\lim_{n \to \infty} I(X_n^A) \leq \lim_{m \to \infty} I(Y_m^A)$. By symmetry the equality holds. Thus the set function μ on $\mathcal{S}$ is well-defined. It is easy to check that μ is indeed a capacity.

We are going to show that μ is upper-continuous. Let (A_n) be a decreasing sequence of sets in $\mathcal{S}$ and $A \in \mathcal{S}$ with $\cap_{n=1}^{\infty} A_n = A$. By definition of $\mathcal{S}$ there are a sequence of functions (X_n) and $X \in \mathcal{F}$ such that $A_n = [X_n \geq c_n]$ and $[X \geq c]$. Since $\mathcal{F}$ is a Stone lattice, we may assume, without loss of generality, that (X_n)

is a decreasing sequence, that $X_n \leq 1$ and $X \leq 1$, and that $A_n = [X_n = 1]$ and $A = [X = 1]$. Let $\epsilon > 0$. By the definition of μ, there are some m and, without loos of generality, an increasing sequence (m_n) of integers with $m_n \geq n$ such that

$$|\mu(A) - I(X_m^A)| < \epsilon, \quad |\mu(A_n) - I(X_{m_n}^{A_n})| < \epsilon, \quad \text{for all } n.$$

We claim that the sequence $(X_{m_n}^{A_n})$ and X_m^A satisfy the condition in Lemma 7.2. If $\omega \in A$, then $X_m^A(\omega) = 1$; and if $\omega \notin A$, since $\cap_{n=1}^\infty A_n = A$, there is an n_ω such that $\omega \notin A_n$ for all $n \geq n_\omega$. By the definition of $X_k^{A_n}$, for any fixed n, if $\omega \notin A_n$ then for sufficiently large k_n one has $X_k^{A_n}(\omega) = 0$ for all $k \geq k_n$. Since $(X_{m_n}^{A_n})$ is a decreasing sequence, for $\omega \notin A$ one has $(X_{m_n}^{A_n}) = 0$ for all $n \geq \max\{n_\omega, k_{n_\omega}\}$. By Lemma 5.6 we obtain

$$\lim_{n \to \infty} I(X_{m_n}^{A_n}) \leq I(X_m^A).$$

Since ϵ is arbitrary, we must have $\lim_{n \to \infty} \mu(A_n) \leq \mu(A)$. By monotonicity of μ this implies upper-continuity of μ.

Now we show that I satisfies conditions (iv) and (v) of Theorem 7.1 for $X \in \mathcal{F}$ with $X \geq 0$. In fact, (iv) follows from $I(X - X \wedge a) = I(X) - I(X \wedge a)$ (due to comonotonic additivity of I) and continuity of I. (v) is trivial, because X is bounded by a constant $K > 0$ and $X \wedge b = X$ for $b \geq K$.

Now assume that $c > 0$, $A = [X \geq c]$ and $Z \leq I_A \leq W$ with $Z, W \in \mathcal{F}, Z, W \geq 0$. We will show that $I(Z) \leq \mu(A) \leq I(W)$. Since $(1 - n(c - X)^+)^+ \geq I_A \geq Z$, we have that $\mu(A) = \lim_{n \to \infty} I((1 - n(c - X)^+)^+) \geq I(Z)$. On the other hand, by Lemma 7.1 we have $\mu(A) = I(X_n^A) \leq I(W)$. Consequently, for any $X \in \mathcal{F}$ with $X \geq 0$, if we approximate X with functions $u_n(X)$, where

$$u_n(x) := \frac{1}{2^n} \sum_{i=1}^{n2^n - 1} I_{x \geq \frac{i}{2^n}},$$

then, using the fact that $I(Y) \leq \mu([X \geq c]) \leq I(Z)$ for $Y \leq I_{[X \geq c]} \leq Z$ with $Y, Z \in \mathcal{F}, Y, Z \geq 0$, the same proof as in Theorem 7.1 gives that $I(X)$ is the Choquet integral of X w.r.t. μ.

Now assume $X \in \mathcal{F}$. Let $K > 0$ such that $K + X \geq 0$. Then by comonotonic additivity of I and Choquet integral we have

$$K + I(X) = I(K + X) = \int (K + X) d\mu = K + \int X d\mu.$$

Thus $I(X) = \int X d\mu$. Using the expression of Choquet integral in terms of $G_{\mu,X}$ and the up-continuity of μ one can show the uniqueness of μ representing of I.

On the other hand, suppose that a functional I is defined by the Choquet integral wr.r.t. an upper-continuous capacity μ. It is clear that I is monotonic and comonotonically additive. So we only have to prove that I is continuous. Let (X_n) be a decreasing sequence in $\mathcal{F}$ with the limit $X \in \mathcal{F}$. Since $[X_n \geq t]$ decreasingly tends to $[X \geq t]$ and μ is upper-continuous, we have $\lim_{n \to \infty} \mu([X_n \geq t]) = \mu([X \geq t])$.